Stability and Control: Theory, Methods and Applications
Volume 21

Asymptotic Methods in Resonance Analytical Dynamics

Stability and Control: Theory, Methods and Applications

A series of books and monographs on the theory of stability and control

Edited by A.A. Martynyuk

Institute of Mechanics, Kiev, Ukraine and V. Lakshmikantham, Florida Institute of Technology, USA

Volume 1
Theory of Integro-Differential Equations
V. Lakshmikantham and M. Rama Mohana Rao

Volume 2
Stability Analysis: Nonlinear Mechanics Equations
A.A. Martynyuk

Volume 3
Stability of Motion of Nonautonomous Systems (Method of Limiting Equations)
J. Kato, A.A. Martynyuk and A.A. Shestakov

Volume 4
Control Theory and its Applications
E. O. Roxin

Volume 5
Advances in Nonlinear Dynamics
Edited by S. Sivasundaram and A.A. Martynyuk

Volume 6
Solving Differential Problems by Multistep Initial and Boundary Value Methods
L. Brugnano and D. Trigiante

Volume 7
Dynamics of Machines with Variable Mass
L. Cveticanin

Volumes 8
Optimization of Linear Control Systems: Analytical Methods and Computational Algorithms
F.A. Aliev and VB. Larin

Volume 9
Dynamics and Control
Edited by G. Leitmann, F.E. Udwadia and A.V Kryazhimskii

Volume 10
Volterra Equations and Applications
Edited by C. Corduneanu and I.W. Sandberg

Volume 11
Nonlinear Problems in Aviation and Aerospace
Edited by S. Sivasundaram

Volume 12
Stabilization of Programmed Motion
E.Ya. Smirnov

Volume 13
Advances in Stability Theory at the End of the 20th Century
A.A. Martynyuk

Volume 14
Dichotomies and Stability in Nonautonomous Linear Systems
Yu A. Mitropolskii, A.M. Samoilenko and V. L Kulik

Volume 15
Almost Periodic Solutions of Differential Equations in Banach Spaces
Y. Hino, T. Naito, Nguyen Van Minh and Jong Son Shin

Volume 16
Functional Equations with Causal Operators
C. Corduneanu

Volume 17
Optimal Control of Growth of Wealth of Nations
E.N. Chukwu

Volume 18
Stability and Stabilization of Nonlinear Systems with Random Structure
I. Ya Kats and A.A Martynyuk

Volume 19
Lyapunov Method & Certain Differential Games
V.I. Zhukovskiv

Volume 20
Stability of Differential Equations with Aftereffect
N.V. Azbelev and P.M. Simonov

Volume 21
Asymptotic Methods in Resonance Analytical Dynamics
E. A. Grebenikov, Yu. A. Mitropolsky and Yu. A. Ryabov

Stability and Control: Theory, Methods and Applications
Volume 21

Asymptotic Methods in Resonance Analytical Dynamics

E. A. Grebenikov
Russian Academy of Sciences
Moscow, Russia

Yu. A. Mitropolsky
National Academy of Sciences of Ukraine
Kiev, Ukraine

Yu. A. Ryabov
Moscow Institute of Auto and Highway Construction
Moscow, Russia

CHAPMAN & HALL/CRC
A CRC Press Company
Boca Raton London New York Washington, D.C.

Library of Congress Cataloging-in-Publication Data

Grebenikov, E. A. (Evgeniæi Aleksandrovich)
Asymptotic methods in resonance analytical dynamics / by Eugeniu Grebenikov, Yu. A. Mitropolsky, and Y. Ryabov.
p. cm. — (Stability and control ; v. 21)
Includes bibliographical references and index.
ISBN 0-415-31008-3 (alk. paper)
1. Averaging method (Differential equations) 2. Differential equations—Asymptotic theory. 3. Resonance—Mathematical models. I. Mitropol§skiæl, ëĩU. A. (ëĩUriæi Alekseevich), 1917- II. Rëiìabov, ëĩU. A. (ëĩUriæi Aleksandrovich) III. Title. IV. Series.

QA372 .G715 2004
515′.35—dc22 2003069583

Direct all inquiries to CRC Press LLC, 2000 N.W. Corporate Blvd., Boca Raton, Florida 33431.

Visit the CRC Press Web site at www.crcpress.com

International Standard Book Number 0-415-31008-3
Library of Congress Card Number 2003069583
Printed in the United States of America 1 2 3 4 5 6 7 8 9 0
Printed on acid-free paper

Contents

Introduction to the Series

The problems of modern society are both complex and interdisciplinary. Despite the apparent diversity of problems, tools developed in one context are often adaptable to an entirely different situation. For example, consider the Lyapunov's well known second method. This interesting and fruitful technique has gained increasing signifycance and has given a decisive impetus for modern development of the stability theory of differential equations. A manifest advantage of this method is that it does not demand the knowledge of solutions and therefore has great power in application. It is now well recognized that the concept of Lyapunov-like functions and the theory of differential and integral inequalities can be utilized to investigate qualitative and quantitative properties of nonlinear dynamic systems. Lyapunov-like functions serve as vehicles to transform the given complicated dynamic systems into a relatively simpler system and therefore it is sufficient to study the properties of this simpler dynamic system. It is also being realized that the same versatile tools can be adapted to discuss entirely different nonlinear systems, and that other tools, such as the variation of parameters and the method of upper and lower solutions provide equally effective methods to deal with problems of a similar nature. Moreover, interesting new ideas have been introduced which would seem to hold great potential.

Control theory, on the other hand, is that branch of application-oriented mathematics that deals with the basic principles underlying the analysis and design of control systems. To control an object implies the influence of its behavior so as to accomplish a desired goal. In order to implement this influence, practitioners build devices that incorporate various mathematical techniques. The study of these devices and their interaction with the object being controlled is the subject of control theory. There have been, roughly speaking, two main lines of work in control theory which are complementary. One is based on the idea that a good model of the object to be controlled is available and that we wish to optimize its behavior, and the other is based on the constraints imposed by uncertainty about the model in which the object operates. The control tool in the latter is the use of feedback in order to correct for deviations from the desired behavior. Mathematically, stability theory, dynamic systems and functional analysis have had a strong influence on this approach.

Volume 1, *Theory of Integro-Differential Equations*, is a joint contribution by V. Lakshmikantham (USA) and M. Rama Mohana Rao (India).
Volume 2, *Stability Analysis: Nonlinear Mechanics Equations*, is by A. A. Martynyuk (Ukraine).
Volume 3, *Stability of Motion of Nonautonomous Systems: The Method of Limiting Equations*, is a collaborative work by J. Kato (Japan), A. A. Martynyuk (Ukraine) and A. A. Shestakov (Russia).

Volume 4, *Control Theory and Its Applications*, is by E. O. Roxin (USA).
Volume 5, *Advances in Nonlinear Dynamics*, is edited by S. Sivasundaram (USA) and A. A. Martynyuk (Ukraine) and is a multiauthor volume dedicated to Professor S. Leela (USA).
Volume 6, *Solving Differential Problems by Multistep Initial and Boundary Value Methods*, is a joint contribution by L. Brugnano (Italy) and D. Trigiante (Italy).
Volume 7, *Dynamics of Machines with Variable Mass*, is by L. Cveticanin (Yugoslavia).
Volume 8, *Optimization of Linear Control Systems: Analytical Methods and Computational Algorithms*, is a joint contribution by F. A. Aliev (Azerbaijan) and V. B. Larin (Ukraine).
Volume 9, *Dynamics and Control*, is edited by G. Leitmann (USA), F. E. Udwadia (USA) and A. V. Kryazhimskii (Russia) and is a multiauthor volume.
Volume 10, *Volterra Equations and Applications*, is edited by C. Corduneanu (USA) and J. W. Sandberg (USA) and is a multiauthor volume.
Volume 11, *Nonlinear Problems in Aviation and Aerospace*, is edited by S. Sivasundaram (USA) and is a multiauthor volume.
Volume 12, *Stabilization of Programmed Motion*, is by E. Ya. Smirnov (Russia).
Volume 13, *Advances in Stability Theory at the End of the 20th Century*, is edited by A. A. Martynyuk (Ukraine) and is a multiauthor volume.
Volume 14, *Dichotomies and Stability in Nonautonomous Linear Systems*, is a collaborative work by Yu. A. Mitropolsky (Ukraine), A. M. Samoilenko (Ukraine) and V. L. Kulik (Ukraine).
Volume 15, *Almost Periodic Solutions of Differential Equations in Banach Spaces*, is a collaborative work by Y. Hino (Japan), T. Naito (Japan), Nguyen Van Minh (Vietnam) and Jong Son Shin (Japan).
Volume 16, *Functional Equations with Causal Operators*, is by C. Corduneanu (USA).
Volume 17, *Optimal Control of the Growth of Wealth of Nations*, is by E. N. Chukwu (USA).
Volume 18, *Stability and Stabilization of Nonlinear Systems with Random Structure*, is a joint contribution by I. Ya. Kats (Russia) and A. A. Martynyuk (Ukraine).
Volume 19, *Lyapunov Functions in Differential Games*, is by V. I. Zhukovskiy (Russia).
Volume 20, *Stability of Differential Equations with Aftereffect*, is a joint contribution by N. V. Azbelev (Russia) and P. M. Simonov (Russia).
Volume 21, *Asymptotic Methods in Resonance Analytical Dynamics*, is a collaborative work by E. A. Grebenikov (Russia), Yu. A. Mitropolsky (Ukraine) and Yu. A. Ryabov (Russia).

Due to the increased interdependency and cooperation among the mathematical sciences across the traditional boundaries, and the accomplishments thus far achieved in the areas of stability and control, there is every reason to believe that many breakthroughs await us, offering existing prospects for these versatile techniques to advance further. It is in this spirit that we see the importance of the 'Stability and Control' series, and we are immensely thankful to Taylor & Francis and Chapman & Hall/CRC for their interest and cooperation in publishing this series.

Preface

Many important systems in analytical dynamics are described by nonlinear mathematical models, and the latter as a rule are represented by differential or integro-differential equations. The absence of exact universal methods for the investigation of nonlinear systems has driven the development of a wide range of approximate analytic and numerical–analytic methods that can be implemented in effective computer algorithms.

Trying to give them a general description, one can assert that practically all approximate methods are constructed by the iteration principle. This means that first an initial approximation is somehow chosen for the problem, and then, by means of iterations, the addition of terms of different infinitesimal order to the initial approximation are found. The term "iteration" is understood here as either successive approximations, or a chain of successive transformations of phase variables, or a functional series with terms decreasing in value. From the historical aspect, the first approximation was considered as a solution of some linear problem (hence the term linearization method), to which some small functions (proportional to the small parameter) determined within the framework of one or another perturbation theory were added. For example, while studying the oscillations of the mathematical pendulum Newton considered the small oscillations $\varphi(t)$ of the pendulum, i.e. a linear problem of the form

$$\ddot{\varphi} + \omega^2 \varphi = A,$$

where ω, A are constants. This linear equation serves as the initial approximation for the nonlinear equation

$$\ddot{\psi} + \omega^2 \sin \psi = A$$

that describes oscillations $\psi(t)$ of a mathematical pendulum, not necessarily small in value.

Lagrange did the same when considering secular perturbations in the planetary three-body problem under small perturbations. It was then that the system of differential equations of the following form was solved:

$$\ddot{x} + \omega^2 x = F(t),$$

where x is a six-, eight- or ten-dimensional vector, ω the vector of constant frequencies, and $F(t)$ a vector-function Ω-periodic with respect to t.

Another approach to nonlinear problem solving is when, as the initial approximation, the solution of a system is taken that is a nonlinear but essentially "simpler" system than the initial one. This simplified system can be obtained by different methods, but since the times of Lagrange and Gauss, in oscillatory models of celestial mechanics the method of averaging (or smoothing) of periodic or quasi-periodic functions being a part of the analytic structure of the equations was most often used. Later, and particularly nowadays, averaging methods in combination with asymptotic representations (in the sense of Poincaré) were used as the basic constructive means of solving intricate problems of analytical dynamics, formalized in the language of differential equations. This became possible thanks to the work of N.N. Bogolyubov in the 1930s, where the problem of nonlinear problem solving was formulated as a problem of transformation of the initial differential equations into new simplified, so-called comparison equations. Since the choice of comparison equations is arbitrary, one can make an optimal choice (e.g. from the standpoint of their solvability, or the proximity of solutions of comparison equations to those of the initial system, or the efficiency of numerical methods applied to them) that determines the corresponding transformation of phase variables. It is easy to show that if the differential equations are written in a normal form in the Cauchy sense, then the functions that perform the change of variables satisfy some system of quasi-linear systems in first-order partial derivatives.

From the aspect of geometric interpretation, the transformation of initial equations into comparison equations can be interpreted as the choice of a phase space most optimal for the given problem, of its metrics and norm. The transformation of equations means the search for the geometry that gives the simplest description to the considered problem. But the simplest description of a problem does not mean that the method of solution has been found. Therefore, as a rule, it is only a reasonable combination of iteration procedures with a successful choice of the initial approximation and with the use of fast algorithms and compact programs that can guarantee the efficiency of the construction of an approximate solution with prescribed accuracy.

So the search for the optimal geometry for each specific problem of analytical dynamics is connected with the finding of some substitution of variables that binds the old phase variables in which the initial description of the problem is made, with new phase variables in which the problem is described by a method that is simplest (or, rather, most convenient for the researcher). If we succeed in solving the equations determining the substitution of variables, we obtain convenient relations binding the initial phase space with the new one.

An excellent illustration of this is the well-known three-body problem of analytical dynamics. Here the famous problem of small denominators occurred for the first time, which is in essence the most attractive mathematical object for research in the field of resonance analytic dynamics.

Small denominators were first discovered in 1784 by Laplace when he studied the motion of Jupiter and Saturn around the Sun. It turned out that small denominators (below we will give their detailed description) led to very important peculiarities in the motion of these planets, which could not be explained by the outstanding mathematicians of the past (Eiler, Lagrange, Lambert and others).

The point is that the motions of planets and satellites (or, rather, their coordinates and velocities) have a very complex nature and can be mathematically represented in the form of multiple Fourier series, i.e. in the form of a combination of a lot (most often an infinite set) of periodic motions with different periods or, which is the same, with different frequencies. Usually it is possible to choose basic or master frequencies that are large quantities compared with the other frequencies. For example, in the two-planet problem (Sun–planet–planet) it is possible to choose two basic frequencies corresponding to the average periods of the planets, revolution around the Sun, being the average angular velocities of their motion along heliocentric orbits. Astronomers call these basic frequencies the average motions. Strictly speaking, average motions of planets or satellites are not constant, but as observations show, they change very slowly, i.e. they are slow functions of time.

If in the two-planet problem the basic frequencies are closely commensurable, it is then that we come across the problem of small denominators. Mathematically the effect of small denominators shows in the fact that in the solutions of the equations of planetary motion, represented by Fourier series, periodic terms appear with coefficients whose denominators are close to zero, or in other words, periodic harmonics with large amplitudes appear.

If there is an exact rational commensurability of the basic frequencies, linear functions of time may also appear in the solutions (instead of periodic functions with very large amplitudes). In planetary motion, effects occur that in physics are called resonance effects, like the resonant oscillations of two pendulums with their points of suspension located on a common horizontal rod.

In the case of frequency resonances, elements of planetary orbits and their angular velocities undergo great changes which in comparatively short time intervals are difficult to distinguish from secular, i.e. linear functions of time. For example, in the case of Jupiter and Saturn their average motion changes so that over 100 years they are almost identical to secular motion. It was this circumstance that prompted Lagrange to make an attempt, however abortive, to construct a theory of the motion of Jupiter and Saturn with the average motions in the form of linear functions of time.

Now we will describe the results obtained by Laplace in more detail. According to the observations made by astronomers in the second half of the 18th century, the average motion of Saturn (heliocentric angular velocity of Saturn per day) is $n_1^{(0)} = 120''.454645$, and of Jupiter is $n_2^{(0)} = 299''.128361$. It is easy to calculate that $n_2^{(0)} : n_1^{(0)} = 2.483326 \approx 5 : 2$, i.e. the average motions of those planets are rationally almost commensurable.

From this relation it follows that

$$2n_2^{(0)} - 5n_1^{(0)} = -4''.016503 = -0.0000194725 \text{ rad.}$$

The corresponding differential equations for the longitude of Saturn (quantity l_1) and Jupiter (quantity l_2) taking account of mutual attractions has the form

$$\frac{dl_1}{dt} = \sum_{k_1,k_2=-N_1}^{N_1} A_{k_1,k_2}\cos(k_1y_1 + k_2y_2),$$
$$\frac{dl_2}{dt} = \sum_{k_1,k_2=-N_2}^{N_2} B_{k_1,k_2}\cos(k_1y_1 + k_2y_2),$$

where y_1, y_2 are functions depending in a complicated way (however, almost linearily) on time t and on the longitudes l_1, l_2 themselves. The analytic structure of the coefficients A_{k_1,k_2}, B_{k_1,k_2} is also very complicated, but in approximate theories they can be considered as constants. If we write a rigorous closed system of differential equations of planetary dynamics, it will take the form of a multifrequency rotary system with slow and fast variables, and consequently A_{k_1,k_2}, B_{k_1,k_2} are represented by power series with respect to slow variables that are here the ratios of the major semiaxes of the planets' orbits. Besides, it is known that A_{k_1,k_2}, B_{k_1,k_2} diminish quickly enough with increase of $|k_1|, |k_2|$, and it is also important to note that the coefficients A_{k_1,k_2} are proportional to the mass of Jupiter, and B_{k_1,k_2} to the mass of Saturn. This means that the coefficients A_{k_1,k_2} are approximately 3 to 5 times larger than B_{k_1,k_2}. Hence it should be expected that the effect of the change of l_1 is greater with time than that of the change of l_2.

In the known theories of planetary motion (both modern and previous ones) it is shown that within a time interval several decades long (i.e. during several revolutions of Jupiter and Saturn around the Sun)

$$\begin{cases} y_1(t) \approx n_1^{(0)}t + y_{1,0}, \\ y_2(t) \approx n_2^{(0)}t + y_{2,0}, \end{cases} \qquad (*)$$

where $n_1^{(0)}, n_2^{(0)}$ almost exactly coincide with the values of average motions obtained directly from observations over centuries. If we assume that A_{k_1,k_2}, B_{k_1,k_2} are constants, and $y_1(t)$, $y_2(t)$ are expressed by formulae $(*)$, then as a result of integration of the equations for the longitude we obtain

$$l_1(t) = A_{0,0}t + \sum_{k_1,k_2=-N_1}^{N_1} \frac{A_{k_1,k_2} \sin\big[(k_1 n_1^{(0)} + k_2 n_2^{(0)})t + d_1\big]}{k_1 n_1^{(0)} + k_2 n_2^{(0)}},$$

$$l_2(t) = B_{0,0}t + \sum_{k_1,k_2=-N_2}^{N_2} \frac{B_{k_1,k_2} \sin\big[(k_1 n_1^{(0)} + k_2 n_2^{(0)})t + d_2\big]}{k_2 n_2^{(0)} + k_1 n_1^{(0)}},$$

d_1, d_2 are constants.

Before Laplace, in the construction of theories of planetary motion around the Sun astronomers confined themselves to terms with the smallest summation indices $k_1, k_2 = 1, 2$ (certainly without a rigorous mathematical justification of such a truncation of the series), implicitly supposing that all the remaining summands are negligibly small (note that this is the situation that occurs in the case of theories of motion of other big planets). However Laplace found that the harmonic

$$\frac{A_{-5,2} \sin\big[(2n_2^{(0)} - 5n_1^{(0)})t + d_1\big]}{2n_2^{(0)} - 5n_1^{(0)}},$$

contrary to former ideas, has a very large amplitude due to the smallness of the denominator $2n_2^{(0)} - 5n_1^{(0)}$ which is, e.g., two orders of magnitude (300 times) smaller than the denominator $2n_2^{(0)} + 5n_1^{(0)}$. If $n_1^{(0)}$ and $n_2^{(0)}$ had not been rationally almost commensurable, then the written harmonic would have had a small amplitude, and it could be ignored in the theory, like Laplace's forerunners had done. So Laplace found that the resonance $n_1^{(0)} : n_2^{(0)} \approx 2 : 5$ results in the appearance of a harmonic with a large amplitude, and when this harmonic was taken into account in the theory of the motion of Saturn and Jupiter it brought the theory and observations into a comparative concord.

Thus, Laplace found that the average angular velocities of Saturn and Jupiter contain long-period perturbations with the period of 833 years and with very large amplitudes. If they are not accounted for, the discrepancies in the longitude between the theory and the observations of Saturn may reach $50''$, and $-20''$ and more for Jupiter. It was such deviations of the theory from observations that stumped researchers, because from the point of view of observational astronomy, these are prohibitive amounts.

Historically the small-denominator problem emerged during the study of the motion in the Sun–Jupiter–Saturn problem, though astronomers now relate it to the problem with "not very sharp" resonance. Imagine for a moment that the average motions of Saturn and Jupiter are equal to $n_1^{(0)} = 119''.843122$, $n_2^{(0)} = 299''.10779541$, respectively. Then $2n_2^{(0)} - 5n_1^{(0)} = -0''.00000019$, and then the above harmonic would have an amplitude 100 times larger. In this hypothetical case the long period would be equal to 83300 years, and the discrepancies in longitudes (because the term $A_{-5,2} \sin\big[(2n_2^{(0)} - 5n_1^{(0)})t + d_1\big]$ was not taken into account in the theory) would reach inconceivable values, around $83°$ and $23°$.

Laplace's investigations into the theory of the motion of Saturn and Jupiter should be regarded as an outstanding achievement of mathematics and celestial mechanics, they are powerful analytical tools. They had put a number of fundamental problems before mathematicians and astronomers; the solving of these problems has enriched not only classical mathematics, but also modern mathematics.

After the discovery of the small-denominator effect by Laplace, and particularly in the 19th century, the two lines of research began to show, clearly expressed but tightly connected with each other.

The first line of investigation can be called astronomical. It lies in the development of methods and direct construction of approximate solutions of those problems of celestial mechanics where there is a resonance of frequencies (the commensurability of average motions in the first place) of celestial bodies' orbital motion. Required of those solutions was a sufficiently precise description of real motions, generally within a limited but large enough time interval. In mathematical terms it is a question of the construction of asymptotic solutions of differential equations of planetary dynamics in an asymptotically large time interval (of order $O(\mu^{-1})$). Astronomers, one might say, have successfully solved this problem of perturbation theory for not very sharp resonances by means of the classical series which we understand to be analytical expressions for the required variables that contain secular, trigonometric and mixed perturbations, i.e. those of the form $B \sin \alpha t$, but as a rule also the secular and mixed functions

$$At^k \quad (k > 0), \qquad Ct^k \sin \alpha t \quad (k > 0).$$

Classical perturbation theory for the description of large planets' orbital motion around the Sun was mainly developed by Euler, Clairaut, Lagrange, Laplace, Gauss, Le Verrier, and Newcomb.

Another line of investigation in the small-denominator problem is mathematical or theoretical; it lies in qualitative analysis of the solutions of differential equations of celestial bodies' motion, and in the construction of such series that are convergent either in an asymptotically large time interval or in an infinite time interval. From the analytical expressions for perturbations constructed by means of classical theory, it follows clearly enough that classical series are usable in a finite time interval outside of which they do not correspond to real planetary motion due to the presence of secular and mixed perturbations in the series. Therefore it seems natural to search for such mathematical methods that would allow us to obtain analytic expressions (above all, for the slow positional variables x) that do not contain terms proportional to t^k $(k > 0)$.

Laplace was the first to pay attention to the drawbacks of classical series and set the objective of finding solutions of planetary motion equations in the form of trigonometric series (if the inevitable secular term of the form ωt always present in expressions for planetary longitudes is not accounted for).

In the 19th century a number of remarkable investigations were accomplished (Delonet, Newcomb, Lindstedt, Gylden), making it possible to represent the solutions of planetary problems as formal trigonometric series. The bright mathematical idea of Poincaré on the applicability of asymptotic representation theory (proposed and developed by himself) to the problems of Hamiltonian dynamics allowed the author to describe in his famous work *New methods of celestial mechanics* from a unified standpoint the methods of the study of analytical dynamics problems that might seem different in essence and form. Thanks to the mathematical genius of Poincaré, not only resonance analytical dynamics reached an unsurpassed level, but in fact a scientific action program was formulated for the analytics of the 20th century.

One of the interesting new directions is the KAM (Kolmogorov–Arnold–Moser) theory allowing us to construct exact (in the sense of convergence) solutions of the Hamiltonian dynamics regular with respect to a small parameter, in spite of the negative impact of the small denominators on the conditions of convergence of an infinite chain of canonical transformations giving exact solutions in the limit. KAM theory made it possible to solve in an exact wording the problem of stability of a hypothetical planetary system (or rather its configuration), which should be recognized as a remarkable achievement of modern mathematics. Unfortunately, these results are so far inapplicable to our solar system because its dynamical parameters (planetary masses, above all) do not satisfy the estimations of KAM theory.

However it would be incorrect to contend that asymptotic methods in resonance analytical dynamics are in a state of completeness. This is especially noticeable in the construction of solutions of differential equations of analytical dynamics, i.e. in constructive equation theory. A modern researcher dreams of such a computer realization of the asymptotic or qualitative theory of differential equations that would allow us to use analytical operations and graphical tools fully. However it turns out that these issues are closely associated with the so-called problem of asymptotic theory bifurcation in the neighborhood of low–order frequency resonances, and with the problem of recalculation of the initial conditions at each step of the iterations. In other words, in resonance analytical dynamics it is possible to formulate some problems that seem to us most urgent:

Problem 1. Let a multifrequency system of differential equations be specified on a torus, as well as the corresponding initial conditions. Is it possible to construct a variant of asymptotic theory such that at each step of the transformation the iterations are minimized by changing the initial conditions? The answer is yes. Moreover, there exists an analytic algorithm allowing us to express the new initial conditions through the old ones, and vice versa.

This algorithm is easily implemented on computer.

Problem 2. In the construction of asymptotic theory by means of successive changes of variables it is inexpedient to specify beforehand the analytic structure of the equations at each step, but this should be determined from some conditions of each iteration norm minimization. We have developed an algorithm of obtaining such minimization conditions for typical problems of resonance analytical dynamics.

Algorithms to solve the above-mentioned problems have called forth the appearance of the concept of the bifurcation of forms of the analytical theory of perturbations of differential equations with a small parameter.

Problem 3. The development of constructive numerical–analytic methods of the construction of periodic and quasi-periodic solutions to problems of resonance analytical dynamics that use converging analytic algorithms and are implemented on computer, using symbolic programming packages.

These problems are actually the subject of this monograph. We are investigating the properties of the solutions of multifrequency regular systems of differential equations of analytical dynamics on the assumption of the presence of frequency resonances in the evolution process (with the change of t). Used for this purpose are the averaging principle, asymptotic representation in the sense of Poincaré, and converging iteration procedures of Lyapunov–Poincaré. A constructive asymptotic theory allows us to obtain in an explicit analytic form iterations of any order with respect to the small parameter, taking into account the above-mentioned minimization considerations.

In this monograph a number of nonlinear oscillations concerned with applications are considered.

Acknowledgments

The authors would like to express their sincere gratitude to the members of the Russian Academy of Sciences S.P. Novikov and D.V. Anosov, member of the National Academy of Sciences of Ukraine A.M. Samoilenko for their advice and comments made in the course of repeated discussions on the monograph's content. We would like to express our sincere gratitude to A.A. Martynyuk for the organization and scientific editing of the text of the book.

Fortunately the authors had the opportunity of discussing fundamental problems of mathematics and mechanics, and their mutual complementarities and dependences with academician L.I. Sedov, and we express our heartfelt gratitude to him.

We would also like to tender thanks to L. Chernetskaja, S. Kostina, S. Rasshivalova and N. Zemtsova for their help in preparing this English edition. Finally, we wish to acknowledge the cooperation of Taylor and Francis, and to thank them for their patience and understanding.

1 Preliminaries

1.0 Introduction

In this chapter the basic theorems of mathematical analysis are given that are necessary for the statement of the contents of the later chapters, as well as the main symbols, properties of series asymptotic in the sense of Poincaré's definition, and the main points of the classical Lyapunov–Poincaré technique for differential equations with small parameter. It seems important to give a full statement of the famous Poincaré theorem on the existence of asymptotic solutions of ordinary differential equations, since in the first place it is seldom found in modern mathematics books and in the second it needs some essential comments when studying multifrequency systems of ordinary differential equations. The matter mainly concerns the value of the time interval within which it is possible to use asymptotic approximations of exact solutions of differential equations, and also the theoretical and practical errors accompanying those approximations.

In the study of the so-called generalized equation of asymptotic perturbation theory (or Krylov–Bogolyubov generalized equation) the method of characteristics for first-order partial differential equations has proved to be very effective because it allows us to find exact solutions of equations of any approximation, and therefore at each step of the iterations we do not introduce an error into the asymptotic approximation of the solution. This assertion is correct at least for rotary multifrequency systems of differential equations with slow and fast variables. In other words, one can say that the matter concerns systems of differential equations set on many-dimensional tori, with their right-hand members expressed by Fourier divisible series, and their frequencies depending on slow state variables only. It is to such systems that the "resonance conditions" are peculiar, or the equivalent "problem of small denominators" that are the main obstacle during construction of their exact solutions by iteration methods.

These are differential equations of the form

$$\begin{aligned} \frac{dx}{dt} &= \mu X(x, y, \mu), \\ \frac{dy}{dt} &= \omega(x) + \mu Y(x, y, \mu), \end{aligned} \tag{1.0.1}$$

where x, X are m-dimensional vectors, y, Y, ω n-dimensional vectors of Euclidean space, and μ is a small positive parameter. We will also assume that the right-hand members of system (1.0.1) are differentiable a sufficient number of times with respect to x and y in some $(m+n)$-dimensional domain G_{m+n}, 2π-periodic with respect to y and regular with respect to μ at the point $\mu = 0$. The vector $\omega(x) = (\omega_1(x), \dots, \omega_n(x))$ is called *the basic frequencies vector.*

The mathematical theory of such systems, due to the high variety of the properties of solutions determined by the nonlinearity of the vector-functions $X(x, y, \mu)$, $Y(x, y, \mu)$, and due to the possible occurrence (in the process of the dynamical evolution of the system) of so-called small denominators, is not complete, though it is essentially advanced. Since Laplace's time the effect of small denominators has been understood as the appearance in formulae for the approximate solution $\tilde{x}(t, \mu)$, $\tilde{y}(t, \mu)$, of a small-denominators system — the scalar product $(k, \omega(x(t, \mu)))$, where k is an integral index vector, and $x(t, \mu)$ is the x-projection of the system solution. The smallness of the function (k, ω) means that for some vectors $k = (k_1, \dots, k_n)$ and some points of the space of solutions, the frequencies $\omega_1(x(t, \mu)), \dots, \omega_n(x(t, \mu))$ are rationally commensurable or almost commensurable, i.e.

$$(k, \omega(x(t_s^*, \mu))) \simeq 0, \quad t_s^* \in [0, T].$$

Obviously for the existence of resonances it is necessary that n should not be less than 2. Why then do small denominators appear?

Let us revert to the system of differential equations (1.0.1) and construct the classical first approximation by the formula

$$x^{(1)}(t, \mu) = x_0 + \mu \int_0^t X(x_0, y_0, \mu)\, d\tau$$

$$= x_0 + \mu a t + \mu \sum_{\|k\| \geq 1} \frac{X_k(x_0, \mu)}{i(k, \omega_0)} \exp\{i(k, \omega_0)t + i(k, y_0)\},$$

where

$$\|k\| = |k_1| + \cdots + |k_n|, \quad k_s = 0, \pm 1, \pm 2, \dots, \quad s = 1, \dots, n.$$

If the initial point (x_0, y_0) is such that there occurs an exact resonance at the initial time $(k, \omega_0) = 0$, then $a \neq 0$, and the function $x^{(1)}(t, \mu)$, besides having periodic summands, also contains the secular term $\mu a t$ proportional to t. If at the initial time $(k,\, \omega_0) \simeq 0$ (i.e. there is an almost exact resonance) for all vectors k from some subset of integers $I_{\rm res}$, then for exactly those vectors the amplitudes of periodic functions on $x^{(1)}$ may become arbitrarily large, and for all other summation vectors k periodic functions will have small amplitudes, since for them the values of $(k,\, \omega_0)$ are not small. Actually the situation is even more complicated, because amplitudes in the trigonometric functions $\exp\{i(k,\, \omega_0)t\}$ depend not only on the values of $(k,\, \omega_0)$, but also on the values of the Fourier coefficients $X_k(x_0, \mu)$ that, generally speaking, decrease with the growing norm $||k||$; therefore finally the amplitude $X_k/(k,\, \omega_0)$ depends both on $X_k(x_0, \mu)$ and on $(k,\, \omega_0)$. Such a complicated process of an increase or decrease of the amplitudes of harmonic functions as early as in the first approximation makes it difficult to study the behavior of the solutions $x(t, \mu)$, $y(t, \mu)$, not to mention higher-order approximations.

Finally, we include in Chapter 1 the iteration variant of the part of the theory of ordinary differential equations, known as the Poincaré–Lyapunov method. The proposed iterations can be easily algorithmized, and this is illustrated by some examples in Chapter 4. Note that most of the problems shown in this chapter are also described by multifunctional systems of differential equations with functions periodic with respect to fast phase variables y, and frequencies depend on slow variables x.

1.1 Main Symbols

1. Let the real Euclidean space of dimension n be denoted by R_n, and the unitary complex space by K_n. Norms of vector spaces R_n and K_n that we are going to use are as follows:

$$||x|| = \sqrt{\sum_{k=1}^{n} |x_k|^2}, \quad ||x|| = \sup_{1 \le k \le n} |x_n|, \quad ||x|| = \sum_{k=1}^{n} |x_k|.$$

2. The spatial domains R_n and K_n will be denoted by G_n or Q_n. The index shows the dimension of a domain or a space.

3. The direct (Cartesian) product of two fields will be denoted by $G_{m+n} = G_m \times G_n$ or $G_{m+n} = \{(x, y) : x \in G_m, y \in G_n\}$.

4. μ everywhere denotes a small nonnegative parameter.

5. The symbol $\rightarrow$ denotes the one-sided transformation (substitution), and the symbol $\leftrightarrow$ denotes two-sided (direct and inverse) substitution, e.g. $x \rightarrow \bar{x}$, $z \rightarrow \bar{z}$.

6. The norm of the n-dimensional integer vector $k = (k_1, ..., k_n)$ is calculated by the formula

$$||k|| = |k_1| + \cdots + |k_n|, \quad k_s = 0, \pm 1, \pm 2, \ldots, \quad s = 1, \ldots, n.$$

The norm of the imaginary part of the n-dimensional complex vector $z \in G_n \subset K_n$ is calculated by the formula

$$\|\operatorname{Im} z\| = |\operatorname{Im} z_1| + \cdots + |\operatorname{Im} z_n|.$$

7. (x, y) will denote as a rule the scalar product of two n-dimensional vectors, and therefore

$$(x, y) = \sum_{k=1}^{n} x_k y_k.$$

The same symbol will denote matrix products and those of matrix and vector.

8. If the n-dimensional vector-function $u = (u_1, \ldots, u_n)$ depends on the n-dimensional argument vector $z = (z_1, \ldots, z_n)$, then $\partial u / \partial z$ is a Jacobian matrix of order $n \times n$

$$\frac{\partial u}{\partial z} = \left(\frac{\partial u_i}{\partial z_k} \right), \quad i, k = 1, \ldots, n.$$

9. The Fourier series of the 2π-periodic function $Z(z)$ of the n-dimensional argument z will be written in complex form:

$$Z(z) = \sum_{\|k\| \geq 0} Z_k \exp\{i(k, z)\}, \tag{1.1.1}$$

$$Z_k = \frac{1}{(2\pi)^n} \int_0^{2\pi} \cdots \int_0^{2\pi} Z(z) \exp\{-i(k, z)\}\, dz_1 \ldots dz_n. \tag{1.1.2}$$

Assume that the function $Z(z)$ satisfies the conditions of Dirichlet's theorem on the expansibility of a periodic function into a Fourier series in some complex domain G_n such that $z \in G_n \subset K_n$ and $\|Im\ z\| \leq \rho$, where ρ is a real number. It will often be convenient to present the series (1.1.1) in the form

$$Z(z) = Z_N(z) + R_N Z(z), \tag{1.1.3}$$

$$Z_N(z) = \sum_{0 \leq \|k\| \leq N} Z_k \exp\{i(k, z)\}, \tag{1.1.4}$$

$$R_N Z(z) = \sum_{\|k\| \geq N+1} Z_k \exp\{i(k, z)\}. \tag{1.1.5}$$

The function $Z_N(z)$ is always a trigonometric polynomial, and $R_N Z(z)$ can be both an infinite trigonometric sequence and a finite polynomial.

10. Function averaging (smoothing) operators will be denoted by the symbol M with the subscript indicating those variables on which the averaging procedure is executed.

Let us introduce the following symbols denoting the averaging operators that are used most often:

$$M_t[Z(z,t,\mu)] = \bar{Z}(z,\mu) = \lim_{T\to\infty}\frac{1}{T}\int_0^T Z(z,t,\mu)\,dt, \tag{1.1.6}$$

$$M_Z[Z(z,\mu)] = \bar{Z}(\mu) = \frac{1}{(2\pi)^n}\int_0^{2\pi}\cdots\int_0^{2\pi} Z(z,\mu)\,dz_1\ldots dz_n. \tag{1.1.7}$$

Operator (1.1.7) is used if the function $Z(z,\mu)$ is 2π-periodic with respect to all components of the vector z in the range of definition, i.e.

$$Z(z+(2\pi),\,\mu) \equiv Z(z,\mu),$$

where $z+(2\pi)$ denotes the vector $(z_1+2\pi,\,\ldots\,,\,z_n+2\pi)$.

In addition, the procedure of averaging over some of the arguments $z_1,\ldots,z_s$ $(s<n)$ is applied:

$$\begin{aligned} M[Z(z,\mu)] &= \bar{Z}(z_{s+1},\ldots,z_n,\mu) \\ &= \frac{1}{(2\pi)^s}\int_0^{2\pi}\cdots\int_0^{2\pi} Z(z_1,\ldots,z_s,z_{s+1},\ldots,z_n)\,dz_1\ldots dz_s. \end{aligned} \tag{1.1.8}$$

11. A vinculum above literal symbols will denote the averaged value of a function or a variable, or a solution of averaged (smoothed) equations. $\overline{Z}$ is an averaged value of Z; if $z(t,\mu)$ is a solution of some equation, then $\bar{z}(t,\mu)$ is the solution of the respective averaged equation.

1.2 Asymptotic Series and their Properties

Consider an infinite power series of the form

$$\sum_{k=0}^{\infty}\mu^k z_k(t,\mu), \tag{1.2.1}$$

with each term defined in $G_2 = \{(t,\mu)\colon\, t\in[0,T],\ \ \mu\in[0,\mu^*]\}$.

Denote its partial sum by

$$S_n(t,\mu) = \sum_{k=0}^{n}\mu^k z_k(t,\mu). \tag{1.2.2}$$

If there exists a function $S(t,\mu)$ in G_2 such that for any $n>0$ there is a limit relationship

$$\lim_{\mu\to\infty}\frac{S(t,\mu)-S_n(t,\mu)}{\mu^n} = 0, \tag{1.2.3}$$

then we will say that series (1.2.1) is an asymptotic representation of the function $S(t,\mu)$ in G_2 (see Poincaré [2]). This is Poincaré's definition, and it has proved to be very efficient in nonlinear equation theory. Poincaré also proposed to use the following symbol of identical equality for the determination of (1.2.3):

$$S(t,\mu) \equiv \sum_{k=0}^{\infty} \mu^k z_k(t,\mu). \tag{1.2.4}$$

He called these relations asymptotic equalities. We will mostly call (1.2.1) an asymptotic series.

Of course, every convergent power series is also an asymptotic series, but the inverse proposition is obviously incorrect. From this definition follows a very important equality:

$$S(t,\mu) = \sum_{k=0}^{n} \mu^k z_k(t,\mu) + o(\mu^n). \tag{1.2.5}$$

Two asymptotic series can be added and multiplied, i.e. if

$$S_1(t,\mu) \equiv \sum_{k=0}^{\infty} \mu^k z_k^{(1)}(t,\mu), \tag{1.2.6}$$

$$S_2(t,\mu) \equiv \sum_{k=0}^{\infty} \mu^k z_k^{(2)}(t,\mu), \tag{1.2.7}$$

then

$$S(t,\mu) = S_1(t,\mu) + S_2(t,\mu) \equiv \sum_{k=0}^{\infty} \mu^k z_k(t,\mu), \tag{1.2.8}$$

$$\prod(t,\mu) = S_1(t,\mu) S_2(t,\mu) \equiv \sum_{k=0}^{\infty} \mu^k \prod_k(t,\mu), \tag{1.2.9}$$

where

$$z_k(t,\mu) = z_k^{(1)}(t,\mu) + z_k^{(2)}(t,\mu), \tag{1.2.10}$$

$$\prod_k(t,\mu) = \sum_{s=0}^{k} z_s^{(1)}(t,\mu) z_{k-s}^{(2)}(t,\mu). \tag{1.2.11}$$

If each term of the asymptotic series (1.2.1) is integrable with respect to t within $[t_1, t_2] \subset [0, T]$, then the following asymptotic equality is true:

$$\int_{t_1}^{t_2} S(t,\mu)\,dt \equiv \sum_{k=0}^{\infty} \mu^k \int_{t_1}^{t_2} z_k(t,\mu)\,dt, \tag{1.2.12}$$

i.e. asymptotic series can be integrated term by term, giving an asymptotic series as a result.

Inverse operations on asymptotic series (division and termwise differentiation) are also possible, but under some additional conditions.

If $z_0^{(2)}(t,\mu) \neq 0$ in G_2, then

$$\frac{S_1(t,\mu)}{S_2(t,\mu)} \equiv D(t,\mu) \equiv \sum_{k=0}^{\infty} \mu^k d_k(t,\mu), \tag{1.2.13}$$

where the coefficients $d_k(t,\mu)$ are defined in series from an infinite system of algebraic equations:

$$\begin{aligned}
d_0 z_0^{(2)} &= z_0^{(1)}, \\
d_1 z_0^{(2)} &= z_1^{(1)} - d_0 z_1^{(2)}, \\
d_2 z_0^{(2)} &= z_2^{(1)} - d_0 z_2^{(2)} - d_1 z_1^{(2)}, \\
&\ldots\ldots\ldots\ldots\ldots\ldots\ldots\ldots \\
d_k z_0^{(2)} &= z_k^{(1)} - d_0 z_k^{(2)} - d_1 z_{k-1}^{(2)} - \cdots - d_{k-1} z_1^{(2)}, \\
&\ldots\ldots\ldots\ldots\ldots\ldots\ldots\ldots
\end{aligned} \tag{1.2.14}$$

It has been shown by Poincaré [1] that generally speaking it is impossible to differentiate the asymptotic series (1.2.4) term by term, even if each term of (1.2.1) is a differentiable function. In other words, the series

$$\sum_{k=0}^{\infty} \mu^k \frac{dz_k(t,\mu)}{dt}$$

and the function dS/dt can be such that

$$\frac{dS(t,\mu)}{dt} \not\equiv \sum_{k=0}^{\infty} \mu^k \frac{dz_k(t,\mu)}{dt}. \tag{1.2.15}$$

At the same time Poincaré [1] showed that asymptotic series could be used for the construction of solutions of differential equations in the following sense. Let the function $z(t,\mu)$, differentiable with respect to t, be the solution of the equation

$$\frac{dz}{dt} = Z(z,t,\mu), \tag{1.2.16}$$

and let the series

$$D(t,\mu) = \sum_{k=0}^{\infty} \mu^k z_k(t,\mu) \tag{1.2.17}$$

formally satisfy equation (1.2.16) (whether it converges or diverges). Now write the asymptotic equality

$$\frac{dD(t,\mu)}{dt} \equiv \sum_{k=0}^{\infty} \mu^k \frac{dz_k(t,\mu)}{dt}. \tag{1.2.18}$$

We can assert that

$$\frac{dz(t,\mu)}{dt} \equiv \frac{dD(t,\mu)}{dt}. \tag{1.2.19}$$

Now formulate an asymptotic equality for a composite function. Let the function $F(z)$ be analytic (see Schwartz [1]) in some neighborhood of the point $z = 0$. Let there also exist the asymptotic equality

$$x(t,\mu) \equiv \sum_{k=0}^{\infty} \mu^k x_k(t,\mu). \tag{1.2.20}$$

Then the following asymptotic equality is true:

$$F(x(t,\mu)) \equiv F\left(\sum_{k=0}^{\infty} \mu^k x_k(t,\mu)\right). \tag{1.2.21}$$

The above properties and relations for asymptotic representations can be easily generalized for the case of multivariable functions when the argument $z = (z_1, \dots, z_n)$ has dimension n.

1.3 Poincaré's Theorem on Asymptotic Approximations of Solutions of Differential Equations

The basis for the applicability of asymptotic series in the theory of ordinary differential equations is the remarkable theorem proved by Poincaré in his famous work *New methods of celestial mechanics.* The theorem is stated below, together with a sufficiently detailed proof. It will henceforth allow us to explicitly write many estimations necessary in specific cases.

Theorem 1.3.1 *Let there be given the equation*

$$\frac{dz}{dt} = Z(z,t,\mu) \tag{1.3.1}$$

and the initial condition

$$z(0,\mu) = z_0. \tag{1.3.2}$$

In addition, let:

(1) *the vector-function $Z(z,t,\mu)$ be defined, continuous and bounded in the $(n+2)$-dimensional domain $G_{n+2} = G_{n+1} \times [0,T]$, analytic with respect to z, μ in the domain $G_{n+1} = z \in G_n,\ \mu \in [0,\mu^*]$;*

(2) *the terms of the series (1.2.17) be analytic with respect to $\mu \in [0,\mu^*]$ and satisfy the conditions*

$$z_k(0,\mu) = 0, \qquad k = 0,1,2,\dots \tag{1.3.3}$$

Then for any $\varepsilon > 0$ there exists $\mu_0 \in [0,\mu^*]$ such that for all $\mu \in [0,\mu_0]$ and $t \in [0, T_0 \le T]$ the following estimation is true

$$||z(t,\mu) - S_p(t,\mu)|| < \varepsilon, \tag{1.3.4}$$

where $z(t,\mu)$ is the exact solution of equation (1.3.1) with initial condition (1.3.2),

$$S_p(t,\mu) = \sum_{k=0}^{p} \mu^k z_k(t,\mu). \tag{1.3.5}$$

Proof Condition (1) ensures the existence and uniqueness of the solution of equation (1.3.1) within the interval $[0,\ T^* \leq T]$. In equation (1.3.1) perform the substitution of variables

$$z(t,\mu) = S_p(t,\mu) + \mu^{p+1} u(t,\mu). \tag{1.3.6}$$

where $u(t,\mu)$ is a new vector-function. Clearly this satisfies the differential equation

$$\mu^{p+1} \frac{du}{dt} = Z(S_p + \mu^{p+1} u,\ t,\ \mu) - \frac{dS_p}{dt}. \tag{1.3.7}$$

Define the vector-function $S_p(t,\mu)$ from the equation

$$\frac{dS_p}{dt} = \bar{Z}(S_p(t,\mu)), \tag{1.3.8}$$

which is obtained from the initial equation if the terms proportional to μ^{p+1}, $\mu^{p+2}, \ldots$ are truncated from its right-hand member.

Taking into account the generalized Cauchy theorem (see Poincaré [2]) on solving differential equations with an analytic right-hand member in the form of power series, for each component $\mu^{p+1} u_k$ of the vector $\mu^{p+1} u$ the following inequality can be written:

$$Z_k - \frac{dS_{p,k}}{dt} \prec \frac{M}{1 - a\mu \left(1 + \mu^p \sum_{s=1}^{m} u_s\right)}, \tag{1.3.9}$$

where a, M are some positive numbers. The symbol $\prec$ means termwise majorization. For $u = 0$ it can be written that $z = S_p$. On the other hand, series (1.2.17) formally satisfies equation (1.3.1), and so the vector difference $Z(S_p + \mu^{p+1} u,\ t,\ \mu) - dS_p/dt$ is divisible by μ^{p+1}. This implies that instead of the majorizing inequality (1.3.9) it is possible to write a "more detailed" inequality

$$Z_k - \frac{dS_{p,k}}{dt} \prec \frac{M \mu^{p+1} a \left(a^p + \sum_{s=1}^{n} u_s\right)}{1 - a\mu \left(1 + \mu^p \sum_{s=1}^{n} u_s\right)}. \tag{1.3.10}$$

Introducing the notation

$$F = \frac{Ma(a^p + f)}{1 - a\mu(1 + \mu^p f)}, \quad \mu^{p+1} R_k = Z_k - \frac{dS_{p,k}}{dt}, \quad f = \sum_{s=1}^{n} u_s$$

it becomes obvious that $R_k \prec F$ $(k = 1, \dots, n)$, and the differential equation (1.3.7) takes the form

$$\frac{du}{dt} = R, \quad R = (R_1, \dots, R_n). \tag{1.3.11}$$

Simultaneously with the vector equation (1.3.7) consider the scalar differential equation

$$\frac{df}{dt} = F(f, \mu) \equiv \frac{Ma(a^p + f)}{1 - a\mu(1 + a^p f)}. \tag{1.3.12}$$

with its general solution expressed by the formula

$$\frac{(a^p + f)^{1 - a\mu + a^{p+1}\mu^{p+1}}}{e^{a\mu^{p+1} f}} = Ce^{aMt}. \tag{1.3.13}$$

If it is assumed that $f(0, \mu) = 0$, then the partial solution of equation (1.3.12) satisfying this initial condition will be nonnegative for sufficiently small denominators μ, and it is defined by the equality

$$\frac{\left[a^{-p}(a^p + f)\right]^{1 - a\mu + a^{p+1}\mu^{p+1}}}{e^{a\mu^{p+1} f}} = e^{aMt}. \tag{1.3.14}$$

Study the behavior of the function $f(t, \mu)$. From (1.3.14) determine the inverse function $t = t(f, \mu)$:

$$aMt = (1 - a\mu + a^{p+1}\mu^{p+1}) \ln(a^p + f) - p(1 - a\mu + a^{p+1}\mu^{p+1}) \ln a - a\mu^{p+1} f. \tag{1.3.15}$$

Since we are interested in the behavior of the solutions of system (1.3.11) with $t \geq 0$, first find the interval within which the function $t = t(f, \mu)$ is monotone decreasing, i.e. the interval within which the derivative $t'_f \geq 0$. It is easy to calculate that if $0 \leq f \leq \bar{f}$, where

$$\bar{f} = \frac{1 - a\mu}{a\mu^{p+1}}, \tag{1.3.16}$$

then $0 \leq t \leq \bar{t}$,

$$\bar{t} = \frac{1}{aM}\left[\ln\left(\frac{1 - a\mu + a^{p+1}\mu^{p+1}}{a^{p+1}\mu^{p+1}}\right)^{1 - a\mu + a^{p+1}\mu^{p+1}} - (1 - a\mu)\right]. \tag{1.3.17}$$

The intervals $[0, \bar{f}]$, $[0, \bar{t}]$ are respectively the intervals of monotone increase of the functions $t = t(f, \mu)$, $f = f(t, \mu)$.

For each specific value of μ the intervals $[0, \bar{f}]$, $[0, \bar{t}]$ are bounded, and consequently the norm of the vector-function $u(t, \mu)$ is also bounded above. Moreover, the maximum interval $[0, t]$ and the norm estimation

$$\|z(t, \mu) - S_p(t, \mu)\| < na(1 - a\mu), \quad t \in [0, \bar{t}], \quad 0 \leq \mu \leq \inf\{a, \bar{\mu}\}. \tag{1.3.18}$$

are known for each value of μ.

From the expression (1.3.17) and inequality (1.3.18) it is clear that the smaller is μ, the bigger is $\bar{t}$, and the value of $na(1 - a\mu)$ tends to the finite value na. From this it follows that for $t \in [0, \bar{t}]$, the norm $\|z - S_p\|$ for any small values μ, generally speaking, may not be arbitrarily small, and therefore we should consider another, smaller interval of the variable t. From (1.3.16) and (1.3.17) it follows that if $\mu \to 0$, then $\bar{f} \to \infty$, $\bar{t} \to \infty$. Intervals of monotone increase of the functions $t = t(f, \mu)$, $f = f(t, \mu)$ become infinite, and consequently the function $f(t, \mu)$ as $\mu \to 0$ becomes arbitrarily large, though $\mu^{p+1} f(t, \mu) \to \bar{a} < a^{-1}$. For sufficiently small μ we have

$$\bar{t} = O(\ln \mu^{p+1}). \tag{1.3.19}$$

In other words, considering such an interval of time, with $\mu \to 0$ one cannot construct a majorant $f(t, \mu)$ by Poincaré's method, which could be effectively used in obtaining ε-estimation of the norm $\|z - S_p\|$. Moreover, it can be shown that for an arbitrary interval $[0, \tilde{t}(\mu)] \subset [0, \bar{t}]$ such that $\tilde{t}(\mu) \to \infty$ with $\mu \to 0$, the majorant $f(t, \mu)$ also increases indefinitely. This implies that in order to build a good bounded majorant it is necessary to consider a limited interval of time $[0, \bar{T}]$ on condition that

$$T(\mu) \to A_0 < \infty \qquad \text{as } \mu \to 0, \tag{1.3.20}$$

where A_0 is a finite positive number. With $t \in [0, T^*]$ and sufficiently small μ the majorant is bounded, and with $\mu \to 0$ it tends to the finite bound

$$\lim_{\mu \to 0} f(t, \mu) = a^p \left(e^{aMt} - 1\right) < a^p \left(e^{aM\bar{T}} - 1\right). \tag{1.3.21}$$

The differential equation (1.3.12) is a majorizing one for each of the equations of system (1.3.11) with the initial conditions $u_1(0, \mu) = \ldots = u_n(0, \mu) = f(0, \mu) = 0$. This implies that for every $k = 1, \ldots, n$ the following relations are true:

$$u_k(t, \mu) < f(t, \mu), \quad k = 1, \ldots, n, \quad t \in [0, T_0], \quad T_0 = \inf\{\bar{T}, T^*\}. \tag{1.3.22}$$

From the boundedness of the functions $u_k(t, \mu)$ by $t \in [0, T_0]$ follows the boundedness of the norm

$$\|u(t, \mu)\| < na^p \left(e^{aMT_0} - 1\right).$$

Now let an arbitrary $\varepsilon > 0$ be set. Choose $\mu_0(\varepsilon)$ from the condition

$$\mu_0^{p+1} na^p \left(e^{aMT_0} - 1\right) = \varepsilon. \tag{1.3.23}$$

Then for every $\mu \in [0, \mu_0(\varepsilon)]$ and $t \in [0, T_0]$ we obtain

$$\mu^{p+1} \|u(t, \mu)\| < \varepsilon,$$

and hence estimation (1.3.4) is true. The theorem is proved.

1.4 Geometric Interpretation of Solutions of Oscillating Systems

The behavior of solutions of multifrequency systems depends not only on whether frequency resonances appear or not, but also on the nature of the norm of the integral vector k for which the resonance relation $k_1 : k_2 = \omega_2 : \omega_1$ holds true in the case of a dual-frequency system. If there are more than two frequencies, then the following equality is called by definition the resonance relationship for frequencies $\omega_1, \ldots, \omega_n$:

$$(k, \omega) = 0, \tag{1.4.1}$$

where $k = (k_1, \ldots, k_n)$ is an n-dimensional integer vector.

Equality (1.4.1) expresses the exact resonance equality, or 0-resonance. But also in real situations, more probable are the cases when instead of (1.4.1) the following approximate equality is true

$$(k^*, \omega) = \alpha, \quad k^* = (k_1^*, \ldots, k_n^*), \tag{1.4.2}$$

where $|\alpha|$ is some sufficiently small quantity. If α in this problem is considered a small quantity, then the (k^*, ω) will be called the α-resonance.

The integer

$$||k^*|| = \sum_{s=1}^{n} |k_s^*| \tag{1.4.3}$$

is called the order of the α-resonance.

Let there exist α-resonances with different integer vectors $k^{(1)}$ and $k^{(2)}$. The α-resonance $(k^{(1)}, \omega)$ is called a lower-order resonance compared with the α-resonance $(k^{(2)}, \omega)$, if $||k^{(1)}|| < ||k^{(2)}||$. And conversely, α resonance $(k^{(2)}, \omega)$ is called a higher-order resonance compared with $(k^{(1)}, \omega)$. If $||k^{(1)}|| = ||k^{(2)}||$, then both α-resonances are of the same order.

There always exists a resonance of least order, with its integer vector k^* defined from the condition

$$||k^*|| = \inf_{s \geq 1} ||k^{(s)}||. \tag{1.4.4}$$

The interval $[-\alpha, \alpha]$ will be called the α-resonance zone.

In the asymptotic theory of differential equations with periodic functions the lowest (with small $||k||$) small (with very small α) α-resonances are of great importance, because the growth of amplitudes of individual harmonics depends both on the α value and on the norm of the vector $||k||$. We give two examples to explain this thesis.

Example 1.4.1 In an n-dimensional domain of G_n Euclidean space let there be given a p times differentiable function $f(y)$ depending on n arguments $y_1, \ldots, y_n$.

In addition, let it be 2π-periodic in y. Then it can be presented by an n-fold Fourier series

$$f(y) = \sum_{\|k\|\geq 0} f_k e^{i(k,y)} \tag{1.4.5}$$

with bounded coefficients

$$|f_k| \leq \frac{V_p}{\pi^k \|k\|^p}, \tag{1.4.6}$$

where V_p is an overall variation $\dfrac{\partial^p f}{\partial y_1^{k_1} \cdots \partial y_n^{k_n}}$ in the parallelepiped of periods. Now assume that

$$y = \omega t, \tag{1.4.7}$$

where $\omega = (\omega_1, \dots, \omega_n)$ is a numeric vector.

After replacement of the argument y by the function ωt in Fourier series (1.4.5) we get either a periodic function t (if the frequencies $\omega_1, \dots, \omega_n$ are rationally commensurable, then there is exact resonance with some integer vectors $k^{(1)}, k^{(2)}, \dots$), or a conditionally periodic function (in the case of rational incommensurability of the frequencies). Integrating with respect to t, we get

$$\int_0^t f(\omega\tau)\, d\tau = a_0 + a_1 t + \sum_{\|k\|\geq 1}'' \frac{f_k e^{i(k,\omega)t}}{i(k,\omega)}, \tag{1.4.8}$$

where the primes on the sum indicate that the summation is performed only with respect to those indices k for which $(k,\omega) \neq 0$, i.e. the index vector k only takes on "nonresonance" values. It is easy to calculate that

$$a_1 = f_0 + A', \tag{1.4.9}$$

where f_0 is an absolute term of Fourier series (1.4.5), and A' only appears when the frequencies $\omega_1, \dots, \omega_n$ are in resonance. In other words,

$$A' = \sum_{\|k\|\geq 1}' f_k, \tag{1.4.10}$$

and in this sum, the index vector k only takes on those "resonance" values for which $(k,\omega) = 0$.

Thus, the resonance terms in the Fourier series lead to the appearance of additional secular terms in the integral (1.4.8) in addition to the typical secular term $f_0 t$ which is always present (except when the average value of $f(y)$ in the period equals zero).

Equality (1.4.8) also shows that the amplitude of the harmonic $\exp\{i(k,y)\}$ equal to $|f_k|/|(k,\omega)|$ depends both on the value of the coefficient f_k and on that of the denominator (k,ω). With the same value of (k,ω) the higher amplitude will be that for which $|f_k|$ is higher, i.e. the greatest influence is exerted by the initial Fourier coefficients, because the larger the norm $||k||$, the smaller is f_k. In particular this is clear from estimation (1.4.6).

This implies that there can be very sharp resonances $(k,\omega) \approx 0$, but due to the large value of $||k||$ their influence is imperceptible, because it is suppressed by the number f_k. On the other hand, a not very sharp resonance with a relatively small value of $||k||$ can provoke strong growth of amplitudes. We are dealing with quite a complicated mechanism of simultaneous growth of harmonic oscillation amplitudes due to the possible appearance of small denominators of the form (k,ω) in representations of functions, and their damping with the increased number of the harmonic k. It is this mechanism that is always seen in the construction of the asymptotic theory of differential equations describing multifrequency oscillatory processes.

Example 1.4.2 Let there be given a Hamiltonian system

$$\begin{aligned} \frac{dx}{dt} &= -\frac{\partial H}{\partial y}, \\ \frac{dy}{dt} &= \frac{\partial H}{\partial x}, \end{aligned} \tag{1.4.11}$$

where x is an n-dimensional pulse vector, y an n-dimensional vector of generalized Lagrange coordinates, and the Hamiltonian has the form

$$H(x,y,\mu) = H_0(x,y) + \mu H_1(x,y,\mu). \tag{1.4.12}$$

Let the undisturbed Hamiltonian $H_0(x,y)$ be calculated by the formula

$$\begin{aligned} H_0(x,y) = h_0 &+ \sum_{k=1}^{n} \omega_k x_k + a_{11}x_1^2 + a_{12}x_1x_2 + a_{22}x_2^2 \\ &+ \frac{1}{2}\sum_{s=3}^{n}\sum_{j=3}^{n} h_{sj}(y_3, y_4, \dots, y_n)x_j x_s, \end{aligned} \tag{1.4.13}$$

where h_0, ω_k, a_{js} are some positive numbers and the functions h_{sj} do not depend on y_1 and y_2.

The undisturbed Hamiltonian system obviously has the form

$$\begin{aligned} \frac{dx_1^{(0)}}{dt} &= 0, \\ \frac{dx_2^{(0)}}{dt} &= 0, \\ \frac{dx_k^{(0)}}{dt} &= -\frac{\partial H_0}{\partial y_k^{(0)}} = -\frac{1}{2}\sum_{j=3}^{n}\sum_{s=3}^{n} \frac{\partial h_{js}}{\partial y_k^{(0)}} x_j^{(0)} x_s^{(0)}, \end{aligned}$$

$$\frac{dy_1^{(0)}}{dt} = \frac{\partial H_0}{\partial x_1^{(0)}} = \omega_1 + 2a_{11}x_1^{(0)} + a_{12}x_2^{(0)},$$
$$\frac{dy_2^{(0)}}{dt} = \frac{\partial H_0}{\partial x_2^{(0)}} = \omega_2 + a_{12}x_1^{(0)} + 2a_{22}x_2^{(0)},$$
$$\frac{dy_k^{(0)}}{dt} = \frac{\partial H_0}{\partial x_k^{(0)}} = \omega_k + \frac{1}{2}\sum_{j=3}^{n} h_{kj}x_j^{(0)}, \qquad k = 3, \dots, n.$$

It admits the partial solution

$$\begin{aligned} x_1^{(0)} &= R_1, \quad x_2^{(0)} = R_2, \quad x_k^{(0)} = 0, \\ y_1^{(0)} &= (\omega_1 + 2a_{11}R_1 + a_{12}R_2)t + y_{10}^{(0)}, \\ y_2^{(0)} &= (\omega_2 + a_{12}R_1 + 2a_{22}R_2)t + y_{20}^{(0)}, \\ y_k^{(0)} &= \omega_k t + y_{k0}^{(0)}, \quad k = 3, \dots, n. \end{aligned} \tag{1.4.14}$$

From (1.4.14) it is clear that in the undisturbed motion all pulses are continuous, and the Lagrangian coordinates are the linear functions of time.

Now consider the initial system (1.4.11) with disturbance of the Hamiltonian

$$\mu H_1 = -\mu \cos y_1.$$

The explicit form of a system with Hamiltonian $H = H_0 - \mu \cos y_1$ is written as

$$\begin{aligned} \frac{dx_1}{dt} &= -\mu \sin y_1, \\ \frac{dx_2}{dt} &= 0, \\ \frac{dx_k}{dt} &= -\frac{1}{2}\sum_{j=3}^{n}\sum_{s=3}^{n} \frac{\partial h_{js}}{\partial y_k} x_j x_s, \\ \frac{dy_1}{dt} &= \omega_1 + 2a_{11}x_1 + a_{12}x_2, \\ \frac{dy_2}{dt} &= \omega_2 + a_{12}x_1 + 2a_{22}x_2, \\ \frac{dy_k}{dt} &= \omega_k + \frac{1}{2}\sum_{j=3}^{n} h_{jk}x_j, \qquad k = 3, \dots, n. \end{aligned} \tag{1.4.15}$$

To equations (1.4.15) add the initial conditions

$$\begin{aligned} &x_1(0) = R_1, \quad x_2(0) = R_2, \quad x_k(0) = 0, \\ &y_s(0) = 0, \quad k = 3, \dots, n, \quad s = 1, \dots, n. \end{aligned} \tag{1.4.16}$$

System (1.4.15) with initial conditions (1.4.16) can be exactly integrated, and its partial solution is expressed by the equalities

$$\begin{aligned}
x_1(t,\mu) &= R_1 - \frac{\omega_1 + 2a_{11}R_1 + a_{12}R_2}{2a_{11}}(1 - \operatorname{dn}\varphi), \\
x_2(t,\mu) &= R_2 - \frac{\omega_2 + a_{12}R_1 + 2a_{22}R_2}{2a_{22}}(1 - \operatorname{dn}\psi), \quad x_k(t,\mu) = 0, \\
y_1(t,\mu) &= 2\operatorname{am}\varphi, \quad y_2(t,\mu) = 2\operatorname{am}\psi, \quad y_k(t,\mu) = \omega_k t, \quad k = 3,\ldots,n, \\
\varphi(t) &= \frac{1}{2}(\omega_1 + 2a_{11}R_1 + a_{12}R_2)t, \\
\psi(t) &= \frac{1}{2}(\omega_2 + a_{12}R_1 + 2a_{22}R_2)t.
\end{aligned} \tag{1.4.17}$$

Formulae (1.4.17) include elliptical Jacobian functions (see Whittaker and Watson [1]) $\operatorname{am}\varphi$, $\operatorname{am}\psi$, $\operatorname{dn}\varphi$, $\operatorname{dn}\psi$ with their moduli equal to

$$\kappa_1 = \frac{2\sqrt{2a_{11}\mu}}{\omega_1 + 2a_{11}R_1 + a_{12}R_2}, \qquad \kappa_2 = \frac{2\sqrt{2a_{22}\mu}}{\omega_2 + a_{12}R_1 + 2a_{22}R_2}.$$

It is known that the functions $\operatorname{dn}\varphi$ and $\operatorname{dn}\psi$ are periodic, therefore $x_1(t,\mu)$ and $x_2(t,\mu)$ are periodic as well. As for the amplitude functions $\operatorname{am}\varphi$ and $\operatorname{am}\psi$, they are semi-linear functions of time. Indeed, if we use the trigonometric expansion (see Duboshin [1])

$$\operatorname{am}\varphi = \frac{\pi\varphi}{2K(\kappa_1)} + 2\sum_{n=1}^{\infty}\frac{1}{n}\frac{q^n}{1+q^{2n}}\sin\frac{n\pi\varphi}{K(\kappa_1)},$$

where $K(\kappa_1)$ is a complete elliptic integral of the first kind,

$$q = \exp\left\{-\frac{\pi K\left(\sqrt{1-\kappa_1^2}\right)}{K(\kappa_1)}\right\},$$

then it is clear that with $\mu \to 0$ the modulus of the elliptic integral $\kappa_1 \to 0$, and under these conditions $K(\kappa_1) \to \pi/2$ and $K\left(\sqrt{1-\kappa_1^2}\right) \to \infty$. So $q \to 0$. Thus with $\mu \to 0$

$$\operatorname{am}\varphi \to \varphi = \frac{1}{2a_{11}}(\omega_1 + 2a_{11}R_1 + a_{12}R_2)t,$$

or

$$\operatorname{am}\varphi = \varphi + O(\mu).$$

For simplicity assume that the Hamiltonian system describes the dynamics of a mechanical or physical system with two degrees of freedom. Then in the solution of (1.4.17) one should truncate the expressions for $x_k(t,\mu)$ and $y_k(t,\mu)$

$(k = 3, \dots, n)$, and obtain

$$
\begin{aligned}
x_1(t,\mu) &= R_1 - \frac{\omega_1 + 2a_{11}R_1 + a_{12}R_2}{2a_{11}}(1 - \operatorname{dn}\varphi),\\
x_2(t,\mu) &= R_2 - \frac{\omega_2 + a_{12}R_1 + 2a_{22}R_2}{2a_{22}}(1 - \operatorname{dn}\psi),\\
y_1(t,\mu) &= 2\operatorname{am}\varphi, \quad y_2(t,\mu) = 2\operatorname{am}\psi.
\end{aligned}
\tag{1.4.18}
$$

Compare this solution with the corresponding undisturbed solution obtained from (1.4.14) by rejecting the formulae for $x_k^{(0)}$ and $y_k^{(0)}$ $(k = 3, \dots, n)$:

$$
\begin{aligned}
x_1^{(0)}(t) &= R_1, \quad x_2^{(0)}(t) = R_2,\\
y_1^{(0)}(t) &= (\omega_1 + 2a_{11}R_1 + a_{12}R_2)t,\\
y_2^{(0)}(t) &= (\omega_2 + a_{12}R_1 + 2a_{22}R_2)t.
\end{aligned}
$$

The disturbed pulses $x_1(t,\mu)$ and $x_2(t,\mu)$ change periodically around their undisturbed values R_1 and R_2. The generalized coordinates $y_1(t,\mu)$ and $y_2(t,\mu)$ with time may differ greatly from the undisturbed functions $y_1^{(0)}(t)$ and $y_2^{(0)}(t)$, since up to μ

$$
\begin{aligned}
y_1(t,\mu) - y_1^{(0)}(t) &= (\omega_1 + 2a_{11}R_1 + a_{12}R_2)t\left[1 - \frac{2a_{11}\mu}{(\omega_1 + 2a_{11}R_1 + a_{12}R_2)^2}\right],\\
y_2(t,\mu) - y_2^{(0)}(t) &= (\omega_2 + a_{12}R_1 + 2a_{22}R_2)t\left[1 - \frac{2a_{22}\mu}{(\omega_2 + a_{12}R_1 + 2a_{22}R_2)^2}\right],
\end{aligned}
$$

and with $t = 1/\mu$ these differences reach a finite value. But besides the secular motion, the difference $y_s(t,\mu) - y_s^{(0)}(t)$ also contains periodic functions of time. Therefore the behavior of the functions $y_s(t,\mu) - y_s^{(0)}(t)$, despite their complexity, suggests that the disturbed coordinates $y_1(t,\mu)$ and $y_2(t,\mu)$ may go arbitrarily far from undisturbed quantities.

The above analysis can be given a clear geometric interpretation (cf. Arnol'd [1], and Poincaré [2]).

Represent a two-dimensional torus by circles of radii $R_1 > 0$ and $R_2 > 0$ (Fig. 1.4.1). In topological terms, a two-dimensional torus (or, more exactly, a toroidal surface or a "hollow bagel") is a direct product of two circles C_1 and C_2 with radii R_1 and R_2. The position of any point P on the surface of the torus is defined by two angular coordinates (longitude and latitude) of the point. We will interpret the variables $y_1^{(0)}(t)$ and $y_2^{(0)}(t)$ as the angular coordinates of the point P on the surface of the torus (or on the torus, for short). Then the initial values of the pulses R_1 and R_2 (if they are positive) define the dimensions of the torus, and the generalized Lagrangian coordinates $y_1^{(0)}(t)$ and $y_2^{(0)}(t)$ describe the trajectory of the point on the torus. The path defined by the two angular coordinates $y_1^{(0)}(t)$

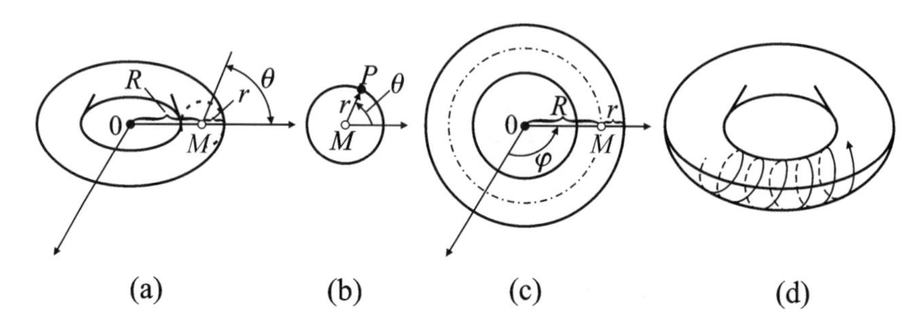

FIG. 1.4.1. Two-dimensional torus (a); cross-section of the torus (b); horizontal (longitudinal) section of the torus (c); winding of a two-dimensional torus (d).

and $y_2^{(0)}(t)$ winds round the torus; depending on the arithmetic properties of the numbers $\omega_1^* = \omega_1 + 2a_{11}R_1 + a_{12}R_2$ and $\omega_2^* = \omega_2 + a_{12}R_1 + 2a_{22}R_2$ which can be considered as some new frequencies, the torus winding may be solid everywhere (if the frequencies ω_1^* and ω_2^* are rationally incommensurable, i.e. if there is no resonance) or will close after a certain number of convolutions by longitude and latitude (if the frequencies are rationally commensurable) (Fig. 1.4.1).

Thus, depending on the arithmetic properties of a frequency vector with components ω_1^* and ω_2^*, the path on the torus can be represented by a periodic function (in the case of resonance) or a conditionally periodic function with two frequencies (in the case of no resonance).

These geometric considerations are true if once the frequencies ω_1^* and ω_2^* are calculated they henceforth remain constant, as for the undisturbed problem. If one considers the disturbed problem (1.4.15) with $n = 2$ and consequently tries to interpret the solution (1.4.18) geometrically, the periodic functions $x_1(t, \mu)$ and $x_2(t, \mu)$ suggest, as it were, pulsation of the torus, and this situation arises when the periods of the functions x_1 and x_2 are rationally commensurable. Then the pulsation is a periodic function, and therefore the initial torus will be repeated in certain time intervals. On the pulsating torus a point will move in accordance with formulae (1.4.18). It should also be noted that the functions x_1 and x_2 are slowly changing (with small values of μ) time functions, therefore the torus pulsation goes on very slowly, and with $\mu \to 0$, $x_1(t, \mu) \to R_1$, $x_2(t, \mu) \to R_2$.

In the case of $n > 2$ degrees of freedom, the paths of Hamiltonian systems can be interpreted in a similar way. It is possible to introduce an n-dimensional tore as a direct product of n circumferences $C_1, C_2, \dots, C_n$ with radii $R_1, R_2, \dots, R_n$, and on the surface of the n-dimensional torus n geographical coordinates $y_1, \dots, y_n$ can be introduced—in the undisturbed version (with $\mu = 0$) of the linear functions of time. If the frequencies $\omega_1^*, \dots, \omega_n^*$ are rationally commensurable (i.e. $(k^*, \omega^*) = 0$), then on the surface of the n-dimensional torus the trajectory of point P will close, and in this case we will obtain a periodic orbit, and if the frequencies are rationally incommensurable with all integer vectors k, we will obtain the everywhere solid winding of the n-dimensional torus. Similar considerations can also be developed for a disturbed Hamiltonian system with n degrees of freedom.

1.5 On the Method of Characteristics for Quasi-Linear First-Order Partial Differential Equations: Method of Characteristics

Since hereinafter we will repeatedly need the Krylov–Bogolyubov equation, we will demonstrate the analytic aspects of the method of characteristics.

The generalized n-dimensional vector equation of Krylov–Bogolyubov considered as the basis of the asymptotic theory of equations using the averaging principle, has the form

$$\frac{\partial u}{\partial t} + \left(\frac{\partial u}{\partial \bar{z}},\ \bar{Z}(\bar{z}, t, \mu)\right) = Z(\bar{z} + u,\ t, \mu) - \bar{Z}(\bar{z}, t, \mu), \tag{1.5.1}$$

where t is a scalar variable, Z, $\bar{Z}$, $\bar{z}$, u are n-dimensional vectors, u is the sought vector-function, and $\bar{z}$ and t are its arguments. The exterior brackets in the left-hand member of equation (1.5.1) stand for the product of the Jacobian matrix of dimensions $n \times n$ and the n-dimensional vector $\bar{Z}$.

Such an equation is called a quasi-linear first-order partial differential equation. The structure of this equation includes the partial derivatives $\partial u/\partial \bar{z}$ in the linear way, but its right-hand member can depend on the function u itself in a nonlinear way.

In conformity with the method of characteristics (see Stepanov [1]), construct a system of ordinary differential equations for characteristics, which can be symbolically written in the form

$$\frac{d\bar{z}}{\bar{Z}} = \frac{du}{Z - \bar{Z}}.$$

This symbolic equality actually stands for the following system of differential equations:

$$\frac{d\bar{z}_1}{\bar{Z}_1} = \cdots = \frac{\bar{z}_n}{\bar{Z}_n} = \frac{du_1}{Z_1 - \bar{Z}_1} = \cdots = \frac{du_n}{Z_n - \bar{Z}_n} = dt, \tag{1.5.2}$$

where $\bar{Z}_1, \ldots, \bar{Z}_n$ are components of the vector $\bar{Z}$; components of vectors $Z - \bar{Z}$, $\bar{z}$, u are denoted in a similar way.

The system (1.5.2) can be written in the so-called normal form

$$
\begin{aligned}
&\frac{d\bar{z}_1}{dt} = \bar{Z}_1(\bar{z}_1, \dots, \bar{z}_n, t, \mu),\\
&\dots\dots\dots\dots\dots\\
&\frac{d\bar{z}_n}{dt} = \bar{Z}_n(\bar{z}_1, \dots, \bar{z}_n, t, \mu),\\
&\frac{du_1}{dt} = Z_1(\bar{z}_1 + u_1, \dots, \bar{z}_n + u_n, t, \mu) - \bar{Z}_1(\bar{z}_1, \dots, \bar{z}_n, t, \mu),\\
&\dots\dots\dots\dots\dots\dots\dots\dots\dots\dots\\
&\frac{du_n}{dt} = Z_n(\bar{z}_1 + u_1, \dots, \bar{z}_n + u_n, t, \mu) - \bar{Z}_n(\bar{z}_1, \dots, \bar{z}_n, t, \mu).
\end{aligned}
\tag{1.5.3}
$$

This has order $2n$, and it is important to note that its subsystem consisting of the first n equations does not contain the functions $u_1, \dots, u_n$, therefore it can be integrated irrespective of the second subsystem. The validity of the applicability of the method of characteristics to quasi-linear first-order partial differential equations is based on the following remarkable assertion.

Let the general integral of system (1.5.3) be known:

$$
\begin{aligned}
&\Psi_1(\bar{z}_1, \dots, \bar{z}_n, u_1, \dots, u_n, t, \mu) = C_1,\\
&\dots\dots\dots\dots\dots\dots\dots\\
&\Psi_{2n}(\bar{z}_1, \dots, \bar{z}_n, u_1, \dots, u_n, t, \mu) = C_{2n}.
\end{aligned}
$$

Then the arbitrary differential function V with its arguments being the functions $\Psi_1, \dots, \Psi_{2n}$ is the solution of the quasi-linear first-order partial differential equation (1.5.1).

Certainly, finding the general integral of a system of ordinary differential equations in normal form is not a simple task, but fortunately for multifrequency systems of differential equations with slow and fast variables, with their right-hand members being multiple Fourier series, such a problem proves to be solvable. As will be shown later, it allows us to advance the asymptotic theory of systems of ordinary differential equations regular with respect to a small parameter arbitrarily far, i.e. it is possible to basically define asymptotic approximations with an arbitrarily large number in the form of explicit functions.

1.6 Iterative Variant of the Poincaré–Lyapunov Small Parameter Method

This method was found for the first time by Lichtenstein [1]. It was developed in more detail by Malkin [1] and applied to different problems of nonlinear oscillations. This section is based mainly on the monograph by Grebenikov and Ryabov [1]. Iteration algorithms are the basis for the numerical–analytic construction of periodic and conditionally periodic solutions of nonlinear systems, which is the subject of Chapter 4.

Let there be given a system of the form

$$\frac{dz}{dt} = P(t)z + \varphi(t) + \mu Z(z,t), \tag{1.6.1}$$

where t is a dimensionless time (or an angular variable), μ a small (positive) parameter, $P(t)$ an $(n \times n)$-matrix, $\varphi(t)$ and $Z(z,t)$ n-dimensional vectors, $P(t)$, $\varphi(t)$, $Z(z,t)$ functions continuously differentiable with respect to their arguments and 2π-periodic with respect to t.

Let the system (1.6.1) for $\mu = 0$ have the 2π-periodic solution $z^0(t)$. Assume $z = z^0(t) + x$ and obtain for x the system

$$\frac{dx}{dt} = P(t)x + \mu Z(z^0(t) + x,\, t), \tag{1.6.2}$$

where the function $Z(z^0(t)+x,\, t)$ in its properties is similar to the function $Z(z,t)$.

1.6.1 Simple iterations

a) The nonresonance (noncritical) case is characterized by the fact that the linear homogeneous system

$$\frac{dx}{dt} = P(t)x \tag{1.6.3}$$

does not have 2π-periodic solutions, except the zero solution. Then the 2π-periodic solution of system (1.6.2) can be found by means of successive approximations

$$x_1(t,\mu),\ \ x_2(t,\mu),\ \ x_3(t,\mu),\ \dots, \tag{1.6.4}$$

satisfying the equations

$$\frac{dx_k}{dt} = P(t)x_k + \mu Z(z^0(t) + x_{k-1},\, t), \quad k = 1,2,\dots, \quad x_0 = 0. \tag{1.6.5}$$

Convergence of the sequence (1.6.4) to the 2π-periodic solution $x(t,\mu)$ of system (1.6.2) occurs at least if μ does not exceed some limit.

b) The resonance (critical) case is characterized by the fact that the homogeneous system (1.6.3) has nonzero 2π-periodic solutions. Then the initial system (1.6.1) has a family of 2π-periodic solutions $z^0(t,c)$, which depends on an arbitrary constant (or an arbitrary constant vector) c if the function $\varphi(t)$ satisfies a condition of the form

$$\int_0^{2\pi} H(t)\varphi(t)\, dt = 0, \tag{1.6.6}$$

where $H(t)$ is some nonsingular 2π-periodic matrix. Then the 2π-periodic solution $x(t,\mu)$ of system (1.6.2), where $z^0(t)$ should be replaced by $z^0(t,c)$, as well as the constant (or the constant vector) c, are found by means of successive approximations

$$x_1(t,\mu),\ x_2(t,\mu),\ \dots, \qquad c_0,\ c_1(\mu),\ c_2(\mu),\ \dots. \tag{1.6.7}$$

These approximations are determined by the following systems of differential and algebraic (transcendent) equations:

$$F_0(c_0) \equiv \int_0^{2\pi} H(t)Z(z^0(t,c_0),t)\,dt = 0, \tag{1.6.8}$$

$$\frac{dx_1}{dt} = P(t)x_1 + \mu Z(z^0(t,c_0),t), \tag{1.6.9}$$

$$F_1(c_1) \equiv \int_0^{2\pi} H(t)Z(z^0(t,c_1) + x_1(t,\mu),\ t)\,dt = 0, \tag{1.6.10}$$

$$\frac{dx_2}{dt} = P(t)x_2 + \mu Z(z^0(t,c_1) + x_1(t,\mu),\ t), \tag{1.6.11}$$

etc.

The simplest situation (first-order critical case) occurs when the system of algebraic equations (1.6.8) has the so-called simple solution $\bar{c}_0$ (or several such solutions), i.e. such that the determinant

$$\det\left(\frac{\partial F_0(c_0)}{\partial c_0}\right)\Bigg|_{c_0=\bar{c}_0} \tag{1.6.12}$$

is nonzero. Then to every such solution $\bar{c}_0$, at least if μ does not exceed a certain limit, there corresponds the singular constant $c_*(\mu)$ (or a singular vector $c_*(\mu)$) and the singular 2π-periodic solution $x_*(t,\mu)$ of system (1.6.2). The series (1.6.7) converges to this solution $x_*(t,\mu)$ and to the constant $\bar{c}_0$ (or to the vector $\bar{c}_0$). When $\mu \to 0$ the solution $x_*(t,\mu)$ approaches zero.

But if the determinant (1.6.12) is zero or equation (1.6.8) is satisfied identically, then different critical cases of second and higher order occur. The existence of 2π-periodic solutions and the construction of successive approximations converging to those solutions are ensured by supplementary conditions (see Grebenikov and Ryabov [4]).

c) Autonomous systems Let the following autonomous system be considered instead of (1.6.1):

$$\frac{dz}{dt} = Az + \mu Z(z), \tag{1.6.13}$$

where A is a constant matrix, $Z(z)$ a vector-function continuously differentiable in a sufficiently large domain D_z, and the matrix A is such that the homogeneous system

$$\frac{dz}{dt} = Az \tag{1.6.14}$$

has a family of 2π-periodic solutions $z^0(t, c)$ depending on the arbitrary constant c (or the constant vector c). Since the period of the sought periodic solution of system (1.6.13) with $\mu \neq 0$ in the general case differs from 2π, the substitution $t = (1 + \mu h)\theta$ is introduced, so system (1.6.13) is rewritten in the form

$$\frac{dz}{d\theta} = (1 + \mu h)Az + \mu(1 + \mu h)Z(z). \tag{1.6.15}$$

With $\mu = 0$ we have a family of 2π-periodic (with respect to θ) solutions $z^0(\theta, c)$. Upon the substitution $z = z^0(\theta, c) + x$, we obtain for x the system

$$\frac{dx}{d\theta} = Ax + \mu hA(z^0(\theta, c) + x) + \mu(1 + \mu h)Z(z^0(\theta, c) + x). \tag{1.6.16}$$

The successive approximations

$$x_1(\theta, \mu),\ x_2(\theta, \mu),\ \ldots,\ h_0,\ h_1(\mu),\ h_2(\mu),\ \ldots,\ c_0,\ c_1(\mu),\ c_2(\mu),\ \ldots \tag{1.6.17}$$

are constructed and these approximations are determined from the following differential and algebraic (transcendent) equations:

$$F_0(c_0, h_0) \equiv \int_0^{2\pi} H(\theta)\left[h_0 Az^0(\theta, c_0) + Z(z^0(\theta, c_0)\right] d\theta = 0, \tag{1.6.18}$$

$$\frac{dx_1}{d\theta} = Ax_1 + \mu h_0 Az^0(\theta, c_0) + \mu Z(z^0(\theta, c_0)), \tag{1.6.19}$$

$$\begin{aligned} F_1(c_1, h_1) \equiv \int_0^{2\pi} H(\theta) & \left[h_1 A(z^0(\theta, c_1) + x_1)\right. \\ & \left. + \mu(1 + \mu h_1)Z(z^0(\theta, c_1) + x_1)\right] d\theta = 0, \end{aligned} \tag{1.6.20}$$

$$\frac{dx_2}{d\theta} = Ax_2 + \mu h_1 A(z^0(\theta, c_1) + x_1) + \mu(1 + \mu h_1)Z(z^0(\theta, c_1) + x_1), \tag{1.6.21}$$

etc.

The simplest situation (the first-order critical case) occurs when the algebraic (transcendent) system (1.6.8) with respect to h_0, c_0 has the simple solution $\bar{c}_0, \bar{h}_0$, such that the determinant

$$\det\left(\frac{\partial F_0(c_0, h_0)}{\partial(c_0, h_0)}\right)\Bigg|_{c_0=\bar{c}_0,\, h_0=\bar{h}_0} \tag{1.6.22}$$

is nonzero. Then in any case, if μ does not exceed a certain limit, then the series (1.6.17) converge to the periodic solution $x_*(\theta, \mu)$ of system (1.6.16), to the constant $h_*(\mu)$ and to the constant (or the constant vector) $c_*(\mu)$ respectively. The constant $h_*(\mu)$ determines the period

$$T = 2\pi(1 + \mu h_*(\mu))$$

of this solution as a function of μ, and the constant (or the constant vector $c_*(\mu)$ determines the solution $z^0(t, c_*(\mu))$ of the family $z^0(t, c)$, to which the T-periodic solution of the initial system (1.6.13) with $\mu \neq 0$ corresponds.

If the determinant (1.6.22) is zero, or equation (1.6.18) is satisfied identically, then we have the critical cases of first and higher order. The analysis of such cases can be found in the book Grebenikov and Ryabov [4].

1.6.2 Iterations with quadratic convergence

a) Nonresonance (noncritical) case Write the system (1.6.2) in the form

$$\frac{dx}{dt} = P(t)x + \mu X(x, t), \tag{1.6.23}$$

where the function $X(x, t)$ is continuously differentiable with respect to its arguments and 2π-periodic with respect to t. Successive 2π-periodic approximations of the form (1.6.4) are determined from the equations

$$\begin{aligned} \frac{dx_k}{dt} &= \left[P(t) + \mu \frac{\partial X(x_{k-1}, t)}{\partial x}\right] x_k \\ &+ \mu \left[X(x_{k-1}, t) - \frac{\partial X(x_{k-1}, t)}{\partial x} x_{k-1}\right], \\ k &= 1, 2, \dots, \quad x_0 \equiv 0. \end{aligned} \tag{1.6.24}$$

The following asymptotic estimations are true (see Grebenikov and Ryabov [1]):

$$|x_1| \sim \mu, \quad |x_{k+1} - x_k| \sim \mu \, |x_k - x_{k-1}|^2, \quad k = 1, 2, \dots \tag{1.6.25}$$

At least, with μ values not exceeding a certain limit, the series $\{x(t, \mu)\}$ converge to the 2π-periodic solution of system (1.6.23).

b) Resonance (critical) cases Replacing $z^0(t)$ by $z^0(t, c)$ in (1.6.2), write this system in the following form:

$$\frac{dx}{dt} = P(t)x + \mu X(x, c, t), \tag{1.6.26}$$

where

$$X(x, c, t) = Z(z^0(t, c) + x, \, t).$$

The approximations

$$x_1(t,\mu),\ x_2(t,\mu),\ \dots,\quad c_0,\ c_1(\mu),\ c_2(\mu),\ \dots \tag{1.6.27}$$

are found successively from the following systems of differential equations:

$$\frac{dx_1}{dt} = P(t)x_1 + \mu X(0, c_0, t), \tag{1.6.28}$$

$$\begin{aligned}\frac{dx_k}{dt} &= \left[P(t) + \mu\frac{\partial X(x_{k-1}, c_{k-2}, t)}{\partial x}\right]x_k \\ &+ \mu\left[X(x_{k-1}, c_{k-1}, t) - \frac{\partial X(x_{k-1}, c_{k-1}, t)}{\partial x}x_{k-1}\right], \\ &k = 2, 3, \dots,\end{aligned} \tag{1.6.29}$$

and from the corresponding algebraic (transcendent) equations with respect to $c_0, c_1, \dots,$ presenting the existence conditions for 2π-periodic solutions of those systems.

The following asymptotic estimations are true:

$$\begin{aligned}&|x_1| \sim \mu, \quad |x_2 - x_1| \sim \mu^2, \\ &|x_{k+1} - x_k| \sim \mu|x_k - x_{k-1}|^2, \quad k = 2, 3, \dots\end{aligned} \tag{1.6.30}$$

In the case of an autonomous system of the form (1.6.13) for the construction of approximations $x_k(t,\mu)$, $k = 1, 2, \dots,$ equations of the form (1.6.28), (1.6.29) are used, with their right-hand members depending on the complementary constant h. The series $\{x_k(t,\mu)\}$, $\{c_k(\mu)\}$, $\{h_k(\mu)\}$ are found simultaneously; their convergence is guaranteed if μ does not exceed a certain limit.

c) Systems with slow and fast variables Let a system be given, of the form

$$\begin{aligned}\frac{dx}{dt} &= \mu X(x, y), \\ \frac{dy}{dt} &= \lambda_0 + \mu Y(x, y),\end{aligned} \tag{1.6.31}$$

where μ is a small positive parameter, x a vector of slow positional variables, λ_0 a constant vector, y a vector of fast angular variables, components of vector-functions $X(x,y)$, $Y(x,y)$ are represented by multiple Fourier series (or polynomials) in angular variables y, and their coefficients are power series (or polynomials) with respect to the components of the variable x. In the general case it is first necessary to perform a Krylov–Bogolyubov transform (see Bogolyubov [1]) (the numerical–analytic implementation of this transform will be described below in Chapter 4). As a result, we obtain the system

$$\begin{aligned}\frac{d\xi}{dt} &= \mu\overline{H}\xi + \mu\overline{F}(\xi, \theta), \\ \frac{d\theta}{dt} &= \lambda + \mu\overline{\Phi}(\xi, \theta),\end{aligned} \tag{1.6.32}$$

where $\overline{H}$ is a constant matrix, λ a constant vector (differing from λ_0 by the elements of order μ), and the functions $\overline{F}$, $\overline{\Phi}$ are of the same nature as the functions X, Y, but

(1) in the function $\overline{F}(\xi,\theta)$ among the terms independent of θ there are no terms linear with respect to ξ, and all the terms depending on θ, including those linear with respect to ξ, have coefficients of infinitesimal order not lower than μ,

(2) in the function $\overline{\Phi}(\xi,\theta)$ the terms depending on θ and independent of ξ, have coefficients of infinitesimal order not lower than μ.

So

$$\mu\|\overline{F}(\xi,\theta)\| \sim \mu\|\xi\|^2 + \mu^2, \quad \mu\overline{\Phi}(\xi,\theta) \sim \mu\|\xi\| + \mu^2. \tag{1.6.33}$$

To simplify the notation of the following formulae, write the system (1.6.32), without showing explicitly the parameter μ, in the form

$$\begin{aligned} \frac{d\xi}{dt} &= H\xi + F(\xi,\theta), \\ \frac{d\theta}{dt} &= \lambda + \Phi(\xi,\theta), \end{aligned} \tag{1.6.34}$$

where the functions F, Φ are expressed by Fourier series (or polynomials),

$$F(\xi,\theta) = \sum_{\|k\|\geq 0} F_k(\xi)\, e^{i(k,\theta)}, \quad \Phi(\xi,\theta) = \sum_{\|k\|\geq 0} \Phi_k(\xi)\, e^{i(k,\theta)} \tag{1.6.35}$$

and their coefficients $F_k(\xi)$, $\Phi_k(\xi)$ are power series (or polynomials) with respect to the components of the vector ξ.

It is supposed that the matrix H is nondegenerate, and its imaginary eigenvalues $\mp i\gamma_1$, $\mp i\gamma_2, \ldots$ are such that all γ_1, $\gamma_2, \ldots$ and the components $\lambda_1, \lambda_2, \ldots, \lambda_n$ of the vector λ are rationally incommensurable.

The solution (approximated) of system (1.6.34) will be calculated in the form of Fourier polynomials of a high enough order N:

$$\xi(t) = \sum_{0\leq\|k\|\leq N} U_k\, e^{i(k,\psi)}, \quad \theta(t) = \psi(t) + \sum_{1\leq\|k\|\leq N} V_k\, e^{i(k,\psi)}, \tag{1.6.36}$$

where $\psi(t) = \omega t + \psi_0$, the vectors U_k, V_k, ω are those we seek, and ψ_0 is an arbitrary vector.

Substituting (1.6.36) into (1.6.34), we arrive at the following relations:

$$\sum_{1\leq\|k\|\leq N} i(k,\omega)U_k\, e^{i(k,\psi)} = H \sum_{0\leq\|k\|\leq N} U_k\, e^{i(k,\omega)} + \tilde{F},$$

$$\omega + \sum_{1\leq\|k\|\leq N} i(k,\omega)V_k\, e^{(k,\omega)} = \lambda + \tilde{\Phi}, \tag{1.6.37}$$

where $\tilde{F}, \tilde{\Phi}$ are found as a result of the substitution of (1.6.36) into the series (or polynomials) (1.6.35). Had we performed such a substitution in an analytic form, we would obtain a Fourier series of the following form:

$$\tilde{F} = \sum_{||k|| \geq 0} D_k(U, V)\, e^{i(k,\psi)}, \quad \tilde{\Phi} = \sum_{||k|| \geq 0} C_k(U, V)\, e^{i(k,\psi)}, \tag{1.6.38}$$

where U, V denote vectors with their components being the sets of components of vectors U_k, $0 \leq ||k|| \leq N$ and V_k, $1 \leq ||k|| \leq N$, respectively, and the coefficients (vectors) D_k, C_k are in the general case certain functions of those vectors.

Restricting ourselves in the series (1.6.36) to harmonics of an order not higher than N, we come to the following finite equations with respect to U_k, V_k, ω:

$$\begin{aligned} \omega &= \lambda + C_0(U, V), \\ [i(k, \omega)E - H]U_k &= D_k(U, V), \quad 0 \leq ||k|| \leq N, \\ i(k, \omega)V_k &= C_k(U, V), \quad 1 \leq ||k|| \leq N, \end{aligned} \tag{1.6.39}$$

where E is a unit matrix.

If the numerical value of μ is given and the right-hand members of equations (1.6.4) are written in explicit form, then the algorithm proposed below allows us to seek U_k, V_k, ω (as vectors with numerical components) by means of successive approximations $U_k^{(j)}$, $V_k^{(j)}$, $\omega^{(j)}$, $j = 1, 2, 3, \ldots$ Newton's method is used, corresponding to the quadratic convergence of iterations. It is essential that here the explicit analytic expressions of the coefficients D_k, C_k as functions of the components of vectors U, V are not necessary.

The first approximation $U_k^{(1)}$, $V_k^{(1)}$, $\omega^{(1)}$ is determined by the formulae

$$\begin{aligned} \omega^{(1)} &= \lambda + C_0(0, 0), \\ [i(k, \omega^{(1)})E - H]\, U_k^{(1)} &= D_k(0, 0), \quad 0 \leq ||k|| \leq N, \\ i(k, \omega^{(1)})\, V_k^{(1)} &= C_k(0, 0), \quad 1 \leq ||k|| \leq N, \end{aligned} \tag{1.6.40}$$

where $D_k(0, 0)$, $C_0(0, 0)$, $C_k(0, 0)$ are equal to the Fourier coefficients for the functions $F(\xi, \theta)$, $\Phi(\xi, \theta)$ with $\xi = 0$, $\theta = \psi$.

The value of $\omega^{(1)}$ is determined immediately, and thanks to the condition $\det H \neq 0$ we then find $U_0^{(1)}$ for any $D_0(0, 0)$.

Due to the incommensurability of $\lambda_1, \ldots, \lambda_n, \gamma_1, \gamma_1, \ldots$ the values

$$\det[i(k, \omega^{(1)}) - H], \quad (k, \omega^{(1)}), \quad ||k|| \leq N,$$

are nonzero, at least with small μ, which allows us to further determine all $U_k^{(1)}$, $C_k^{(1)}$, $1 \leq ||k|| \leq N$.

Note that for some vectors $k = k_*$ the above-mentioned quantities may be small, and in the corresponding formulae for $U_k^{(1)}$, $V_k^{(1)}$ so-called small denominators will appear. If those small denominators are not accounted for and μ is retained in the formulae as a literal component, we would obtain the following asymptotic estimations:

$$\|U_k^{(1)}\| \sim \mu, \quad \|V_k^{(1)}\| \sim \mu. \tag{1.6.41}$$

The first approximation $\xi^{(1)}(t)$, $\theta^{(1)}(t)$ to the sought solution of equations (1.6.34) will be obtained in the following form:

$$\begin{aligned} \xi^{(1)}(t) &= \sum_{0\le\|k\|\le N} U_k^{(1)} e^{i(k,\psi^{(1)})}, \\ \theta^{(1)}(t) &= \psi^{(1)}(t) + \sum_{1\le\|k\|\le N} V_k^{(1)} e^{i(k,\psi^{(1)})}, \end{aligned} \tag{1.6.42}$$

where $\psi^{(1)}(t) = \omega^{(1)}t + \psi_0$.

For the calculation of the further approximation Newton's procedure is used; viz, in the calculation of the second approximation $U_k^{(2)}$, $V_k^{(2)}$, $\omega^{(2)}$ we assume in (1.6.39)

$$U_k = U_k^{(1)} + b_k^{(1)}, \quad V_k = V_k^{(1)} + v_k^{(1)}, \quad \omega = \omega^{(1)} + \nu^{(1)} \tag{1.6.43}$$

and construct equations with respect to $b_k^{(1)}$, $v_k^{(1)}$, $\nu_k^{(1)}$ (corrections to the first approximation), discarding terms of infinitesimal order higher than 1 with respect to $\|b_k^{(1)}\|$, $\|v_k^{(1)}\|$, $\|\nu_{(1)}\|$. Solving those (algebraic linear heterogeneous) equations and substituting the obtained $b_k^{(1)}$, $v_k^{(1)}$, $\omega^{(1)}$ into (1.6.43), we obtain the second approximation for the vectors U_k, V_k, ω, as well as the corresponding second approximation for $\xi(t)$, $\theta(t)$. In a similar way we obtain the third approximation, etc. (This method of construction of the solution of equations (1.6.34) will be described in more detail in Section 4.6 below.)

1.7 Comments and References

Section 1.2 The correct definition of asymptotic series is given in Poincaré [2].

Section 1.3 The detailed proof of the theorem on asymptotic representations of solutions of differential equations solutions can be found in Poincaré [2] and Grebenikov [3].

Section 1.4 The material for this Section is taken from Grebenikov [1], Grebenikov and Ryabov [3], Duboshin [1], and Arnold [2].

Section 1.5 The statement of the method of characteristics is taken from Stepanov [1].

Section 1.6 A more detailed statement of this Section can be found in Grebenikov and Ryabov [4], Lichtenstein [1], and Malkin [1].

2 Averaging Principle for Multifrequency Systems of Differential Equations

2.0 Introduction

We have mentioned above that the constructive theory of nonlinear equations based on the application of iteration methods is the principal means of solving problems in nonlinear analysis. Two approaches can be distinguished clearly enough: *classical* and *modern*. We will analyze their peculiar properties, taking the class of ordinary differential equations regular with respect to a small parameter.

Let there be given an n-dimensional differential equation with a small parameter

$$\frac{dz}{dt} = Z(z, t, \mu), \quad z(0) = z_0, \tag{2.0.1}$$

where the vector-function $Z(z, t, \mu)$ regular with respect to a small parameter μ is defined and has properties ensuring the existence and uniqueness of the solution to the Cauchy problem in an $(n+1)$-dimensional field, $z \in G_n$, $t \in R = [0, \infty)$. Our problem lies in finding this solution.

Along with equation (2.0.1) write an equivalent equation

$$\frac{dz}{dt} = \bar{Z}(z, t, \mu) + Z(z, t, \mu) - \bar{Z}(z, t, \mu), \quad z(0) = z_0, \tag{2.0.2}$$

where $\bar{Z}(z, t, \mu)$ is a so far arbitrary function of its arguments. Then write the linear relation

$$z(t, \mu) = \bar{z}(t, \mu) + u(t, \mu), \tag{2.0.3}$$

where $\bar{z}, u$ are some new unknown functions. The solution of Cauchy problem (2.0.1) with the help of relation (2.0.3) reduces to the solution of two Cauchy problems for the differential equations

$$\frac{d\bar{z}}{dt} = \bar{Z}(\bar{z}, t, \mu), \quad \bar{z}(0) = \bar{z}_0 \in G_n \tag{2.0.4}$$

and

$$\frac{du}{dt} = Z(\bar{z} + u, t, \mu) - \bar{Z}(\bar{z}, t, \mu), \quad u(0) = z_0 - \bar{z}_0. \tag{2.0.5}$$

Equation (2.0.4) determines the selection of the initial approximation $\bar{z}(t, \mu)$ to the exact solution $z(t, \mu)$ of problem (2.0.1), and equation (2.0.5) determines the so-called full disturbance $u(t, \mu)$. From (2.0.5) it is clear that the determination of the function $u(t, \mu)$ depends on the selection of the function $\bar{Z}(\bar{z}, t, \mu)$ and the initial point $\bar{z}_0$. Moreover, finding this is only possible upon solving equation (2.0.4), i.e. upon calculation of the initial approximation $\bar{z}(t, \mu)$. Thus, for the Cauchy problem (2.0.1) several variations of the theory of disturbances with parameters $\bar{Z}$ and $\bar{z}_0$ can be constructed, so together they can be called *a perturbation theory generator* for problem (2.0.1), and equation (2.0.4) itself can be called *a generator equation.*

It is more common to use a generator $A(t)\bar{z}$ linear with respect to $\bar{z}$ with initial point $\bar{z}_0 = z_0$, i.e. problem (2.0.1) is replaced by the problems

$$\begin{aligned} \frac{d\bar{z}}{dt} &= A(t)\bar{z}, & \bar{z}(0) &= z_0, \\ \frac{du}{dt} &= Z(\bar{z} + u, t, \mu) - A(t)\bar{z}, & u(0) &= 0, \end{aligned} \tag{2.0.6}$$

or

$$\begin{aligned} \frac{d\bar{z}}{dt} &= Z(\bar{z}, t, 0), & \bar{z}(0) &= z_0, \\ \frac{du}{dt} &= Z(\bar{z} + u, t, \mu) - Z(\bar{z}, t, 0), & u(0) &= 0, \end{aligned} \tag{2.0.7}$$

or

$$\begin{aligned} \frac{d\bar{z}}{dt} &= \bar{Z}(\bar{z}, t, \mu), & \bar{z}(0) &= z_0, \\ \frac{du}{dt} &= Z(\bar{z} + u,\, t, \mu) - \bar{Z}(\bar{z}, t, \mu), & u(0) &= 0. \end{aligned} \tag{2.0.8}$$

Equations (2.0.6) express the familiar method of linearization, equations (2.0.7) express the small-parameter method, and equations (2.0.8) express the averaging method if generator $\bar{Z}$ is constructed by means of any averaging operator.

The essence of the classical iteration method is the replacement of problem (2.0.1) by problems (2.0.4) and (2.0.5), using relation (2.0.3) where the function u depends solely on time t and does not depend on spatial values. To solve problem (2.0.1) completely, first one should find the solution of the generator equation

(2.0.4) and only after that start solving the equation for disturbances (2.0.5) by some iteration method that we denoted symbolically by

$$\begin{aligned} \frac{du_k}{dt} &= Z\left(\bar{z}(t,\mu) + u_{k-1}(t,\mu),\, t,\mu\right) - \bar{Z}\left(\bar{z}(t,\mu),t,\mu\right), \\ u_k(0) &= z_0 - \bar{z}_0, \qquad k = 1,2,\ldots \end{aligned} \tag{2.0.9}$$

The above statement has an obvious geometric interpretation. In the phase space $\{z\}$ there exists an integral curve $z(t,\mu,z_0)$ passing through the initial point z_0. In some n-dimensional cylinder with its axis being the sought integral curve, construct a family of integral curves $\bar{z}(t,\mu)$, $\bar{z}(t,\mu) + u_k(t,\mu)$, $k = 1,2,\ldots$ which has a limit integral curve $z(t,\mu,z_0)$ (if the iteration process is convergent). We see that from the geometrical point of view all classical iteration methods are seen in the initial phase space $\{z\}$.

Now assume that the function u depends not only on t and μ, but also on the spatial coordinates of $\bar{z}$, i.e. instead of (2.0.3) we write the equality

$$z(t,\mu) = \bar{z}(t,\mu) + u(\bar{z},t,\mu). \tag{2.0.10}$$

Geometrically relation (2.0.10) expresses the transition of the spatial value $\{z\}$ to the new spatial value $\{\bar{z}\}$ and vice versa, if the Jacobian matrix is nonsingular. There is an obvious differential equality

$$\frac{dz}{dt} = \frac{d\bar{z}}{dt} + \left(\frac{\partial u}{\partial \bar{z}}, \frac{d\bar{z}}{dt}\right) + \frac{\partial u}{\partial t}, \tag{2.0.11}$$

where the symbol $\left(\frac{\partial u}{\partial \bar{z}}, \frac{d\bar{z}}{dt}\right)$ denotes the product of the matrix $\left(\frac{\partial u}{\partial \bar{z}}\right)$ and the vector $\left(\frac{d\bar{z}}{dt}\right)$.

Therefore here instead of the two Cauchy problems (2.0.4) and (2.0.5) we will have the following Cauchy problems:

$$\frac{d\bar{z}}{dt} = \bar{Z}(\bar{z},t,\mu), \qquad \bar{z}(0) = \bar{z}_0, \tag{2.0.4}$$

$$\begin{aligned} \frac{\partial u}{\partial t} + \left(\frac{\partial u}{\partial \bar{z}}, \bar{Z}(\bar{z},t,\mu)\right) &= Z(\bar{z}+u,t,\mu) - \bar{Z}(\bar{z},t,\mu), \\ u(0) &= z_0 - \bar{z}_0. \end{aligned} \tag{2.0.12}$$

The methods using equations (2.0.4) and (2.0.12) differ from the former ones at least in one essential point: the determination of the multivariable function $u(\bar{z},t,\mu)$ from the first-order partial differential equation (2.0.12) does not require preliminary solution of the generator equation (2.0.4), since in equation (2.0.12) $\bar{z}$ and t are peer arguments of the function u. This allows us to calculate the functions $u(\bar{z},t,\mu)$ and $\bar{z}(t,\mu)$ independently of each other, which is impossible in the case of classical iteration methods.

Equation (2.0.12) is a Cauchy problem for quasi-linear n-dimensional systems of first-order partial differential equations with respect to n-dimensional disturbance vector u. In principle its solution can be found by the method of characteristics (see Stepanov [1]). A particular case of this equation was considered for the first time by N.N. Bogolyubov when he studied the issue of the applicability of the averaging method to one special class of ordinary differential equations (see Bogolyubov [1]). That is why we call equation (2.0.12) the generalized Krylov–Bogolyubov equation.

So in this chapter the reader's attention will be called to the main results of the asymptotic theory of vector equation (2.0.12) with reference to multifrequency systems of differential equations defined on multidimensional tori, typical of many nonlinear problems in dynamics. In particular, we will consider some modern iteration methods for construction of asymptotic solutions of Hamiltonian systems with action—angle type variables and with a Hamiltonian periodic with respect to slow angle variables.

2.1 Mathematical Averaging Principle

Now assume that the generator $\bar{Z}$ is constructed by means of some averaging operator. Then it is more convenient to consider the generator equation (2.0.4) not as an equation defining the first approximation for the asymptotic theory of equation (2.0.1), but as a final equation into which the initial equation (2.0.1) should be transformed with the function u used as a function of such a transform.

Thus, it is also mathematically correct to state the problem as the finding of a substitution of variables that transforms equation (2.0.1) into (2.0.4). Such a statement was formulated by N.N. Bogolyubov in the 1930s in nonlinear equation theory.

As is well known, the method of substitution of variables often offers real opportunities for obtaining specific results. As applied to ordinary differential equations, it can be used, in particular, for transformation of initial equations into equations with a predetermined analytic structure.

Of course, it makes sense to state the problem of the transformation of equations when the transformed equations admit if not the complete solution, then in any case a fuller study than the initial ones do.

Problem statement Find a substitution of variables $z \to \bar{z}$ that will transform equation (2.0.1) into (2.0.4). It is natural to seek the substitution in the class of nondegenerate substitutions, i.e. such that if there exists a substitution $z \to \bar{z}$, then there exists an inverse one $\bar{z} \to z$.

Let the desired substitution of variables be represented by the equalities

$$z = u(\bar{z}, t, \mu), \quad z_0 = u(\bar{z}_0, 0, \mu). \tag{2.1.1}$$

Obviously, the following identity is true:

$$\frac{dz}{dt} \equiv \left(\frac{\partial u}{\partial \bar{z}}, \frac{d\bar{z}}{dt} \right) + \frac{\partial u}{\partial t}. \tag{2.1.2}$$

Replacing in this identity the variables dz/dt and $d\bar{z}/dt$ by the corresponding functions $Z(z,t,\mu)$ and $\bar{Z}(z,t,\mu)$, and taking account of the substitution of variables (2.1.1) we obtain a quasi-linear first-order partial differential equation, where the unknown vector-function is the conversion function $u(\bar{z},t,\mu)$:

$$\frac{\partial u}{\partial t} + \left(\frac{\partial u}{\partial \bar{z}}, \bar{Z}, t, \mu)\right) = Z(u(\bar{z},t,\mu),t,\mu). \tag{2.1.3}$$

By analogy with (2.0.12) we will call equation (2.1.3) *a generalized equation of asymptotic perturbation theory.* In coordinate form equation (2.1.3) is as follows:

$$\begin{aligned} \frac{\partial u_k}{\partial t} &+ \sum_{s=1}^{n} \frac{\partial u_k}{\partial \bar{z}_s} \bar{Z}_s(\bar{z}_1,\dots,\bar{z}_n,t,\mu) \\ &= Z_k(u_1,\dots,u_n,t,\mu), \qquad k=1,\dots,n. \end{aligned} \tag{2.1.4}$$

To equation (2.1.3) or (2.1.4) the following initial conditions should be added:

$$u(\bar{z}_0,t,\mu) = z_0. \tag{2.1.5}$$

Thus, the substitution of variables (2.1.1) is defined by a system of quasi-linear first-order partial differential equations with sought functions $u(\bar{z},t,\mu)$ and initial conditions (2.1.5). If transform (2.1.1) exists and is not degenerate in the domain $G^*_{n+2} \subset G_{n+2}$ (i.e. the Jacobian matrix is nondegenerate in G^*_{n+2}), then in this domain equations (2.0.1) and (2.0.4) are equivalent. Here instead of the terms generator and generator equation it is more appropriate to use the terms *system* or *comparison equation.* In other words, equation (2.1.3) is the *comparison equation* for the initial system (2.0.1).

In principle, finding exact solutions of equations (2.0.1), (2.0.4) and (2.1.3) is equally difficult, but for the construction of approximate solutions (and this is what asymptotic perturbation theory deals with) of initial equations (2.0.1) the use of comparison equations (2.0.4) and variables substitution (2.1.1) may be highly effective, as numerous applications show.

To understand the sense of this discourse, consider two extreme cases that denote the bounds of applicability of the variable substitution method.

A. *Identical transformation.* Let the comparison equation (2.0.4) coincide with the initial equation (2.0.1), i.e.

$$\frac{d\bar{z}}{dt} = Z(\bar{z},t,\mu). \tag{2.1.6}$$

Then the generalized equation (2.1.3) takes the form

$$\frac{\partial u}{\partial t} + \left(\frac{\partial u}{\partial \bar{z}}, Z(\bar{z},t,\mu)\right) = Z(u,t,\mu). \tag{2.1.7}$$

Obviously equation (2.1.7) admits the simplest solution represented by the identical transform

$$z = \bar{z}. \tag{2.1.8}$$

In this case it is very easy to solve the equation defining the substitution of variables, but this solution would be no good, because we have not gained ground towards solving the initial equations.

B. *Transformation reducing to the simplest comparison equation.* In this case we will assume that the comparison vector-function $\bar{Z}(\bar{z}, t, \mu)$ is continuous: $\bar{Z} = a$. Then the comparison equation

$$\frac{d\bar{z}}{dt} = a \tag{2.1.9}$$

can be easily integrated, and its general solution has the form

$$\bar{z} = at + C, \tag{2.1.10}$$

and equation (2.1.3) will be as follows:

$$\frac{\partial u}{\partial t} + \left(\frac{\partial u}{\partial \bar{z}}, a\right) = Z(u, t, \mu). \tag{2.1.11}$$

From (2.1.11) one can write the corresponding system for characteristics:

$$\begin{aligned} \frac{d\bar{z}}{dt} &= a, \\ \frac{du}{dt} &= Z(u, t, \mu). \end{aligned} \tag{2.1.12}$$

Hence it is clear that the transformation of the initial equation (2.0.1) into the simplest comparison equation (2.1.9) is equivalent to the solving of the initial equations themselves, and consequently the problem of transforming the equations turns out to be equivalent to the initial problem. This should have been expected. If there existed a substitution of variables that transformed an initial nonlinear problem into an integrable one, there would exist a universal method of solving nonlinear problems, which is clearly impossible.

After the discourse on limiting situations it seems natural to find an intermediate, compromise approach both to the selection of comparison equations and to the study of the properties of solving quasi-linear partial differential equations, determining the corresponding substitution of variables. This constructive approach is as follows.

1. The comparison vector-function $\bar{Z}(\bar{z}, t, \mu)$ should be chosen so that its analytical structure should be as simple as possible compared with $Z(z, t, \mu)$, but at the same time inherent to it should be the main, characteristic properties of the latter, defining the properties of the solution of the initial equations. Only when the comparison function $\bar{Z}(\bar{z}, t, \mu)$ is chosen in that way should it be expected that the behavior of the solutions of the comparison equations may be analogous to that of the solutions of the initial equations.

2. Comparison functions should be selected so that even if the quasi-linear first-order partial differential equations for variables substitutions cannot be integrated

exactly, then in any case they would allow for some analytical or qualitative analysis of the behavior of their solutions. Among all possible properties of the function $u(\bar{z}, t, \mu)$ we will be interested in its boundedness and smoothness in some domain variation of the arguments.

Items 1 and 2 characterize, as one might say, the mathematical *averaging principle*—the essence of the Krylov–Bogolyubov transform (see Bogolyubov and Mitropolsky [1], Grebenikov [3], and Grebenikov and Ryabov [1]) as applied to systems of ordinary differential equations.

Now from these general observations we turn to the fundamentals of the constructive part of asymptotic perturbation theory for multifrequency systems of differential equations with slow and fast variables:

$$\begin{aligned} \frac{dx}{dt} &= \mu X(x, y, t), \\ \frac{dy}{dt} &= \omega(x, y) + \mu Y(x, y, t). \end{aligned} \tag{2.1.13}$$

Here x, X are m-dimensional variables, and y, ω, Y are n-dimensional variables. It is better to make the selection of comparison functions X and Y (here it should be stressed that, as a rule, the frequencies $\omega(x, y)$ remain unchanged in comparison equations as well) using the averaging operators. If vector-functions X, ω, Y are analytical with respect to their arguments, then it seems possible to apply the formalism of power series both for the construction of comparison equations and for finding solutions of equations defining variable substitution. For example, write a comparison system for (2.1.13) in the form

$$\begin{aligned} \frac{d\bar{x}}{dt} &= \mu \bar{X}(\bar{x}, \bar{y}) + \sum_{k \geq 2} \mu^k A_k(\bar{x}, \bar{y}), \\ \frac{d\bar{y}}{dt} &= \omega(\bar{x}, \bar{y}) + \mu \bar{Y}(\bar{x}, \bar{y}) + \sum_{k \geq 2} \mu^k B_k(\bar{x}, \bar{y}), \end{aligned} \tag{2.1.14}$$

where $\bar{X}$, $\bar{Y}$ are some "average" values of the functions X, Y, and the functions A_k, B_k will be selected in each particular case, using some additional considerations.

The substitution of variables $(x, y) \to (\bar{x}, \bar{y})$, transforming the multifrequency system (2.1.13) into system (2.1.14) will be sought in the form of power series of a small parameter:

$$\begin{aligned} x &= \bar{x} + \sum_{k \geq 1} \mu^k u_k(\bar{x}, \bar{y}, t), \\ y &= \bar{y} + \sum_{k \geq 1} \mu^k v_k(\bar{x}, \bar{y}, t). \end{aligned} \tag{2.1.15}$$

The analytic representations (2.1.15) allow us to rearrange the generalized equation (2.1.3) (which, as is known, is quasi-linear) into derivatives for the unknown functions u_k and v_k:

$$\begin{aligned}
&\frac{\partial u_k}{\partial t} + \left(\frac{\partial u_k}{\partial \bar{y}}, \omega(\bar{x}, \bar{y})\right) \\
&\qquad = F_k(\bar{x}, \bar{y}, t, u_1, v_1, \dots, u_{k-1}, v_{k-1}, A_2, B_2, \dots, A_k), \\
&\frac{\partial v_k}{\partial t} + \left(\frac{\partial v_k}{\partial \bar{y}}, \omega(\bar{x}, \bar{y})\right) \\
&\qquad = \Psi_k(\bar{x}, \bar{y}, t, u_1, v_1, \dots, v_{k-1}, u_k, A_2, B_2, \dots, A_k, B_k), \\
&\qquad k = 2, 3, \dots
\end{aligned} \tag{2.1.16}$$

With $k = 1$ we have

$$\begin{aligned}
&\frac{\partial u_1}{\partial t} + \left(\frac{\partial u_1}{\partial \bar{y}}, \omega(\bar{x}, \bar{y})\right) = X(\bar{x}, \bar{y}, t) - \bar{X}(\bar{x}, \bar{y}), \\
&\frac{\partial v_1}{\partial t} + \left(\frac{\partial v_1}{\partial \bar{y}}, \omega(\bar{x}, \bar{y})\right) = \left(\frac{\partial \omega(\bar{x}, \bar{y})}{\partial \bar{x}}, u_1\right) + Y(\bar{x}, \bar{y}, t) - \bar{Y}(\bar{x}, \bar{y}).
\end{aligned} \tag{2.1.17}$$

The infinite system (2.1.16) has a remarkable property. It can be integrated in series: first from the first equation of system (2.1.17) the vector-function $u_1(\bar{x}, \bar{y}, t)$ is found, then the second equation of system (2.1.17) is integrated; as a result, the vector-function $v_1(\bar{x}, \bar{y}, t)$ is found. Now we pass to system (2.1.16) with $k = 2$. The analytic structure of the first equation of system (2.1.16) with $k = 2$ includes the unknown function $A_2(\bar{x}, \bar{y})$ that can be selected quite arbitrarily. Usually it is defined so as to destroy in the second member of the first equation of system (2.1.16), i.e. in function $F_2(\bar{x}, \bar{y}, t, u_1, v_1, A_2)$, the summands that can give the largest increment in the solution $u_2(\bar{x}, \bar{y}, t)$. Having determined from any conditions the vector-function $B_2(\bar{x}, \bar{y})$, we define the vector-function $v_2(\bar{x}, \bar{y}, t)$ from the second equation of system (2.1.16) with $k = 2$. This process can be continued for any value of k, and therefore, at least in principle, we can advance arbitrarily far in the formal definition of series (2.1.15).

If we could define all the terms of these series and, moreover, prove their convergence as well as the convergence of the series in (2.1.14) in some $(m + n + 2)$-dimensional domain of the Euclidean space G^*_{m+n+2} of the substitution of arguments $\bar{x}$, $\bar{y}$, t, μ, this would mean that the initial system is equivalent to system (2.1.14) in G^*_{m+n+2}. In this situation we would deal with exact transforms but not with asymptotic theory. Unfortunately, for complex nonlinear systems it is practically impossible to do this, so in such cases asymptotic methods of problem solving seems to be most suitable. Here a lot of variations are possible. Let us consider two of them.

Instead of the substitution of variables (2.1.15) that contains an infinite number of functions $u_k(\bar{x}, \bar{y}, t)$ and $v_k(\bar{x}, \bar{y}, t)$ consider the finite expressions

$$x = \bar{x} + \sum_{k=1}^{r} \mu^k u_k(\bar{x}, \bar{y}, t), \qquad y = \bar{y} + \sum_{k=1}^{r} \mu^k v_k(\bar{x}, \bar{y}, t) \tag{2.1.18}$$

and demand that the substitution (2.1.18) should transform the initial equations into the comparison system (2.1.14) containing an infinite number of functions $A_k(\bar{x}, \bar{y})$ and $B_k(\bar{x}, \bar{y})$. Such a substitution of variables $(x, y) \to (\bar{x}, \bar{y})$ can be found in principle, but unlike the previous case, not all functions $A_k(\bar{x}, \bar{y})$ and $B_k(\bar{x}, \bar{y})$ can be chosen at one's own discretion.

Indeed, the equations determining $2r$ of the functions $u_1, v_1, \dots, u_r, v_r$, exactly coincide with the first $2r$ equations of the finite system (2.1.16), where the choice of functions A_2, $B_2, \dots, A_r$, B_r can be performed as before. As regards the functions A_{r+1}, $B_{r+1}, \dots$, they are determined successively and uniquely as quantities depending on similar quantities and functions with smaller indices, i.e.

$$A_k, B_k = \theta_k(\bar{x}, \bar{y}, t, u_1, \dots, u_r, v_r, A_2, \dots, B_{k-1}),$$
$$k = r+1, r+2, \dots$$

Hence, the use of a substitution of variables of the type (2.1.18) in principle solves the problem of the initial system transformation into the comparison system (2.1.14), but the latter in this case only contains $2r-2$ (but not an infinite number) arbitrary functions A_2, $B_2, \dots, A_r$, B_r.

But one can put the following question: is it possible to find a substitution of the type (2.1.18), which would transform the initial equations into the comparison system

$$\begin{aligned} \frac{d\bar{x}}{dt} &= \mu \bar{X}(\bar{x}, \bar{y}) + \sum_{k=2}^{r} \mu^k A_a(\bar{x}, \bar{y}), \\ \frac{d\bar{y}}{dt} &= \omega(\bar{x}, \bar{y}) + \mu \bar{Y}(\bar{x}, \bar{y}) + \sum_{k=2}^{r} \mu^k B_k(\bar{x}, \bar{y})? \end{aligned} \tag{2.1.19}$$

A detailed analysis shows that this problem has only an approximate solution even in the framework of formal transforms. In other words, substitution of variables (2.1.18) allows us to transform initial equations into system (2.1.19) of order $O(\mu^{r+1})$. Indeed, let us suppose that the functions X and Y are representable by the equalities

$$\begin{aligned} X(x, y, t, \mu) &= \sum_{k=0}^{r-1} \mu^k X_k(x, y, t) + O(\mu^{r+1}), \\ Y(x, y, t, \mu) &= \sum_{k=0}^{r-1} \mu^k Y_k(x, y, t) + O(\mu^{r+1}) \end{aligned} \tag{2.1.20}$$

in the neighborhood of the point $(\bar{x}, \bar{y}, t, 0)$. Then the functions u_k, v_k, A_k, B_k can be found from the first $2r$ equations of system (2.1.16) by means of the above procedure. But if for all that we substitute the second members of (2.1.20) and (2.1.19) into

$$\frac{dx}{dt} = \frac{d\bar{x}}{dt} + \sum_{k=1}^{r} \mu^k \left[\left(\frac{\partial u_k}{\partial \bar{x}}, \frac{d\bar{x}}{dt} \right) + \left(\frac{\partial u_k}{\partial \bar{y}}, \frac{d\bar{y}}{dt} \right) + \frac{\partial u_k}{\partial t} \right],$$

$$\frac{dy}{dt} = \frac{d\bar{y}}{dt} + \sum_{k=1}^{r} \mu^k \left[\left(\frac{\partial v_k}{\partial \bar{x}}, \frac{d\bar{x}}{dt} \right) + \left(\frac{\partial v_k}{\partial \bar{y}}, \frac{d\bar{y}}{dt} \right) + \frac{\partial v_k}{\partial t} \right]$$

then the difference "the first member minus the second member" will have order $O(\mu^{r+1})$. It is exactly in this sense that one can use finite transforms (from the point of view of a finite number of items) to obtain comparison equations. Here it is no longer necessary to investigate the convergence of series, but at the same time the transforms are performed with a certain error of an order that can be determined.

2.2 Classification of Systems of Differential Equations, where Resonances are Possible

Poincaré's theorem on asymptotic approximation of solutions says specifically that in asymptotic theory the ε-proximity of solutions of exact and smoothed equations in one of the following time intervals is determined:

(1) $t \in [0, a]$, $a > 0$ and if $\mu \to 0$ $a < \infty$;
(2) $t \in [0, \mu^{\delta-1}]$, $1 > \delta > 0$;
(3) $t \in [0, \mu^{-1}]$;
(4) $t \in [0, \mu^{-\beta}]$, $\beta > 1$;
(5) $t \in [0, \infty)$.

In the case (2) $\delta = 0.5$ is often taken, and then the time interval is equal to $[0, 1/\sqrt{\mu}]$.

With sufficiently small values of μ we have the inclusions

$$[0, a] \subset [0, \mu^{\delta-1}] \subset [0, \mu^{-1}] \subset [0, \mu^{-\beta}] \subset [0, \infty).$$

Note that the above time intervals do not include the trivial domain $[0, a(\mu)]$, where $a(\mu) \to 0$ as $\mu \to 0$.

In Section 1.4 the classification of resonances was given for the numeric vector of frequencies $\omega = (\omega_1, \dots, \omega_n)$. In multifrequency systems of differential equations with variable frequencies of the form (2.1.13) resonance phenomena show up in a far more complicated way, because in the course of time some resonances may disappear, and other resonances may appear, and this process may occur a great, even an infinite number of times irrespective of the length of the time interval. The analysis given in Section 1.4 allows us to conclude that resonances of lower orders are most influential. Therefore during the development of a classification of resonance systems it seems advisable to bound above the norm of vector k that determines the resonance relations.

So, let there be given:

(1) a multifrequency system of the form (2.1.13);
(2) the range of definition of the right-hand sides of the system $G_{m+n+2} = G_{m+n} \times [0, T] \times [0, \bar{\mu}]$, and the initial point $(x_0, y_0) \in G_{m+n}$;
(3) the positive number N determining the inequality

$$||k|| \leq N, \tag{2.2.1}$$

and amongst them there are resonance vectors k^* determining the α-resonance zone, i.e. such that

$$|(k^*, \omega(x, y))| \leq \alpha. \tag{2.2.2}$$

According to the behavior of the partial solution $x(t, x_0, y_0, \mu)$, $y(t, x_0, y_0, \mu)$ (hereinafter denoted by $(x(t, \mu), y(t, \mu)))$ proceeding from the initial point $(x_0, y_0) \in G_{m+n}$, any multifrequency system can be related to one of the following classes of equations.

I. Class of locally nonresonance systems.
II. Class of systems with the property of the solution $(x(t, \mu), y(t, \mu))$ remaining in one of the α-resonance zones, or in other words, in the neighborhood of one of the resonances with $||k^*|| \leq N$.
III. Class of systems with the property of the solution remaining in the neighborhoods of several resonances.
IV. Class of systems not having the property of the solution remaining in resonance zones.
V. Class of degenerate resonance systems.

To the class of locally nonresonance systems we will also relate multifrequency systems for which

$$|(k, \omega(x(t, \mu), y(t, \mu)))| > \alpha \tag{2.2.3}$$

with all $t \in [0, T]$ and $||k|| \leq N$. From this it does not in the least follow that in systems belonging to class I there are no α-resonances of frequencies. They may occur, but for them $||k|| > N$ or $t \notin [0, T]$.

Class II includes those multifrequency systems with their partial solution $(x(t, \mu), y(t, \mu))$ corresponding to the integer vector $k^{(1)}$ with its norm less than N and satisfying the condition

$$|(k^{(1)}, \omega(x(t, \mu), y(t, \mu)))| \leq \alpha \tag{2.2.4}$$

for $t \in [a, b] \subset [0, T]$, $0 \leq a < b \leq T$ and $b - a = O(T)$. To Class II we will relate systems of differential equations that have the property of remaining in the

α-resonance zone with the resonance vector $k^{(1)}$ within the time interval of length commensurable with the given domain $[0, T]$.

Now we will describe the systems that we relate to class III.

Let there be some resonance vectors $k^{(1)}, \ldots, k^{(s)}$ for which the conditions of remaining in the α-resonance zone are satisfied:

$$\begin{gathered}
|(k^{(r)}, \omega(x(t,\mu), y(t,\mu))| \leq \alpha, \quad r = 1, \ldots, s, \\
t \in [a_r, b_r], \quad 0 \leq a_1 \leq \ldots \leq a_s < T, \\
0 < b_1 \leq \ldots \leq b_s \leq T, \quad a_r < b_r, \\
b_r - a_r = O(T), \quad \sum_{r=1}^{s} (b_r - a_r) = O(T).
\end{gathered} \tag{2.2.5}$$

Analyzing the conditions (2.2.5) we see that there are s resonance time intervals, the length of each of them is commensurable with the total domain (in the sense of the order of magnitude), and the sum of their lengths is also commensurable with T.

To class IV we relate those multifrequency systems with their partial solutions satisfying the conditions of (2.2.5) on the one hand, and the conditions of the form

$$0 < \gamma < \left| \left(k^{(r)}, \frac{d\omega(x(t,\mu), y(t,\mu))}{dt} \right) \right|, \tag{2.2.6}$$

on the other, where γ is positive. This shows the minimum rate of change of the function $(k^{(r)}, \omega)$, guaranteeing the withdrawal from the α-resonance zone within $[0, T]$.

All the other behaviors of solutions of multifrequency systems will be related to class V. For instance, such a situation is possible when with a finite T the number of α-resonances s approaches infinity.

We have not hitherto ascertained the width of the α-zone and the value of γ. If α does not depend on the small parameter μ, and γ does, then the intersection of classes II, III and IV is possible. The opposite situation is also possible: $\gamma(\mu) \to 0$ as $\mu \to 0$, and $\alpha > 0$ as $\mu \to 0$, and the solution may remain in the α-resonance zone of one of the resonance points.

Classes III and IV are most difficult for study, because their solutions are actually both resonance and nonresonance, and we do not know any conditions allowing us to a priori point out the resonance and nonresonance time intervals and consequently the evolution of the system towards the resonance or nonresonance state. This assertion is even stronger when the matter concerns degenerate multifrequency systems (class V).

In conclusion we note that at present some results on ε-proximity of exact and smoothed solutions are known in the case when $\alpha(\mu) \to 0$ as $\mu \to 0$, and $t \in [0, \mu^{\delta-1}]$ or $t \in [0, \mu^{-1}]$, and some of them will be described below.

2.3 The Basis of the Asymptotic Theory for Locally Nonresonance Systems

In Section 2.2 we noted that one of the parameters of multifrequency system classification is an integral number N giving the number of those harmonics that should be preserved in the expansions for functions $X(x,y,t,\mu)$ and $Y(x,y,t,\mu)$, in order to ensure the construction of approximate asymptotic solutions with a given error. In short, the number N makes a division between the retained finite expressions in the expansions of functions X and Y and the truncated infinite residual series. For Fourier series there exist effective estimations for the number N, which will be given below.

Now we will decide an issue that is very important for further study. Consider two rotary systems of order $(m+n)$:

$$\begin{aligned} \frac{dx}{dt} &= \mu X(x,y,\mu), \\ \frac{dy}{dt} &= \omega(x) + \mu Y(x,y,\mu); \end{aligned} \tag{2.3.1}$$

$$\begin{aligned} \frac{d\tilde{x}}{dt} &= \mu X_N(\tilde{x},\tilde{y},\mu), \\ \frac{d\tilde{y}}{dt} &= \omega(\tilde{x}) + \mu Y_N(\tilde{x},\tilde{y},\mu) \end{aligned} \tag{2.3.2}$$

with the same initial conditions

$$x(0,\mu) = \tilde{x}(0,\mu), \quad y(0,\mu) = \tilde{y}(0,\mu), \tag{2.3.3}$$

with their right-hand members defined in $G_{m+n} = \{(x,y): \; x \in P_m, \; y \in Q_n\}$. A rotary system will be understood as a multifrequency system with its frequencies $\omega(x)$ only depending on slow variables x, and vector-functions $X(x,y,\mu)$, $Y(x,y,\mu)$ being 2π-periodic functions with respect to y in the domain $Q_n = \|\operatorname{Im} y\| < \varrho$.

Assume that:

(1) the solution of system (2.3.1) with the initial conditions (2.3.3) exists and is unique;

(2) functions X_N, Y_N, ω in the domain G_{m+n} satisfy the Lipschitz condition

$$\begin{aligned} \|X_N(x_1,y_1) - X_N(x_2,y_2)\| &\le L_x\|x_1 - x_2\| + L_y\|y_1 - y_2\|, \\ \|\omega(x_1) - \omega(x_2)\| &\le L_\omega\|x_1 - x_2\|, \\ \|Y_N(x_1,y_1) - Y_N(x_2,y_2)\| &\le l_x\|x_1 - x_2\| + l_y\|y_1 - y_2\|, \end{aligned} \tag{2.3.4}$$

where L_x, L_y, L_ω, l_x, l_y are Lipschitz constants;

(3) in the domain G_{m+n} the following inequalities hold:

$$\|X(x,y) - X_N(x,y)\| \le \varepsilon_1, \quad \|Y(x,y) - Y_N(x,y)\| \le \varepsilon_2, \tag{2.3.5}$$

where $\varepsilon_1, \varepsilon_2$ are given positive numbers.

Under these conditions the following estimations for solutions of rotary systems (2.3.1) and (2.3.2) are true:

$$\begin{aligned}
&\|y(t,\mu)-\tilde{y}(t,\mu)\| < C_1^{(x)}e^{\lambda_1 t} + C_2^{(x)}e^{\lambda_2 t} + C_3,\\
&\|x(t,\mu)-\tilde{x}(t,\mu)\| \leq |L_\omega + \mu l_x|^{-1}\Big[C_1^x\lambda_1 e^{\lambda_1 t} + C_2^{(x)}\lambda_2 e^{\lambda_2 t}\\
&\qquad - \mu l_y\|y(t,\mu)-\tilde{y}(t,\mu)\| - \mu\varepsilon_2\Big],
\end{aligned} \tag{2.3.6}$$

where

$$\begin{aligned}
C_1^{(x)} &= \frac{C_3\lambda_2 + \mu\varepsilon_1}{\lambda_1 - \lambda_2}, \qquad C_2^{(x)} = C_3 - C_1^{(x)},\\
C_3 &= \frac{\varepsilon_1(L_\omega + \mu l_x) - \mu L_x\varepsilon_2}{\mu L_x l_y - L_y L_\omega - \mu L_y l_x},\\
\lambda_{1,2} &= \frac{\mu(L_x + l_y)}{2} \pm \frac{1}{2}\big[\mu^2(L_x + l_y)^2\\
&\qquad - 4(\mu^2 L_y l_x - \mu^2 L_x l_y - \mu L_y L_\omega)\big]^{1/2}.
\end{aligned} \tag{2.3.7}$$

Inequalities (2.3.6) can be used for solving an inverse problem.

Let $t \in [0, T = C\mu^{-1}]$, and let there be given deviations of solutions of the system (2.3.1) and (2.3.2) from each other:

$$\|x - \tilde{x}\| \leq \delta_1, \quad \|y - \tilde{y}\| \leq \delta_2. \tag{2.3.8}$$

It is necessary to estimate the norms $\|X(x,y,\mu) - X_N(x,y,\mu)\|$, $\|Y(x,y,\mu) - Y_N(x,y,\mu)\|$, i.e. calculate the values of ε_1 and ε_2.

If we solve the algebraic equations

$$\begin{aligned}
&C_1^{(x)}\lambda_1 e^{\lambda_1 T} + C_2^{(x)}\lambda_2 e^{\lambda_2 T} - \mu\varepsilon_2 = \delta_1[L_\omega + \mu l_x] + \mu l_y\delta_2,\\
&C_1^{(x)} + C_2^{(x)}e^{\lambda_2 T} + C_3 = \delta_2,
\end{aligned} \tag{2.3.9}$$

with respect to ε_1 and ε_2, we obtain expressions of the form

$$\begin{aligned}
\varepsilon_1 &= a_1(\mu,T)\delta_1 + a_2(\mu,T)\delta_2,\\
\varepsilon_2 &= b_1(\mu,T)\delta_1 + b_2(\mu,T)\delta_2,
\end{aligned} \tag{2.3.10}$$

where $a_k(\mu,T)$, $b_k(\mu,T)$ are composite functions of μ and T. However it is possible to show that with sufficiently small μ and $T = C\mu^{-1}$

$$a_k(\mu,T) = O(1), \quad b_k(\mu,T) = O(1). \tag{2.3.11}$$

So, from given values of δ_1, δ_2, μ and T one can compute ε_1 and ε_2 used for truncation of infinite expansions in the second members of equations (2.3.1). Indeed, let the functions $X(x,y)$, $Y(x,y)$ be l times $(l > n)$ differentiable with respect to y in the domain G_{m+n}. Then we have the estimation (see Grebenikov and Ryabov [1])

$$\Big\|X(x,y) - \sum_{0\leq\|k\|\leq N} X_k(x)e^{i(k,y)}\Big\| < \frac{C(n,N)}{l-n}\sum_{s=1}^{m} C_l^{(s)}(N+0.5)^{n-l}, \tag{2.3.12}$$

with the following agreed notation:

$$C(n,N)=\begin{cases} 2n^{l+1}, & \text{if } N\geq 1,\\ \dfrac{n^{l+2}}{2^{n-2}}, & \text{if } N\geq 2,\\ \dfrac{3n^l}{\sqrt{\pi}(4n-3)}\left(\dfrac{6N}{n_1}\right)^{n-1}, & \text{if } n\geq 2,\ N\geq n-1. \end{cases}$$

$C_l^{(s)}$ is a positive constant bounding above the norm of the s-th component of the vector-function X in the domain G_{m+n}:

$$\left\|X^{(s)}\right\|_{\|\varrho\|}=\max_{0\leq\|\alpha\|\leq\|\varrho\|}\ \sup_{(x,y)\in G_{m+n}}\left|\frac{\partial^{(\alpha)}X^{(s)}}{\partial y_1^{\alpha_1}\cdots\partial y_n^{\alpha_n}}\right|\leq C_l^{(s)}. \tag{2.3.13}$$

A similar estimation can also be made for the components $Y(x,y)$. If X and Y depend on μ, then $C_l^{(s)}=C_l^{(s)}(\mu)$.

Then construct two equations (or inequalities) of the form

$$\frac{C(n,N)}{l-n}(N+0.5)^{n-l}\sum_{s=1}^{m,n}C_l^{(s)}\leq\varepsilon_k,\quad k=1,2, \tag{2.3.14}$$

where ε_k are computed from the formulae (2.3.10). Each inequality is solved independently, the minimum N is found that satisfies the inequality, and then from the two minimum values of N the higher one is chosen. This is the value of N for which the infinite expansions for the functions X and Y are truncated.

Thus, we have solved three problems. If a number N is given, then, depending on the degree of smoothness of the functions X and Y, it is possible to calculate ε_1 and ε_2 characterizing the deviations in norm of the Fourier polynomials from the given functions. With these estimations we are able to estimate the norms of differences of solutions of two systems (formulae (2.3.6)) via known time functions. If the deviations of solutions and the value of the time interval T are given, then the number N can be found. In other words, at the first stage of the asymptotic theory of differential equations we learned to truncate infinite expansions in the right-hand side of equations in order to further deal with finite expressions.

Now instead of the rotary system (2.3.1) we will consider the multifrequency system

$$\begin{aligned} \frac{dx}{dt}&=\mu X_N(x,y,\mu)\equiv\mu\sum_{0\leq\|k\|\leq N}X_k(x,\mu)e^{i(k,y)},\\ \frac{dy}{dt}&=\omega(x)+\mu Y_N(x,y,\mu)\equiv\omega(x)+\mu\sum_{0\leq\|k\|\leq N}Y_k(x,\mu)e^{i(k,y)},\\ &x(0,\mu)=x_0,\quad y(0,\mu)=y_0, \end{aligned} \tag{2.3.15}$$

with its second members being trigonometric polynomials (with respect to fast variables y), but not infinite series.

Assume that the solution $x(t, x_0, y_0, \mu)$, $y(t, x_0, y_0, \mu)$ of the system (2.3.15) is such that for all $1 \leq ||k|| \leq N$ and for any $t \in [0, T]$ the following inequality holds:

$$|(k,\, \omega(x(t, x_0, y_0, \mu)))| > \alpha, \tag{2.3.16}$$

where α is a predefined number.

Under these conditions, in accordance with resonance system classifications, multifrequency systems of the type (2.3.15) are locally nonresonant. To the right-hand side of the system (2.3.15) we apply the averaging operator(1.1.7) with respect to angle variables:

$$\begin{aligned} \bar{X}_N(x, \mu) &= \frac{1}{(2\pi)^n} \int_0^{2\pi} \cdots \int_0^{2\pi} X_N(x, y, \mu)\, dy = X_0(x, \mu), \\ \bar{Y}_N(x, \mu) &= Y_0(x, \mu). \end{aligned} \tag{2.3.17}$$

Then the comparison system of the first approximation for (2.3.15) will be written as

$$\begin{aligned} \frac{d\bar{x}}{dt} &= \mu X_0(\bar{x}, \mu), \\ \frac{d\bar{y}}{dt} &= \omega(\bar{x}) + \mu Y_0(\bar{x}, \mu), \end{aligned} \tag{2.3.18}$$

the system of the r-th approximation will be written in the form

$$\begin{aligned} \frac{d\bar{x}}{dt} &= \mu X_0(\bar{x}, \mu) + \sum_{k=2}^{r} \mu^k A_k(\bar{x}, \mu), \\ \frac{d\bar{y}}{dt} &= \omega(\bar{x}) + \mu Y_0(\bar{x}, \mu) + \sum_{k=2}^{r} \mu^k B_k(\bar{x}, \mu), \end{aligned} \tag{2.3.19}$$

and the comparison system of the general form (or of any approximation) will be written as

$$\begin{aligned} \frac{d\bar{x}}{dt} &= \mu X_0(\bar{x}, \mu) + \sum_{k\geq 2} \mu^k A_k(\bar{x}, \mu), \\ \frac{d\bar{y}}{dt} &= \omega(\bar{x}) + \mu Y_0(\bar{x}, \mu) + \sum_{k\geq 2} \mu^k B_k(\bar{x}, \mu). \end{aligned} \tag{2.3.20}$$

The vector-functions A_k and B_k are so far unknown.

Comparison systems (2.3.18), (2.3.19), and (2.3.20) have an important property. The averaging operator with respect to y results in splitting of the $(m+n)$-order equation system into two subsystems that are integrated independently: one subsystem determining the slow averaged variables $\bar{x}$ has order m, the other subsystem determining $\bar{y}$ has order n. If the subsystem of the first approximation

$$\frac{d\bar{x}}{dt} = \mu X_0(\bar{x}, \mu) \tag{2.3.21}$$

can be integrated, then the calculation of fast averaged variables y adds up to the simple quadrature:

$$\bar{y}(t,\mu) = y_0 + \int_0^t [\omega(\bar{x}(\tau,\mu)) + \mu Y_0(\bar{x}(\tau,\mu),\mu)]\, d\tau. \tag{2.3.22}$$

It remains unclear how the functions $\bar{x}$ and $\bar{y}$ relate to the exact solution of x, y of rotary system (2.3.15). For the study of this question we will use the averaging method, because we will need asymptotic representations of these functions. We will seek for the substitution of variables

$$\begin{aligned} x(t,\mu) &= \bar{x}(t,\mu) + \sum_{k\geq 1} \mu^k u_k(\bar{x},\bar{y},\mu), \\ y(t,\mu) &= \bar{y}(t,\mu) + \sum_{k\geq 1} \mu^k v_k(\bar{x},\bar{y},\mu), \end{aligned} \tag{2.3.23}$$

that transforms system (2.3.15) into one of the comparison systems (2.3.18), (2.3.19), (2.3.20), and we will see that in each case we get different substitutions of variables (i.e. different functions u_k and v_k).

Upon differentiation of (2.3.23) and the execution of the corresponding calculations for transform (2.3.15) → (2.3.18) we get an infinite-dimensional system of linear partial differential equations:

$$\begin{aligned}
\left(\frac{\partial u_1}{\partial \bar{y}}, \omega(\bar{x})\right) &= X_N(\bar{x},\bar{y},\mu) - X_0(\bar{x},\mu), \\
\left(\frac{\partial v_1}{\partial \bar{y}}, \omega(\bar{x})\right) &= Y_N(\bar{x},\bar{y},\mu) - Y_0(\bar{x},\mu) + \left(\frac{\partial \omega(\bar{x})}{\partial \bar{x}}, u_1\right), \\
\left(\frac{\partial u_2}{\partial \bar{y}}, \omega(\bar{x})\right) &= \left(\frac{\partial X_N(\bar{x},\bar{y},\mu)}{\partial \bar{x}}, u_1\right) + \left(\frac{\partial X_N(\bar{x},\bar{y},\mu)}{\partial \bar{y}}, v_1\right) \\
&\quad - \left(\frac{\partial u_1}{\partial \bar{x}}, X_0(\bar{x},\mu)\right) - \left(\frac{\partial u_1}{\partial \bar{y}}, Y_0(\bar{x},\mu)\right), \\
\left(\frac{\partial v_2}{\partial \bar{y}}, \omega(\bar{x})\right) &= \left(\frac{\partial Y_N(\bar{x},\bar{y},\mu)}{\partial \bar{x}}, u_1\right) + \left(\frac{\partial Y_N(\bar{x},\bar{y},\mu)}{\partial \bar{y}}, v_1\right) \\
&\quad + \left(\frac{\partial \omega(\bar{x})}{\partial \bar{x}}, u_2\right) + \frac{1}{2!}\left(\left(\frac{\partial^2 \omega(\bar{x})}{\partial \bar{x}^2}, u_1\right), u_1\right) \\
&\quad - \left(\frac{\partial v_1}{\partial \bar{x}}, X_0(\bar{x},\mu)\right) - \left(\frac{\partial v_1}{\partial \bar{y}}, Y_0(\bar{x},\mu)\right), \\
\left(\frac{\partial u_r}{\partial \bar{y}}, \omega(\bar{x})\right) &= U_r(\bar{x},\bar{y},\mu,u_1,v_1,\dots,u_{r-1},v_{r-1}), \\
\left(\frac{\partial v_r}{\partial \bar{y}}, \omega(\bar{x})\right) &= V_r(\bar{x},\bar{y},\mu,u_1,v_1,\dots,u_{r-1},v_{r-1},u_r),
\end{aligned} \tag{2.3.24}$$

For the transform (2.3.15) → (2.3.19) we have

$$
\begin{aligned}
\left(\frac{\partial u_1}{\partial \bar{y}}, \omega(\bar{x})\right) &= X_N(\bar{x},\bar{y},\mu) - X_0(\bar{x},\mu), \\
\left(\frac{\partial v_1}{\partial \bar{y}}, \omega(\bar{x})\right) &= Y_N(\bar{x},\bar{y},\mu) - Y_0(\bar{x},\mu) + \left(\frac{\partial \omega(\bar{x})}{\partial \bar{x}}, u_1\right), \\
\left(\frac{\partial u_2}{\partial \bar{y}}, \omega(\bar{x})\right) &= \left(\frac{\partial X_N(\bar{x},\bar{y},\mu)}{\partial \bar{x}}, u_1\right) + \left(\frac{\partial X_N(\bar{x},\bar{y},\mu)}{\partial \bar{y}}, v_1\right) \\
&\quad - \left(\frac{\partial u_1}{\partial \bar{x}}, X_0(\bar{x},\mu)\right) - \left(\frac{\partial u_1}{\partial \bar{y}}, Y_0(\bar{x},\mu)\right) - A_2(\bar{x},\mu), \\
\left(\frac{\partial v_2}{\partial \bar{y}}, \omega(\bar{x})\right) &= \left(\frac{\partial Y_N(\bar{x},\bar{y},\mu)}{\partial \bar{x}}, u_1\right) + \left(\frac{\partial Y_N(\bar{x},\bar{y},\mu)}{\partial \bar{y}}, v_1\right) \\
&\quad + \left(\frac{\partial \omega(\bar{x})}{\partial \bar{x}}, u_2\right) + \frac{1}{2!}\left(\left(\frac{\partial^2 \omega(\bar{x})}{\partial \bar{x}^2}, u_1\right), u_1\right) \\
&\quad - \left(\frac{\partial v_1}{\partial \bar{x}}, X_0(\bar{x},\mu)\right) - \left(\frac{\partial v_1}{\partial \bar{y}}, Y_0(\bar{x},\mu)\right) - B_2(\bar{x},\mu), \\
&\cdots\cdots\cdots\cdots\cdots\cdots\cdots\cdots\cdots\cdots\cdots\cdots \\
\left(\frac{\partial u_r}{\partial \bar{y}}, \omega(\bar{x})\right) &= U_r(\bar{x},\bar{y},\mu,u_1,v_1,\ldots,u_{r-1},v_{r-1},A_2,B_2,\ldots,A_r), \\
\left(\frac{\partial v_r}{\partial \bar{y}}, \omega(\bar{x})\right) &= V_r(\bar{x},\bar{y},\mu,u_1,\ldots,v_{r-1},u_r,A_2,B_2,\ldots,A_r,B_r), \\
\left(\frac{\partial u_{r+1}}{\partial \bar{y}}, \omega(\bar{x})\right) &= U_{r+1}(\bar{x},\bar{y},\mu,u_1,v_1,\ldots,u_r,v_r,A_2,B_2,\ldots,A_r), \\
\left(\frac{\partial v_{r+1}}{\partial \bar{y}}, \omega(\bar{x})\right) &= V_{r+1}(\bar{x},\bar{y},\mu,u_1,v_1,\ldots,v_r,u_{r+1},A_2,B_2,\ldots,A_r,B_r), \\
&\cdots\cdots\cdots\cdots\cdots\cdots\cdots\cdots\cdots\cdots\cdots\cdots
\end{aligned} \tag{2.3.25}
$$

And finally, for the transform (2.3.15) → (2.3.20) we have

$$
\begin{aligned}
\left(\frac{\partial u_1}{\partial \bar{y}}, \omega(\bar{x})\right) &= X_N(\bar{x},\bar{y},\mu) - X_0(\bar{x},\mu), \\
\left(\frac{\partial v_1}{\partial \bar{y}}, \omega(\bar{x})\right) &= Y_N(\bar{x},\bar{y},\mu) - Y_0(\bar{x},\mu) + \left(\frac{\partial \omega(\bar{x})}{\partial \bar{x}}, u_1\right), \\
&\cdots\cdots\cdots\cdots\cdots\cdots\cdots\cdots\cdots\cdots\cdots\cdots \\
\left(\frac{\partial u_r}{\partial \bar{y}}, \omega(\bar{x})\right) &= U_r(\bar{x},\bar{y},\mu,u_1,v_1,\ldots,u_{r-1}v_{r-1},A_2,\ldots,A_r), \\
\left(\frac{\partial v_r}{\partial \bar{y}}, \omega(\bar{x})\right) &= V_r(\bar{x},\bar{y},\mu,u_1,v_1,\ldots,v_{r-1},u_r,A_2,\ldots,A_r,B_r), \\
\left(\frac{\partial u_{r+1}}{\partial \bar{y}}, \omega(\bar{x})\right) &= U_{r+1}(\bar{x},\bar{y},\mu,u_1,\ldots,v_r,A_2,\ldots,A_{r+1}), \\
\left(\frac{\partial v_{r+1}}{\partial \bar{y}}, \omega(\bar{x})\right) &= V_{r+1}(\bar{x},\bar{y},\mu,u_1,\ldots,v_r,u_{r+1},A_2,\ldots,A_{r+1},B_{r+1}), \\
&\cdots\cdots\cdots\cdots\cdots\cdots\cdots\cdots\cdots\cdots\cdots\cdots
\end{aligned} \tag{2.3.26}
$$

The infinite systems (2.3.24), (2.3.25), and (2.3.26) can be integrated successively (to any value of the index) in analytic form. Indeed, substituting X_N, Y_N into (2.3.24) – (2.3.26) for u_1 we obtain the equation

$$\left(\frac{\partial u_1}{\partial \bar{y}}, \omega(\bar{x})\right) = \sum_{1\le\|k\|\le N} X_k(\bar{x},\mu)e^{i(k,\bar{y})}, \tag{2.3.27}$$

and its general solution can be written in the form

$$u_1(\bar{x},\bar{y},\mu) = \sum_{1\le\|k\|\le N} \frac{X_k(\bar{x},\mu)e^{i(k,\bar{y})}}{i(k,\omega(\bar{x}))} + \varphi_1(\bar{x},\mu), \tag{2.3.28}$$

where φ_1 is an arbitrary vector-function differentiable with respect to $\bar{x}$. Due to the locally nonresonance nature of system (2.3.15) and the boundedness of $\|k\| \le N$, the denominators $(k,\, \omega(\bar{x}))$ in (2.3.28) with $t \in [0,\, T]$ do not vanish.

Now we will make two essential remarks.

Remark 1 Comparing the first two equations of the infinite-dimensional systems (2.3.24) – (2.3.26) we see that they exactly coincide, and consequently, the first approximation of asymptotic perturbation theory has the same analytic structure irrespective of the analytical form of the comparison equations.

Remark 2 In the substitution of variables (2.3.23), in equations (2.3.24) – (2.3.27), and in the solution (2.3.28) the dependence of functions on the small parameter μ as an argument is shown. It is this parameter that is included as a symbol into (2.3.23) and into the functions from the comparison equations.

If we now substitute the expression for u_1 (2.3.28) into the second equation of any of the comparison systems (2.3.24) – (2.3.26) and construct the general solution for v_1 by the method of characteristics, we will obtain

$$\begin{aligned} v_1(\bar{x},\bar{y},\mu) = &\sum_{1\le\|k\|\le N} \frac{Y_k(\bar{x},\mu)e^{i(k,\bar{y})}}{i(k,\omega(\bar{x}))} \\ &+ \left(\frac{\partial \omega(\bar{x})}{\partial \bar{x}}, \sum_{1\le\|k\|\le N} \frac{X_k(\bar{x},\mu)e^{i(k,\bar{y})}}{i^2(k,\omega(\bar{x}))^2}\right)\left(\left(\frac{\partial \omega(\bar{x})}{\partial \bar{x}}, \varphi_1(\bar{x},\mu)\right), \bar{y}\right) \\ &+ \psi_1(\bar{x},\mu), \end{aligned} \tag{2.3.29}$$

where ψ_1 is also an arbitrary function differentiable with respect to $\bar{x}$.

Study the behavior of the functions $u_1(\bar{x},\bar{y},\mu)$ and $v_1(\bar{x},\bar{y},\mu)$. If $\varphi_1 \not\equiv 0$, then in the function v_1 the following item appears:

$$\left(\left(\frac{\partial \omega(\bar{x})}{\partial \bar{x}}, \varphi_1(\bar{x},\mu)\right), \bar{y}\right),$$

which has the nature of a secular term, since it is proportional to the fast vector $\bar{y}$, and the latter (or rather, its norm) is a semilinear time function t. If we assume that $\psi_1 \neq 0$, then the function v_1 will generate a secular term when finding the function u_2 from the third equation of comparison systems (2.3.24) – (2.3.26). Consequently, to build an asymptotic theory of the first approximation for the rotary system (2.3.15) in the form of trigonometric items, it is necessary to assume

$$\varphi_1(\bar{x},\mu) \equiv 0, \quad \psi_1(\bar{x},\mu) \equiv 0. \tag{2.3.30}$$

Under this condition we obtain

$$u_1(\bar{x},\bar{y},\mu) = \sum_{1\le\|k\|\le N} \frac{X_k(\bar{x},\mu)e^{i(k,\bar{y})}}{i(k,\omega(\bar{x}))}, \tag{2.3.31}$$

$$\begin{aligned} v_1(\bar{x},\bar{y},\mu) = &\sum_{1\le\|k\|\le N} \frac{Y_k(\bar{x},\mu)e^{i(k,\bar{y})}}{i(k,\omega(\bar{x}))} \\ &+ \left(\frac{\partial\omega(\bar{x})}{\partial\bar{x}}, \sum_{1\le\|k\|\le N} \frac{X_k(\bar{x},\mu)e^{i(k,\bar{y})}}{i^2(k,\omega(\bar{x}))^2}\right). \end{aligned} \tag{2.3.32}$$

So, from the infinite number of solutions of the first two equations of the comparison system (2.3.24) – (2.3.26) we have chosen solutions 2π-periodical with respect to the vector $\bar{y}$, for u_1 and v_1.

Now consider the second approximation of asymptotic theory, i.e. determine the functions $u_2(\bar{x},\bar{y},\mu)$, $v_2(\bar{x},\bar{y},\mu)$ of the substitution of variables (2.3.23). First of all consider the third equation of the comparison system (2.3.24) that determines the function u_2. Having calculated the partial derivatives $\partial u_1/\partial\bar{x}$ and $\partial u_1/\partial\bar{y}$ and having substituted them into the right-hand member of the third equation with u_1 and v_1, together we will get the following equation:

$$\begin{aligned} \left(\frac{\partial u_2}{\partial\bar{y}}, \omega(\bar{x})\right) = &\left(\sum_{0\le\|k\|\le N} \frac{\partial X_k(\bar{x},\mu)}{\partial\bar{x}}, \sum_{1\le\|s\|\le N} \frac{X_s(\bar{x},\mu)e^{i(k+s,\bar{y})}}{i(s,\omega(\bar{x}))}\right) \\ &+ \sum_{1\le\|k\|\le N}\ \sum_{1\le\|s\|\le N} \frac{(kX_k, Y_s)e^{i(k+s,\bar{y})}}{(s,\omega(\bar{x}))} \\ &+ \left(\frac{\partial\omega(\bar{x})}{\partial\bar{x}}, \sum_{1\le\|k\|\le N}\ \sum_{1\le\|s\|\le N} \frac{(X_s,\ (X_k,(k,e)))e^{i(k+s,\bar{y})}}{i(s,\omega(\bar{x}))^2}\right) \\ &- \left(\sum_{1\le\|k\|\le N} \frac{(\partial X_k(\bar{x},\mu)/\partial\bar{x})e^{i(k,\bar{y})}}{i(k,\omega(\bar{x}))}, X_0(\bar{x},\mu)\right) \\ &- \left(\sum_{1\le\|k\|\le N} \frac{X_k(\bar{x},\mu)e^{i(k,\bar{y})}}{(k,\omega(\bar{x}))}(k,\ Y_0(\bar{x},\mu))\right), \end{aligned} \tag{2.3.33}$$

where the symbol (k,e) signifies $\sum_{s=1}^{r} k_s$. In the right-hand member of equation

(2.3.33) there are terms that do not depend on $\bar{y}$. We will write them separately, because it is these terms that generate the secular terms in the solution u_2:

$$
\begin{aligned}
U_{2,\text{sec}} = &- \sum_{1\le\|k\|\le N} \frac{(\partial X_k(\bar{x},\mu)/\partial x,\ X_{-k}(\bar{x},\mu))}{i(k,\omega(\bar{x}))} \\
&- \sum_{1\le\|k\|\le N} \frac{(kX_k(\bar{x},\mu),\ Y_{-k}(\bar{x},\mu))}{(k,\ \omega(\bar{x}))} \\
&- \left(\frac{\partial\omega(\bar{x})}{\partial\bar{x}},\ \sum_{1\le\|k\|\le N} \frac{(X_{-k}(\bar{x},\mu),\ (X_k(\bar{x},\mu),\ (k,e)))}{i(k,\omega(\bar{x}))^2}\right)
\end{aligned}
\tag{2.3.34}
$$

with $k+s=0$.

Now if we again write an equation for $u_2(\bar{x},\bar{y},\mu)$, it will take the form

$$
\begin{aligned}
\left(\frac{\partial u_2}{\partial\bar{y}},\omega(\bar{x})\right) = U_{2,\text{sec}} &+ \sum_{\substack{0\le\|k\|\le N\ 1\le\|s\|\le N \\ k+s\ne 0}} \frac{(\partial X_k/\partial\bar{x},\ X_s)\, e^{i(k+s,\bar{y})}}{i(s,\omega(\bar{x}))} \\
&+ \sum_{\substack{1\le\|k\|\le N 1\le\|s\|\le N \\ k+s\ne 0}} \frac{(kX_k,\ Y_s)\, e^{i(k+s,\bar{y})}}{(s,\omega(\bar{x}))} \\
&+ \left(\frac{\partial\omega(\bar{x})}{\partial\bar{x}},\ \sum_{\substack{1\le\|k\|\le N 1\le\|s\|\le N \\ k+s\ne 0}} \frac{(X_s,\ (X_k,\ (k,e)))e^{i(k+s,\bar{y})}}{i(s,\omega(\bar{x}))^2}\right) \\
&- \left(\sum_{1\le\|k\|\le N} \frac{(\partial X_k/\partial\bar{x})e^{i(k,y)}}{i(k,\omega(\bar{x}))},\ X_0\right) \\
&- \left(\sum_{1\le\|k\|\le N} \frac{X_k e^{i(k,\bar{y})}}{(k,\omega(\bar{x}))}\ (k,\ Y_0)\right).
\end{aligned}
\tag{2.3.35}
$$

Integration of equation (2.3.35) by the method of characteristics gives

$$
u_2(\bar{x},\bar{y},\mu) = U_{2,\text{sec}} \times \frac{\bar{y}}{\omega(\bar{x})} + S_1 + S_2, \tag{2.3.36}
$$

where

$$
\begin{aligned}
S_1 = &\sum_{\substack{0\le\|k\|\le n\ 1\le\|s\|\le N \\ k+s\ne 0}} \frac{(\partial X_k/\partial\bar{x},\ X_s)e^{i(k+s,\bar{y})}}{i^2(k+s,\omega(\bar{x}))(s,\omega(\bar{x}))} \\
&+ \sum_{\substack{1\le\|k\|\le N\ 1\le\|s\|\le N \\ k+s\ne 0}} \frac{(kX_k,\ Y_s)e^{i(k+s,\bar{y})}}{i(k+s,\omega(\bar{x}))(s,\omega(\bar{x}))},
\end{aligned}
$$

$$S_2 = \left(\frac{\partial \omega(\bar{x})}{\partial \bar{x}}, \sum_{1 \le \|k\| \le N} \sum_{\substack{1 \le \|s\| \le N \\ k+s \ne 0}} \frac{(X_s, (X_k, (k,e)))e^{i(k+s,\bar{y})}}{i^2(s,\omega(\bar{x}))^2(k+s,\omega(\bar{x}))} \right)$$

$$- \left(\sum_{1 \le \|k\| \le N} \frac{(\partial X_k/\partial \bar{x}) e^{i(k,\bar{y})}}{i^2(k,\omega(\bar{x}))^2}, X_0(\bar{x},\mu) \right)$$

$$- \left(\sum_{1 \le \|k\| \le N} \frac{X_k e^{i(k,\bar{y})}}{i(k\omega(x))^2} (k, Y_0(\bar{x},\mu)) \right) + \varphi_2(\bar{x},\mu).$$

Here φ_2 is an arbitrary vector-function differentiable with respect to $\bar{x}$, and the symbol $U_{2,\mathrm{sec}} \times \bar{y}/\bar{\omega}$ denotes a vector with components

$$\frac{\bar{y}_1}{\bar{\omega}_1} U^{(1)}_{2,\mathrm{sec}}, \quad \frac{\bar{y}_2}{\bar{\omega}_2} U^{(2)}_{2,\mathrm{sec}}, \quad \dots, \quad \frac{\bar{y}_n}{\bar{\omega}_n} U^{(n)}_{2,\mathrm{sec}},$$

with their upper index coinciding with the number of the component of vector u_2, $\bar{\omega} = \omega(\bar{x})$.

So if we find a transformation of the form (2.3.23) transforming (2.3.15) into (2.3.18), then in the first approximation the indestructible secular terms ($U_{2,\mathrm{sec}} \times \bar{y}/\bar{\omega} \sim t$) already appear in the function $u_2(\bar{x},\bar{y},\mu)$. If we integrate the equation for the function v_2, it will be easy to see that the latter will also contain secular terms proportional to t.

Hence it can be concluded that there is no trigonometric substitution of variables (2.3.23) that would actualize the transform (2.3.15) $\to$ (2.3.18). Indestructible secular terms are present among the second-order terms with respect to μ in the relations (2.3.23).

Now consider the transform (2.3.15) – (2.3.19). This is achieved through the use of the infinite-dimensional system (2.3.25). As we have mentioned above, the first two equations of systems (2.3.24) and (2.3.25) coincide, therefore the functions of the first approximation u_1 and v_1 are expressed by the same formulae (2.3.28), (2.3.29), or (2.3.31), (2.3.32) as for the transform (2.3.15) $\to$ (2.3.18). But the equations of the second approximation already differ in the fact that system (2.3.25) still includes indefinite functions $A_2(\bar{x},\mu)$, $B_2(\bar{x},\mu)$. The presence of such functions in the second members of equations for u_2, v_2 allows us to use them at our discretion. The most expedient method seems to be to use an approach when with the help of A_2, B_2 we attempt to exclude from those equations items non-periodic with respect to $\bar{y}$, that have generated the secular terms in u_2, v_2 (see formula (2.3.36)).

In other words, the functions A_2, B_2 should be chosen in such a way that the following equalities will be true

$$\begin{gathered} \int_0^{2\pi} \cdots \int_0^{2\pi} U_2(\bar{x},\bar{y},\mu,u_1,v_1,A_2)\, d\bar{y}_1 \cdots d\bar{y}_n = 0, \\ \int_0^{2\pi} \cdots \int_0^{2\pi} V_2(\bar{x},\bar{y},\mu,u_1,v_1,u_2,A_2,B_2)\, d\bar{y}_1 \cdots d\bar{y}_n = 0, \end{gathered} \tag{2.3.37}$$

or explicitly,

$$A_2(\bar{x}, \mu) = U_{2,\mathrm{sec}}(\bar{x}, \mu), \tag{2.3.38}$$

$$B_2(\bar{x}, \mu) = V_{2,\mathrm{sec}}(\bar{x}, \mu). \tag{2.3.39}$$

Here $V_{2,\mathrm{sec}}(\bar{x}, \mu)$ is an item independent of $\bar{y}$ in the fourth equation of system (2.3.25) and results from the substitution of the previously found functions u_1, v_1, u_2, A_2 into it, and the execution of algebraic operations on trigonometric polynomials.

Upon such a choice of the function A_2 the third equation of the system (2.3.25) takes the form

$$\begin{aligned}
\left(\frac{\partial u_2}{\partial \bar{y}}, \omega(\bar{x})\right) &= \sum_{\substack{1\le\|k\|\le N\\ k+s\neq 0}} \sum_{1\le\|s\|\le N} \frac{(\partial X_k/\partial \bar{x}, X_s)\, e^{i(k+s,y)}}{i(s,\omega(\bar{x}))} \\
&+ \sum_{\substack{1\le\|k\|\le N\\ k+s\neq 0}} \sum_{1\le\|s\|\le N} \frac{(kX_k, Y_s)\, e^{i(k+s,\bar{y})}}{(s,\omega(\bar{x}))} \\
&+ \left(\frac{\partial \omega(\bar{x})}{\partial \bar{x}}, \sum_{\substack{1\le\|k\|\le N\\ k+s\neq 0}} \sum_{1\le\|s\|\le N} \frac{(X_s, (X_k, (k,e)))\, e^{i(k+s,\bar{y})}}{i(s,\omega(\bar{x}))^2}\right) \\
&- \left(\sum_{1\le\|k\|\le N} \frac{(\partial X_k/\partial \bar{x}) e^{i(k,\bar{y})}}{i(k,\omega(\bar{x}))}, X_0(\bar{x}, \mu)\right) \\
&- \left(\sum_{1\le\|k\|\le N} \frac{X_k e^{i(k,\bar{y})}}{(k,\omega(\bar{x}))} (k, Y_0)\right).
\end{aligned} \tag{2.3.40}$$

The method of characteristics gives

$$\begin{aligned}
u_2(\bar{x}, \bar{y}, \mu) &= \sum_{\substack{0\le\|k\|\le N\\ k+s\neq 0}} \sum_{1\le\|s\|\le N} \frac{(\partial X_k/\partial \bar{x}, X_s) e^{i(k+s,\bar{y})}}{i^2(k+s,\omega(\bar{x}))(s,\omega(\bar{x}))} \\
&+ \sum_{\substack{1\le\|k\|\le N\\ k+s\neq 0}} \sum_{1\le\|s\|\le N} \frac{(kX_k, Y_s) e^{i(k+s,\bar{y})}}{i(k+s,\omega(\bar{x}))(s,\omega(\bar{x}))} \\
&+ \left(\frac{\partial \omega(\bar{x})}{\partial \bar{x}}, \sum_{\substack{1\le\|k\|\le N\\ k+s\neq 0}} \sum_{1\le\|s\|\le N} \frac{(X_s, (X_k, (k,e)))\, e^{i(k+s,\bar{y})}}{i^2(k+s,\omega(\bar{x}))(s,\omega(\bar{x}))^2}\right) \\
&- \left(\sum_{1\le\|k\|\le N} \frac{(\partial X_k/\partial \bar{x}) e^{i(k,\bar{y})}}{i^2(k,\omega(\bar{x}))^2}, X_0\right) - \sum_{1\le\|k\|\le N} \frac{X_k e^{i(k,\bar{y})}}{i(k,\omega(\bar{x}))^2} (k, Y_0).
\end{aligned} \tag{2.3.41}$$

In a similar way one can find the function $v_2(\bar{x}, \bar{y}, \mu)$; if B_2 is determined from the condition (2.3.39), then the function will not contain secular terms but will only be expressed via functions periodic with respect to $\bar{y}$. Of course, this is true on the assumption that the arbitrary functions φ_2 and ψ_2 emerging in the integration process are considered as identically equal to zero.

So the obtained analytic expressions (2.3.31), (2.3.32) show that the functions u_1, v_1, u_2, v_2 constituting the Krylov–Bogolyubov transform (2.3.23), to an accuracy of $O(\mu^2)$ contain only 2π-periodic functions with respect to the fast variables $\bar{y}$. This property is also valid for each $s = 2, 3, \dots, r$, if A_s, B_s are determined from the conditions

$$\begin{aligned} &\int_0^{2\pi} \cdots \int_0^{2\pi} U_s(\bar{x}, \bar{y}, \mu, u_1, v_1, \dots, v_{s-1}, A_2, B_2, \dots, A_s)\, d\bar{y} = 0, \\ &\int_0^{2\pi} \cdots \int_0^{2\pi} V_s(\bar{x}, \bar{y}, \mu, u_1, v_1, \dots, v_{s-1}, u_s, A_2, B_2, \dots, A_s, B_s)\, d\bar{y} = 0. \end{aligned} \tag{2.3.42}$$

As regards the functions u_{r+1}, v_{r+1}, in accordance with (2.3.25) the defining equations do not contain any arbitrary functions, and the destruction of secular terms at the $(r+1)$ step is impossible. The functions u_{r+1}, v_{r+1} will contain both secular terms and items periodic with respect to $\bar{y}$, and can be symbolically written as

$$\begin{aligned} u_{r+1}(\bar{x}, \bar{y}, \mu) &= U_{r+1,\mathrm{sec}}(\bar{x}, \mu)\, \frac{\bar{y}}{\omega(\bar{x})} + \{\text{terms periodic with respect to } \bar{y}\}, \\ v_{r+1}(\bar{x}, \bar{y}, \mu) &= V_{r+1,\mathrm{sec}}(\bar{x}, \mu)\, \frac{\bar{y}}{\omega(\bar{x})} + \{\text{terms periodic with respect to } \bar{y}\}, \end{aligned} \tag{2.3.43}$$

where the secular coefficients only depend on $\bar{x}$ and μ.

Hence a conclusion can be drawn, which is very important for the asymptotic theory of differential equations of the form (2.3.15). The transformation (2.3.15) $\to$ (2.3.19) can be achieved in the class of trigonometric substitutions of variables to an accuracy of $O(\mu^r)$ only. There is no exact substitution of variables (2.3.23) in the class of trigonometric functions, that would transform the initial rotary system (2.3.15) into the comparison system of the r-th approximation (2.3.25).

The situation fundamentally changes in case of the mapping (2.3.15) $\to$ (2.3.20), i.e. when the comparison system is a system of any approximation. For the mapping (2.3.15) $\to$ (2.3.20) substitution of variables (2.3.23) is determined by the infinite-dimensional system of linear partial differential equations (2.3.26). The analysis of this system shows that at any step of s (not only for $s \le r$) the equations contain arbitrary functions of the form A_s, B_s that can be chosen in such a way that they would destroy secular terms at each step. Application of the method of characteristics allows us to find transfer functions u_s, v_s in an analytical form for each s. For this purpose it is only necessary to integrate equations (2.3.26) at the previous steps $1, 2, \dots, s-1$. Thus, for u_s, v_s we obtain expressions of the form

$$\begin{aligned} u_s(\bar{x}, \bar{y}, \mu) &= \{\text{terms periodic with respect to } \bar{y}\}, \\ v_s(\bar{x}, \bar{y}, \mu) &= \{\text{terms periodic with respect to } \bar{y}\} \\ s &= 1, 2, \dots \end{aligned} \tag{2.3.44}$$

Conclusion. The mapping of the initial rotary system (2.3.15) into the comparison system of any approximation (2.3.20) can be performed with any prescribed accuracy $O(\mu^s)$ in the class of trigonometric substitutions of variables of the form (2.3.23). With $s \to \infty$ there exist formal trigonometric substitutions of variables of the form (2.3.23), transforming equations (2.3.15) into equations (2.3.20) that also contain infinite power series of μ. Convergence of such series is problematic.

Having constructed the substitution of variables (2.3.23), i.e. having found the functions u_s, v_s, it is possible to analyze the values $\|x(t, \mu) - \bar{x}(t, \mu)\|$ and $\|y(t, \mu) - \bar{y}(t, \mu)\|$. To an accuracy of $O(\mu^2)$ it can be concluded that

$$\|x(t, \mu) - \bar{x}(t, \mu)\| \leq \mu \|u_1(\bar{x}, \bar{y}, \mu)\|,$$

where $\bar{x}(t, \mu)$ are slow variables of one of the comparison systems (2.3.18) – (2.3.20). For systems locally nonresonant on $t \in [0, T]$ the norm of the function u_1 is easily calculated from formula (2.3.31):

$$\|u_1(\bar{x}, \bar{y}, \mu)\| < C/\alpha, \tag{2.3.45}$$

where C is a constant bounding the initial function $X_N(x, y, \mu)$ in the definitional domain G_{m+n+1}. Then, obviously,

$$\|x(t, \mu) - \bar{x}(t, \mu)\| < \mu C/\alpha \tag{2.3.46}$$

and with $\mu \to 0$, $\|x - \bar{x}\| \to 0$ for all $t \in [0, T]$, where the condition of the local nonresonance character of the system (2.3.15) is fulfilled.

For the norm $\|y - \bar{y}\|$ the matter becomes far more complicated. Indeed, with the mapping (2.3.15) $\to$ (2.3.18) the function u_2 contains a secular term proportional to t. When finding the function v_2 this term generates a secular term proportional to t^2, so $\|y(t, \mu) - \bar{y}(t, \mu)\| > 0$, if, e.g., $T = C\mu^{-1}$. Hence it follows that without some essential additional conditions put on the initial functions ω, X_N, Y_N it is impossible to obtain the ε-proximity of $y(t, \omega)$ and $\bar{y}(t, \omega)$.

2.4 Initial Conditions for Comparison Equations

In Section 2.3 we described the algorithm for finding the functions u_k, v_k for three classes of comparison equations: comparison equations of the first approximation (2.3.18), those of the r-th approximation (2.3.19) and those of any (including the infinite) approximation (2.3.20). Under certain conditions the Krylov–Bogolyubov transform (2.3.23) has a trigonometric form, but it turns out that in this case the comparison equations should be solved with the initial conditions differing from those given for the initial equations (2.3.15). We would like to emphasize this important circumstance, because when justifying Krylov–Bogolyubov method for different classes of ordinary differential equations the overwhelming majority of authors formulate theorems on ε-proximity of exact and averaged solutions for the same initial conditions.

So we will look for such initial conditions for comparison equations that would guarantee the existence of trigonometric substitutions of variables of the form (2.3.23). We know that the necessary condition for the existence of 2π-periodic substitutions of variables is the equality of functions $\varphi_k(\bar{x},\mu)$, $\psi_k(\bar{x},\mu)$ $(k=1,2,\dots)$ to zero. Under these conditions the functions u_k and v_k are expressed by functions (2.3.31), (2.3.32), and if we substitute $t=0$ in them, then

$$u_1(\bar{x}(0,\mu),\,\bar{y}(0,\mu),\mu) = \sum_{1\le\|k\|\le N} \frac{X_k(\bar{x}(0,\mu),\mu)\,e^(i(k,\bar{y}(0,\mu)))}{i(k,\omega(\bar{x}(0,\mu)))} \neq 0, \tag{2.4.1}$$

$$\begin{aligned} v_1(\bar{x}(0,\mu),\,\bar{y}(0,\mu),\mu) = {} & \sum_{1\le\|k\|\le N} \frac{Y_k(\bar{x}(0,\mu),\mu)\,e^{i(k,\bar{y}(0,\mu))}}{i(k,\omega(\bar{x}(0,\mu)))} \\ & + \left(\frac{\partial\omega(\bar{x}(0,\mu))}{\partial\bar{x}},\ \sum_{1\le\|k\|\le N} \frac{X_k(\bar{x}(0,\mu),\mu)e^{i(k,\bar{y}(0,\mu))}}{i^2(k,\omega(\bar{x}(0,\mu)))^2}\right) \neq 0. \end{aligned} \tag{2.4.2}$$

For the above reason, in the first approximation the vectors $\|x(0,\mu)-\bar{x}(0,\mu)\|$, $\|y(0,\mu)-\bar{y}(0,\mu)\|$ are not equal to zero vectors because

$$\begin{aligned} x(0,\mu)-\bar{x}(0,\mu) &= \mu u_1(\bar{x}(0,\mu),\,\bar{y}(0,\mu),\mu), \\ y(0,\mu)-\bar{y}(0,\mu) &= \mu v_1(\bar{x}(0,\mu),\,\bar{y}(0,\mu),\mu). \end{aligned} \tag{2.4.3}$$

These norms have order $O(\mu)$ which is retained in any s-th approximation.

Thus, it is possible to find a Krylov–Bogolyubov transform of the form (2.3.23) in a trigonometric form (in the form of functions 2π-periodic with respect to $\bar{y}$) in case one calculates the solution of comparison equations with the initial conditions defined from the following relations: for the mapping (2.3.15) $\to$ (2.3.18)

$$\begin{aligned} \bar{x}(0,\mu)+\mu u_1(\bar{x}(0,\mu),\,\bar{y}(0,\mu),\mu) &= x_0, \\ \bar{y}(0,\mu)+\mu v_1(\bar{x}(0,\mu),\,\bar{y}(0,\mu),\mu) &= y_0; \end{aligned} \tag{2.4.4}$$

for the mapping (2.3.15) $\to$ (2.3.19)

$$\begin{aligned} \bar{x}(0,\mu)+\sum_{k=1}^{r}\mu^k u_k(\bar{x}(0,\mu),\,\bar{y}(0,\mu),\mu) &= x_0, \\ \bar{y}(0,\mu)+\sum_{k=1}^{r}\mu^k v_k(\bar{x}(0,\mu),\,\bar{y}(0,\mu),\mu) &= y_0; \end{aligned} \tag{2.4.5}$$

and for the mapping (2.3.15) → (2.3.20)

$$\begin{aligned}\bar{x}(0,\mu) + \sum_{k\geq 1} \mu^k u_k(\bar{x}(0,\mu), \bar{y}(0,\mu), \mu) &= x_0,\\ \bar{y}(0,\mu) + \sum_{k\geq 1} \mu^k v_k(\bar{x}(0,\mu), \bar{y}(0.\mu), \mu) &= y_0.\end{aligned} \tag{2.4.6}$$

Equations (2.4.4) – (2.4.6) are function-transcendent equations that include $\bar{x}(0,\mu)$ and $\bar{y}(0,\mu)$ as unknowns; x_0, y_0 are the initial values for the initial system (2.3.15). When solving such equations, as a rule one should use different iteration methods (cf. Bakhvalov [1]), including that of simple iterations.

For the mapping (2.3.15) → (2.3.18)

$$\begin{aligned}\bar{x}^{(k)}(0,\mu) &= x_0 - \mu u_1(\bar{x}^{(k-1)}(0,\mu), \bar{y}^{(k-1)}(0,\mu), \mu),\\ \bar{y}^{(k)}(0,\mu) &= y_0 - \mu v_1(\bar{x}^{(k-1)}(0,\mu), \bar{y}^{(k-1)}(0,\mu), \mu).\end{aligned} \tag{2.4.7}$$

At the first step

$$\begin{aligned}\bar{x}^{(1)}(0,\mu) &= x_0 - \mu u_1(x_0, y_0, \mu),\\ \bar{y}^{(1)}(0,\mu) &= y_0 - \mu v_1(x_0, y_0, \mu).\end{aligned} \tag{2.4.8}$$

For the mapping (2.3.15) → (2.3.19) we obtain

$$\begin{aligned}\bar{x}^{(k)}(0,\mu) &= x_0 - \sum_{s=1}^{r} \mu^s u_s(\bar{x}^{(k-1)}(0,\mu), \bar{y}^{(k-1)}(0,\mu), \mu),\\ \bar{y}^{(k)}(0,\mu) &= y_0 - \sum_{s=1}^{r} \mu^s v_s(\bar{x}^{(k-1)}(0,\mu), \bar{y}^{(k-1)}(0,\mu), \mu),\\ \bar{x}^{(0)}(0,\mu) &= x_0, \quad \bar{y}^{(0)}(0,\mu) = y_0, \quad k = 1, 2, \dots ;\end{aligned} \tag{2.4.9}$$

and for the mapping (2.3.15) → (2.3.20)

$$\begin{aligned}\bar{x}^{(k)}(0,\mu) &= x_0 - \sum_{s\geq 1} \mu^s u_s(\bar{x}^{(k-1)}(0,\mu), \bar{y}^{(k-1)}(0,\mu), \mu),\\ \bar{y}^{(k)}(0,\mu) &= y_0 - \sum_{s\geq 1} \mu^s v_s(\bar{x}^{(k-1)}(0,\mu), \bar{y}^{(k-1)}(0,\mu), \mu),\\ \bar{x}^{(0)}(0,\mu) &= x_0, \quad \bar{y}^{(0)}(0,\mu) = y_0, \quad k = 1, 2, \dots .\end{aligned} \tag{2.4.10}$$

Now we will give some totals of this asymptotic theory of locally nonresonance systems with slow and fast variables of the form (2.3.1). We write them in the form of an algorithm.

Step 1. If we are given the errors δ_1 and δ_2 for the solution of equation (2.3.1) within the time interval $t \in [0, T]$, then we will first define the values ε_1 and ε_2 characterizing the deviations of the right-hand members of the approximate equations (2.3.2) from the right-hand members of the given equations (2.3.1). When finding ε_1 and ε_2, in formulae (2.3.10) one should substitute, e.g. $\delta_1/2$ and $\delta_2/2$ instead of δ_1 and δ_2. The compensation of the deviation of the solutions, equal to $\delta_1/2$ and $\delta_2/2$, will be found through asymptotic theory.

Step 2. Knowing the values ε_1 and ε_2, we find N giving the place of "truncation" of the Fourier series for the functions $X(x, y, \mu)$, $Y(x, y, \mu)$. Now instead of the rotary system (2.3.1) we will consider the multifrequency system

$$\begin{aligned}
\frac{dx}{dt} &= \mu X_N(x, y, \mu) \equiv \mu \sum_{0 \le \|k\| \le N} X_k(x, \mu) e^{i(k,y)}, \\
\frac{dy}{dt} &= \omega(x) + \mu Y_N(x, y, \mu) \equiv \omega(x) + \mu \sum_{0 \le \|k\| \le N} Y_k(x, \mu) e^{i(k,y)}, \\
& x(0, \mu) = x_0, \quad y(0, \mu) = y_0,
\end{aligned} \tag{2.4.11}$$

with its right-hand members being trigonometric polynomials with respect to y, but not infinite series.

Step 3. We apply the operator M_y of averaging with respect to phase variables y, to the right-hand members of equations (2.4.11) i.e. we find the comparison functions

$$\begin{aligned}
\bar{X}_N(\bar{x}, \mu) &= M_y[X_N(\bar{x}, y, \mu)] \\
&\equiv \frac{1}{(2\pi)^n} \int_0^{2\pi} \cdots \int_0^{2\pi} X_N(\bar{x}, y, \mu)\, dy_1 \cdots dy_n = X_0(\bar{x}, \mu), \\
\bar{Y}_N(\bar{x}, \mu) &= Y_0(\bar{x}, \mu)
\end{aligned} \tag{2.4.12}$$

and construct the comparison equations of the first, r-th or any approximation (2.3.18), (2.3.19), or (2.3.20). For example, write the averaged system of the r-th approximation for (2.4.11):

$$\begin{aligned}
\frac{d\bar{x}}{dt} &= \mu X_0(\bar{x}, \mu) + \sum_{s=2}^{r} \mu^s A_s(\bar{x}, \mu), \\
\frac{d\bar{y}}{dt} &= \omega(\bar{x}) + \mu Y_0(\bar{x}, \mu) + \sum_{s=2}^{r} \mu^s B_s(\bar{x}, \mu).
\end{aligned} \tag{2.4.13}$$

Step 4. To write the equations (2.4.13) uniquely it is necessary to find the transfer functions

$$\begin{aligned}
x &= \bar{x} + \sum_{s=1}^{r} \mu^s u_s(\bar{x}, \bar{y}, \mu), \\
y &= \bar{y} + \sum_{s=1}^{r} \mu^s v_s(\bar{x}, \bar{y}, \mu).
\end{aligned} \tag{2.4.14}$$

They are found in an analytical form by the method described in Section 2.3, and of course, this method seems extremely promising from the point of view of the use of computers for the execution of analytical (literal) operations. Expressions for A_s, B_s $(s = 2, 3, \dots, r)$ are written in analytic form as well, i.e. the right-hand members of comparison equations are written uniquely.

Step 5. If we consider the transformation formulae (2.4.14) as asymptotic transformations of the initial unknown variables $x(t,\mu)$, $y(t,\mu)$, it is necessary to find the solution of the comparison equations $\bar{x}(t,\mu)$, $\bar{y}(t,\mu)$. First we will calculate the initial values of $\bar{x}(0,\mu)$, $\bar{y}(0,\mu)$ from the functional equations

$$\bar{x}(0,\mu) + \sum_{s=1}^{r} \mu^s u_s(\bar{x}(0,\mu), \bar{y}(0,\mu), \mu) = x_0,$$
$$\bar{y}(0,\mu) + \sum_{s=1}^{r} \mu^s v_s(\bar{x}(0,\mu), \bar{y}(0,\mu), \mu) = y_0.$$

Then we will solve the Cauchy problem for the comparison equations (2.4.13) with the initial values $\bar{x}(0,\mu)$, $\bar{y}(0,\mu)$. However, since the averaging operator M_y results in a separation of movements and the subsystem of m-th order differential equations for $\bar{x}$ does not contain the vector $\bar{y}$, the Cauchy problem should be solved for the subsystem of slow variables only:

$$\frac{d\bar{x}}{dt} = \mu X_0(\bar{x},\mu) + \sum_{s=1}^{r} \mu^s A_s(\bar{x},\mu), \qquad \bar{x}(t,\mu)\big|_{t=0} = \bar{x}(0,\mu). \tag{2.4.15}$$

If $\bar{x}$ are found, then the fast variables $\bar{y}$ will be found by simple integration:

$$\bar{y}(t,\mu) = \bar{y}(0,\mu) + \int_0^t \Big[\omega(\bar{x}(\tau,\mu)) + \mu Y_0(\bar{x}(\tau,\mu),\mu) + \sum_{s=2}^{r} \mu^s B_s(\bar{x}(\tau,\mu),\mu) \Big] \, d\tau. \tag{2.4.16}$$

Now substituting the functions $\bar{x}(t,\mu)$ and $\bar{y}(t,\mu)$ into the transformation formulae (2.4.14) we will find the solution of the initial system (2.3.15) $x(t,\mu)$, $y(t,\mu)$ with error $O(\mu^r)$.

2.5 Averaging Operator for Time-Independent Disturbances

The most widely used averaging operators $M_t, M_z, M_z(s)$ described in Section 1.1 have a certain drawback: in the smoothing process the dependence of the solution on time is not accounted for. Therefore when smoothing it is very important to consider, at least partially, the information on the dynamic properties of the solutions.

Such smoothing operators are *the operator of averaging along the generating solution* and *the operator of averaging for time-independent disturbances.* Apparently, the former was first used in the theory of lunar motion by Charles Delaunay (see Delaunay [1]), and the mathematical aspects of its applicability in nonresonance multifrequency systems were first considered by Volosov [1].

To construct this operator, substitute $\mu = 0$ into the initial differential equation (1.3.1) and obtain, in accordance with Poincaré's terms, the generating equation

$$\frac{dz^{(0)}}{dt} = Z(z^{(0)}, t, 0) \equiv Z^{(0)}(z^{(0)}, t). \tag{2.5.1}$$

Assume that the general solution of this generating equation is known:

$$z^{(0)}(t) = \varphi(t, C), \tag{2.5.2}$$

where C is an n-dimensional vector of arbitrary constants. Replacing in $Z(z, t, \mu)$ the variable z by the function φ, we obtain a new function depending on the parameters C, μ and on time:

$$Z^*(t, C, \mu) \equiv Z(\varphi(t, C),\, t, \mu). \tag{2.5.3}$$

To this function we can apply the time averaging operator $M_t[Z^*]$ determined from formula (1.1.6):

$$M_t[Z^*] = \bar{Z}^*(C, \mu) = \lim_{T\to\infty} \frac{1}{T} \int\limits_0^T Z^*(t, C, \mu)\, dt. \tag{2.5.4}$$

In this case the comparison equation of the first approximation will be written in the form

$$\frac{d\bar{z}}{dt} = \bar{Z}^*(C, \mu), \tag{2.5.5}$$

and evidently it can be integrated right away:

$$\bar{z}(t, \mu) = \bar{Z}^*(C, \mu)t + z_0. \tag{2.5.6}$$

It is hard to expect that the solutions of the system (1.3.1), $z(t, \mu)$ and $\bar{z}(t, \mu)$, would be ε-close in norm; therefore some workers recommend replacing the vector C (or the vector of initial values z_0) in comparison equations by the function $\bar{z}$. After that the comparison equation of the first approximation for (1.3.1) takes the form

$$\frac{d\bar{z}}{dt} = \bar{Z}^*(\bar{z}, \mu), \tag{2.5.7}$$

and this was the subject of inquiry in the work of many mathematicians. But it should be noted that the procedure of changing C to $\bar{z}$ is not strongly valid.

To eliminate this drawback to a certain extent, another averaging operator was proposed by Grebenikov [2, 3] which can be called *an averaging operator for time-independent disturbances.*

In (1.3.1) perform the substitution of variables

$$p(t,\mu) = z(t,\mu) - \varphi(t,C). \tag{2.5.8}$$

Then the vector-function p is defined by the differential equation

$$\frac{dp}{dt} = P(p,t,C,\mu) \equiv Z(p+\varphi,t,\mu) - Z^{(0)}(\varphi(t,C),t), \tag{2.5.9}$$

equivalent to the initial equation (1.3.1). Now apply the procedure of averaging with respect to t to the second member of (2.5.9), in the integration process regarding p, C, μ as constants:

$$\bar{P}(p,C,\mu) = \lim_{T\to\infty} \frac{1}{T} \int_0^T P(p,t,C,\mu)\,dt. \tag{2.5.10}$$

It is the formula (2.5.10) that defines the averaging operator for time-independent disturbances.

The comparison equation of the first approximation for (2.5.9) takes the form

$$\frac{d\bar{p}}{dt} = \bar{P}(\bar{p},C,\mu). \tag{2.5.11}$$

If we write a comparison equation of the first approximation for (1.3.1), constructed by means of an averaging operator for time-independent disturbances, we will obtain

$$\frac{d\bar{z}}{dt} = \bar{P}(\bar{z}-\varphi,\, C,\mu) + Z^{(0)}(\varphi,t). \tag{2.5.12}$$

The rule of construction of such an operator automatically saves, in the second members of the comparison equations (2.5.11) and (2.5.12), the unknown $\bar{p}(t,\mu)$ and $\bar{z}(t,\mu)$; therefore there is no uncertainty like in the case of the use of an operator of averaging along the generating solution.

Now we will give an example of an application of an averaging operator for time-independent disturbances to show its main advantage.

Let $X(x,y)$ be a 2π-periodic function with respect to y and satisfying in some domain $G_{m+n} = \{(x,y)\colon\ x \in P_m,\ y \in Q_n\}$ the conditions of the Jordan–Dirichlet theorem on Fourier series expansibility. In the domain G_{m+n} we will have

$$X(x,y) = \sum_{\|k\|\geq 0} X_k(x) e^{i(k,y)}. \tag{2.5.13}$$

Applying the operator of averaging with respect to y (Gauss's operator) to (2.5.13), we obtain

$$M_y[X] = \frac{1}{(2\pi)^n} \int_0^{2\pi} \ldots \int_0^{2\pi} X(x,y)\,dy_1 \ldots dy_n = X_0(x). \tag{2.5.14}$$

This very operator was used for the construction of the asymptotic theory of locally nonresonant systems (see sections 2.3 and 2.4). Now suppose that

$$y(t) = \omega t + p \tag{2.5.15}$$

and the numeric vector ω has rationally incommensurable components, i.e. $(k, \omega) \neq 0$ for all integer vectors k, except the vector with zero norm. If the substitution (2.5.15) is performed in (2.5.13), we obtain the conditionally periodic function of time

$$X(x, \omega t + p) = \sum_{||k|| \geq 0} X_k(x) \exp\{i(k, \omega)t + i(k, p)\}. \tag{2.5.16}$$

To (2.5.16) we apply an averaging operator for time-independent disturbances, in the integration process regarding p as a constant vector:

$$M_t[X(x, \omega t + p)] = \lim_{T \to \infty} \frac{1}{T} \int_0^T X(x, \omega t + p)\, dt = X_0(x). \tag{2.5.17}$$

Now let the frequency vector ω have rationally commensurable components, i.e. the equality $(k, \omega) = 0$ is true for some set of integer vectors k^*. Then

$$\begin{aligned} X(x, \omega t + p) = &\sum_{k \in \{k^*\}}{}' X_k(x) \exp\{i(k, p)\} \\ &+ \sum{}'' X_k(x) \exp\{i(k, \omega)t + i(k, p)\}, \end{aligned} \tag{2.5.18}$$

where two primes denote summation over the set of integers apart those for which $(k, \omega) = 0$. In other words, the first sum includes resonance harmonics, and the other nonresonance harmonics.

After application of an averaging operator for time-dependent disturbances we obtain

$$M_t[X(x, \omega t + p)] = \bar{X}(x, p) = \sum_{k \in \{k^*\}}{}' X_k(x) e^{i(k,p)}, \tag{2.5.19}$$

hence it is clear that after the smoothing procedure not only do the "resonance" summands of expansion remain, but implicitly the dependence of the averaged function on p was kept as well.

Though the averaging operator for time-dependent disturbances is denoted by the same symbol as the operator of averaging with respect to the time-averaging operator, there is an essential difference between them. First, the averaging operator for time-dependent disturbances is also applied when the initial functions do not depend explicitly on time, and second, time is introduced through the dependence of the generating solution on it, but not arbitrarily.

2.6 Asymptotic Theory of Systems with Their Paths Passing through Resonance Points

Revert to rotary system (2.3.15)

$$\begin{aligned} \frac{dx}{dt} &= \mu X_N(x, y, \mu), \\ \frac{dy}{dt} &= \omega(x) + \mu Y_N(x, y, \mu), \\ x(0, \mu) &= x_0, \quad y(0, \mu) = y_0, \end{aligned} \tag{2.6.1}$$

where the vector-functions X_N, Y_N are represented by n-fold Fourier polynomials with respect to y

$$\begin{aligned} X_N(x, y, \mu) &= \sum_{0 \le \|k\| \le N} X_k(x, \mu) e^{i(k,y)}, \\ Y_N(x, y, \mu) &= \sum_{0 \le \|k\| \le N} Y_k(x, \mu) e^{i(k,y)}. \end{aligned} \tag{2.6.2}$$

Instead of (2.6.1) we introduce a system for disturbances, using the substitution of variables

$$p = x - x_0, \quad q = y - \omega_0 t - y_0, \quad \omega_0 = \omega(x_0), \tag{2.6.3}$$

and obtain

$$\begin{aligned} \frac{dp}{dt} &= \mu X_N(p + x_0,\, q + \omega_0 t + y_0,\, \mu), \\ \frac{dq}{dt} &= \omega(p + x_0) - \omega_0 + \mu Y_n(p + x_0,\, q + \omega_0 t + y_0,\, \mu); \end{aligned} \tag{2.6.4}$$

$$\begin{aligned} X_N &= \sum_{0 \le \|k\| \le N} X_k(p + x_0,\, \mu) \exp\{i(k, \omega_0)t + i(k, q + y_0)\}, \\ Y_N &= \sum_{0 \le \|k\| \le N} Y_k(p + x_0,\, \mu) \exp\{i(k, \omega_0)t + i(k, q + y_0)\}. \end{aligned} \tag{2.6.5}$$

The equivalence of (2.6.1) and (2.6.4) is obvious.

Let the initial vector of frequencies ω_0 satisfy, for $k \in \{k^*\}$, $\|k\| \neq 0$, the resonance condition

$$(k, \omega_0) = 0. \tag{2.6.6}$$

Apply Gauss's operator M_y to the right-hand members of equation (2.6.1). Then the averaged equations of the first, r-th and any approximation will exactly coincide with the comparison equations (2.3.18) – (2.3.20). As an example see below the comparison equations of any approximation

$$\begin{aligned} \frac{d\bar{x}}{dt} &= \mu X_0(\bar{x}, \mu) + \sum_{k \ge 2} \mu^k A_k(\bar{x}, \mu), \\ \frac{d\bar{y}}{dt} &= \omega(\bar{x}) + \mu Y_0(\bar{x}, \mu) + \sum_{k \ge 2} \mu^k B_k(\bar{x}, \mu). \end{aligned} \tag{2.6.7}$$

It seems that for the construction of the mapping (2.6.1) $\to$ (2.6.7) one should use the substitution of variables (2.3.23) and the system (2.3.26) since there is no external difference between (2.3.15) and (2.6.1). However the solution of system (2.3.26) for (2.6.1) with consideration for the resonance conditions (2.6.6) will differ from the solution given in Section 2.3. Indeed, the equation for u_1 in this case can be written in the form

$$\left(\frac{\partial u_1}{\partial \bar{y}}, \omega(\bar{x})\right) = \sum_{1\le\|k\|\le N}' X_k(\bar{x},\mu)e^{i(k,\bar{y})} + \sum_{1\le\|k\|\le N}'' X_k(\bar{x},\mu)e^{i(k,\bar{y})}, \tag{2.6.8}$$

where the first sum contains terms for which $(k, \omega_0) = 0$ (resonance sum), and the second sum contains nonresonant terms. In locally nonresonant systems the first sum is lacking.

Due to the presence of the resonance sum the solution of equation (2.6.8) is a complicated procedure.

First assume that the paths of system (2.6.1) have the property of remaining in a resonance point, i.e. the equality

$$(k,\, \omega(\bar{x})) = 0 \tag{2.6.9}$$

is true not only at the initial point x_0, but also for some interval of t. In equality (2.6.9) the vector k takes the same values as in the first sum in (2.6.8). Then, using the method of characteristics, we can write the general solution for (2.6.8) in the form

$$\begin{aligned} u_1(\bar{x},\bar{y},\mu) &= \frac{1}{n}\sum_{s=1}^{n}\frac{\bar{y}_s}{\omega_s(\bar{x})}\sum_{1\le\|k\|\le N}' X_k(\bar{x},\mu)e^{i(k,\bar{y})} \\ &\quad + \sum_{1\le\|k\|\le N}'' \frac{X_k(\bar{x},\mu)e^{i(k,\bar{y})}}{i(k,\omega(\bar{x}))} + \varphi_1(\bar{x},\mu), \end{aligned} \tag{2.6.10}$$

where $\varphi_1(\bar{x},\mu)$ is an arbitrary vector-function differentiable with respect to $\bar{x}$, like it was before. The first summand in (2.6.10) actually contains a secular term multiplied by $e^{i(k,\bar{y})}$, because

$$\frac{1}{n}\sum_{s=1}^{n}\frac{\bar{y}_s}{\omega_s(\bar{x})} = t + \{\text{small function}\}, \tag{2.6.11}$$

i.e. the first summand consists of terms of the form $t\sin\alpha t$, $t\cos\alpha t$ which Poincaré called *mixed disturbances.*

So, if the averaging of the right-hand members is performed by means of the operator M_y and for some vectors k the resonance relation (2.6.9) holds, then already at the first step in the Krylov–Bogolyubov transform (2.3.23) secular (or, more exactly, mixed) terms appear, and consequently an asymptotic theory of disturbances of rotary systems of the form (2.6.1) in trigonometric form cannot be constructed.

Now assume that the resonance relation (2.6.9) holds true on some discrete set of points $\{\bar{x}_{\text{res}}\}$. In this case we have two analytic representations for the function u_1:

$$u_1(\bar{x},\bar{y},\mu) = \frac{1}{n}\sum_{s=1}^{n}\frac{\bar{y}_s}{\omega_s(\bar{x})}\sum_{1\le\|k\|\le N}{}' X_k(\bar{x},\mu)e^{i(k,\bar{y})} + \sum_{1\le\|k\|\le N}{}'' \frac{X_k(\bar{x},\mu)e^{i(k,\bar{y})}}{i(k,\omega(\bar{x}))} + \varphi_1(\bar{x},\mu), \tag{2.6.12}$$

if $(k,\omega(\bar{x}_{\text{res}})) = 0$;

$$u_1(\bar{x},\bar{y},\mu) = \sum_{1\le\|k\|\le N} \frac{X_k(\bar{x},\mu)e^{i(k,\bar{y})}}{i(k,\omega(\bar{x}))} + \varphi_1(\bar{x},\mu), \tag{2.6.13}$$

if $(k,\omega(\bar{x})) \neq 0$.

So for the transform (2.3.23) we will have two variations: (2.6.12) for the resonance case (it contains secular functions) and (2.6.13) for nonresonance points. Of course, the double expression for the function $u_1(\bar{x},\bar{y},\mu)$ generates a double expression for the function v_1.

In the first case we obtain

$$\begin{aligned} v_1&(\bar{x},\bar{y},\mu) \\ &= \frac{1}{n}\sum_{s=1}^{n}\frac{\bar{y}_s}{\omega_s(\bar{x})}\sum_{1\le\|k\|\le N}{}' Y_k(\bar{x},\mu)e^{i(k,\bar{y})} + \sum_{1\le\|k\|\le N}{}'' \frac{Y_k(\bar{x},\mu)e^{i}(k,\bar{y})}{i(k,\omega(\bar{x}))} \\ &+ \frac{\bar{y}_1^2}{2n\omega_1(\bar{x})}\left(\frac{\partial\omega(\bar{x})}{\partial\bar{x}}, \left(\frac{1}{\omega_1(\bar{x})} + \sum_{s=2}^{n}\omega_s(\bar{x})\sum_{1\le\|k\|\le N}{}' X_k(\bar{x},\mu)e^{i(k,\bar{y})}\right)\right) \\ &+ \frac{\bar{y}_1}{n\omega_1(\bar{x})}\left(\frac{\partial\omega(\bar{x})}{\partial\bar{x}}, \left(\sum_{s=2}^{n}\frac{\bar{y}_s}{\omega_s(\bar{x})} - \frac{n\bar{y}_1}{\omega_1(\bar{x})}\right)\sum_{1\le\|k\|\le N}{}' X_k(\bar{x},\mu)e^{i(k,\bar{y})}\right) \\ &+ \frac{1}{\omega_1(\bar{x})}\left(\frac{\partial\omega(\bar{x})}{\partial\bar{x}}, \sum_{1\le\|k\|\le N}{}'' \frac{X_k(\bar{x},\mu)e^{i(k,\bar{y})}}{i^2(k,\omega(\bar{x}))^2}\right) \\ &+ \frac{\bar{y}_1}{\omega_1(\bar{x})}\left(\frac{\partial\omega(\bar{x})}{\partial\bar{x}}, \varphi_1(\bar{x},\mu)\right) + \psi_1(\bar{x},\mu). \end{aligned} \tag{2.6.14}$$

In the second case,

$$\begin{aligned} v_1&(\bar{x},\bar{y},\mu) \\ &= \sum_{1\le\|k\|\le N} \frac{Y_k(\bar{x},\mu)e^{i(k,\bar{y})}}{i(k,\omega(\bar{x}))} + \left(\frac{\partial\omega(\bar{x})}{\partial\bar{x}}, \sum_{1\le\|k\|\le N} \frac{X_k(\bar{x},\mu)e^{i(k,\bar{y})}}{i^2(k,\omega(\bar{x}))^2}\right) \\ &+ \frac{\bar{y}_1}{\omega_1(\bar{x})}\left(\frac{\partial\omega(\bar{x})}{\partial\bar{x}}, \varphi_1(\bar{x},\mu)\right) + \psi_1(\bar{x},\mu). \end{aligned} \tag{2.6.15}$$

The expression (2.6.14) shows that already in the first approximation in the resonance case the function v_1 contains a secular term of the form t^2 $(\bar{y}_1/\omega_1 \sim t)$, and in the nonresonance case, a secular term of the form t, if $\varphi_1(\bar{x},\mu) \equiv 0$.

So we obtain two variations of the asymptotic theory of disturbances of first order: one for the resonance case, another for the nonresonance case. It is easy to infer that this conclusion remains true for the theory of disturbances of a higher order as well.

Now let the condition of initial resonance (2.6.6) be kept, and a procedure of averaging of functions X_N, Y_N by means of an averaging operator for time-independent disturbance be performed. Then the comparison equations of any approximation will be written in the form

$$\begin{aligned}\frac{d\bar{p}}{dt} &= \mu \sum_{0\leq\|k\|\leq N}{}' X_k(\bar{p}+x_0,\mu)e^{i(k,y_0+\bar{q})} + \sum_{k\geq 2}\mu^k A_k(\bar{p},\bar{q},\mu),\\ \frac{d\bar{q}}{dt} &= \omega(\bar{p}+x_0) - \omega(x_0) + \mu \sum_{0\leq\|k\|\leq N}{}' Y_k(\bar{p}+x_0,\mu)e^{i(k,y_0+\bar{q})}\\ &\quad + \sum_{k\geq 2}\mu^k B_k(\bar{p},\bar{q},\mu).\end{aligned} \tag{2.6.16}$$

Our main objective is to show that the mapping (2.6.4) $\rightarrow$ (2.6.16) can be achieved by means of trigonometric substitution of variables of the form (2.3.23).

Find the Krylov–Bogolyubov transform in the form

$$p = \bar{p} + \sum_{k\geq 1}\mu^k u_k(\bar{p},\bar{q},t,\mu), \qquad q = \bar{q} + \sum_{k\geq 1}\mu^k v_k(\bar{p},\bar{q},t,\mu). \tag{2.6.17}$$

Then the system determining the transformation functions u_s, v_s will have the following analytic form:

$$\begin{aligned}&\frac{\partial u_1}{\partial t} + \left(\frac{\partial u_1}{\partial \bar{q}}, \omega(\bar{p}+x_0) - \omega_0\right)\\ &\quad = \sum_{1\leq\|k\|\leq N}{}'' X_k(\bar{p}+x_0\mu)\exp\{i(k,\, \bar{q}+\omega_0 t + y_0)\}.\\ &\frac{\partial v_1}{\partial t} + \left(\frac{\partial v_1}{\partial \bar{q}}, \omega(\bar{p}+x_0) - \omega_0\right)\\ &\quad = \sum_{1\leq\|k\|\leq N}{}'' Y_k(\bar{p}+x_0,\mu)\exp\{i(k,\bar{q}+\omega_0 t + y_0)\} + \left(\frac{\partial\omega(\bar{p}+x_0)}{\bar{p}}, u_1\right),\\ &\cdots\cdots\cdots\cdots\cdots\cdots\cdots\cdots\cdots\cdots\\ &\frac{\partial u_s}{\partial t} + \left(\frac{\partial u_s}{\partial \bar{q}}, \omega(\bar{p}+x_0) - \omega_0\right) =\\ &\quad = U_s(\bar{p},\bar{q},t,\mu,u_1,v_1,\ldots,u_{s-1},v_{s-1},A_2,B_2,\ldots,A_s),\\ &\frac{\partial v_s}{\partial t} + \left(\frac{\partial v_s}{\partial \bar{q}}, \omega(\bar{p}+x_0) - \omega_0\right)\\ &\quad = V_s(\bar{p},\bar{q},t,\mu,u_1,\ldots,v_{s-1},u_s,A_2,B_2,\ldots,A_s,B_s).\end{aligned} \tag{2.6.18}$$

In sums with two primes k only takes nonresonance values. The system (2.6.18) may be successively solved in analytic form. For example, we have

$$
\begin{aligned}
u_1(\bar{p},\bar{q},t,\mu) = & \sum_{1\le\|k\|\le N}'' \frac{X_k(\bar{p}+x_0,\mu)\exp\{i(k,\bar{q}+\omega_0 t+y_0)\}}{i(k,\omega(\bar{p}+x_0))} \\
& + \varphi_1(\bar{p},\mu), \\
v_1(\bar{p},\bar{q},t,\mu) = & \sum_{1\le\|k\|\le N}'' \frac{Y_k(\bar{p}+x_0,\mu)\exp\{i(k,\bar{q}+\omega_0 t+y_0)\}}{i(k,\omega(\bar{p}+x_0))} \\
& + \left(\frac{\partial\omega(\bar{p}+x_0)}{\partial\bar{p}}, \sum_{1\le\|k\|\le N}'' \frac{X_k(\bar{p}+x_0,\mu)\exp\{i(k,\bar{q}+\omega_0 t+y_0)\}}{i^2(k,\omega(\bar{p}+x_0))^2}\right) \\
& + \psi_1(\bar{p},\mu).
\end{aligned}
\tag{2.6.19}
$$

Expressions (2.6.19) show that in the case of averaging with respect to t it is possible to determine the first terms of the Krylov–Bogolyubov transform in a trigonometric form, assuming that

$$\varphi_1(\bar{p},\mu) \equiv \psi_1(\bar{p},\mu) \equiv 0.$$

If at each step of the iterations the functions $A_s(\bar{p},\bar{q},\mu)$, $B_s(\bar{p},\bar{q},\mu)$ are chosen as average values of the right-hand members of equations (2.6.18) with respect to t, i.e. if they are determined from the relations

$$\lim_{T\to\infty}\frac{1}{T}\int_0^T U_s\,dt = 0, \quad \lim_{T\to\infty}\frac{1}{T}\int_0^T V_s\,dt = 0, \tag{2.6.20}$$

and the functions φ_s, ψ_s are equated to zero, then it is possible to construct the Krylov–Bogolyubov transform in a trigonometric form for higher approximations as well. Then, to obtain the asymptotic representations of the solutions of the initial system (2.6.1) it is necessary to substitute the solution of comparison equations (2.6.16) and the solutions of the form (2.6.19) into transform (2.6.17). The transition from the variables (disturbances) p, q to the initial variables x, y is trivial.

Conclusion If the solution of the multifrequency system (2.6.1) passes through one or several resonance points, then on averaging of the right-hand members by means of the Gauss operator M_y we obtain a sufficiently complicated asymptotic perturbation theory with secular terms, with different analytical representations for resonance and nonresonance sections of the path. By means of the averaging operator for time-independent disturbances the analytical form for the disturbances is noticeably simplified, but solution of the problem of the first, second and other approximations becomes more complicated, because in the presence of the resonance of initial frequencies, comparison equations of the first, second and any approximations do not split into subsystems separately determining the slow variables $\bar{x}(t,\mu)$

and the fast variables $\bar{y}(t,\mu)$. If frequency resonances are present, there is no split of motions in phase space.

2.7 The Algorithm of Joining of Resonance and Nonresonance Path Sections

Formulae (2.6.12) – (2.6.15) show that for approximate solutions of the initial multi-frequency system (2.6.1) we will have different analytic representations for different points of time. These analytic differences generate analytic formulae for solutions, quasi-independent from each other, but since the integral path of initial equations is a continuous function, we should join the different sections.

For this purpose we will split the closed interval $[0,\, T]$ into two sets:

(a) the time resonance set $I_{\text{res}}(T)$ consisting of intervals for which

$$|(k, \omega(\bar{x}(t,\mu)))| \leq \alpha, \quad t \in \big[t_{\text{l,res}}^{(s)},\, t_{\text{r,res}}^{(s)}\big], \tag{2.7.1}$$

where α is the width of the resonance zone, and $t_{\text{l,res}}^{(s)}$ and $t_{\text{r,res}}^{(s)}$ are the left and the right ends of the s-th resonance interval;

(b) the time nonresonance set $I_{\text{nonres}}(T)$ consisting of intervals for which

$$|(k, \omega(\bar{x}(t,\mu)))| > \alpha, \quad t \in \big[t_{\text{l,nonres}}^{(s)},\, t_{\text{r,nonres}}^{(s)}\big], \tag{2.7.2}$$

where $t_{\text{l,nonres}}^{(s)}$, $t_{\text{r,nonres}}^{(s)}$ are the left and the right ends of the s-th nonresonance interval ($t_{\text{l,nonres}}^{(s)} = t_{\text{r,res}}^{(s)}$, because the first interval is a resonance one, $(k,\omega_0) = 0$, $t_{\text{l,res}}^{(1)} = 0$, $t_{\text{r,nonres}}^{(s)} = t_{\text{l,res}}^{(s+1)}$).

It is obvious that

$$I_{\text{res}}(T) \cup I_{\text{nonres}}(T) = [0,\, T]. \tag{2.7.3}$$

On the set $I_{\text{res}}(T)$ asymptotic perturbation theory is represented by formulae with secular disturbances, if the averaging was made by means of the operator M_y, and on the set $I_{\text{nonres}}(T)$ by formulae of the form (2.6.13). For both variants to represent the solution $x(t,\mu)$, $y(t,\mu)$, it is necessary to join these expressions at the points $t_{\text{l,res}}^{(s+1)} = t_{\text{r,nonres}}^{(s)}$, $t_{\text{r,res}}^{(s)} = t_{\text{l,nonres}}^{(s)}$. It is also necessary to join the expressions for the derivatives dx/dt, dy/dt, because the right-hand members of the initial equations are continuous functions. To perform the joining it is necessary that $2(m+n)$ satisfies the conditions of the form

$$\begin{aligned} &x\big(t_{\text{l,res}}^{(s+1)}, \mu\big) = x\big(t_{\text{r,nonres}}^{(s)}, \mu\big), \qquad y\big(t_{\text{l,res}}^{(s+1)}, \mu\big) = y\big(t_{\text{r,nonres}}^{(s)}, \mu\big), \\ &\frac{dx\big(t_{\text{l,res}}^{(s+1)}, \mu\big)}{dt} = \frac{dx\big(t_{\text{r,nonres}}^{(s)}, \mu\big)}{dt}, \qquad \frac{dy\big(t_{\text{l,res}}^{(s+1)}, \mu\big)}{dt} = \frac{dy\big(t_{\text{r,nonres}}^{(s)}, \mu\big)}{dt}. \end{aligned} \tag{2.7.4}$$

The joining conditions (2.7.4) constitute a system of $2(m+n)$ functional equations with the arbitrary functions φ_1, ψ_1 and their partial derivatives $d\varphi_1/dt$,

$d\psi_1/dt$ to be considered as the unknown functions, if the theory of the first approximation is being constructed. Since φ_1, ψ_1 are arbitrary functions differentiable with respect to $\bar{x}$, then for system (2.7.4) to be compatible we should introduce $2(m+n)$ unknown parameters. For example, as $\varphi_1^{(s)}$, $\psi_1^{(r)}$ we can take functions linear relative to $\bar{x}$

$$\begin{aligned} \varphi_1^{(s)}(\bar{x}, \mu) &= a_s \bar{x}_s + \mu b_s, & \psi_1^{(r)}(\bar{x}, \mu) &= c_r \bar{x}_r + \mu d_r, \\ s &= 1, \dots, m; & r &= 1, \dots, n, \end{aligned} \tag{2.7.5}$$

and then system (2.7.4) adds up to two subsystems of linear algebraic equations. (One is of order $2m$ with the unknowns a_s, b_s $(s = 1, \dots, m)$, the other is of order $2n$ with the unknowns c_r, d_r $(r = 1, \dots, n)$). Systems of the form (2.7.4) have to be solved each time when passing from a resonance interval to a nonresonance one and vice versa.

Now it remains to define the width of the resonance zone—the number α determining the moments $t_{1,\mathrm{res}}^{(s)}$, $t_{\mathrm{r,res}}^{(s)}$ in particular. The latter essentially depend on T, i.e. in the time interval in which the asymptotic theory of multifrequency systems is constructed. If $T = O(\mu^{-1})$, then it is expedient to take $O(\mu)$ as α, e.g.

$$\alpha = \mu, \tag{2.7.6}$$

i.e. we think that the width of the resonance zone is equal to 2μ. To obtain $t_{\mathrm{r,res}}^{(1)}$ $\big(t_{1,\mathrm{res}}^{(1)} = 0$ owing to the fact that $(k, \omega_0) = 0\big)$, it is necessary to solve the following equation with respect to t:

$$(k^*, \, \omega(\bar{x}(t, \mu))) = \mu \tag{2.7.7}$$

where k^* is the resonance vector with the lowest norm.

Let the solutions of the equation (2.7.7) be denoted by

$$0 < t_1^* < t_2^* < \ldots \leq T. \tag{2.7.8}$$

Then it is clear that $t_{\mathrm{r,res}}^{(1)} = t_1^*$, $t_{1,\mathrm{res}}^{(2)} = t_2^*$, etc. Of course, the situation can arise when the equation (2.7.7) has no solution $t_k^* \leq T$. This means that within the interval $[0, T]$ the path does not leave the α-zone of the initial resonance $(k, \omega_0) = 0$.

2.8 Periodic and Quasi-Periodic Oscillations in the Van der Pol Oscillator System

In electric circuit theory (see Van der Pol [1]) mathematical models are used that are called autonomous and nonautonomous Van der Pol oscillators. The former is the second-order differential equation

$$\frac{d^2x}{dt^2} + \omega^2 x = \mu(1 - x^2)\frac{dx}{dt}, \tag{2.8.1}$$

and the latter is the equation

$$\frac{d^2x}{dt^2} + \omega^2 x = \mu(1 - x^2)\frac{ddx}{dt} + \mu\kappa\lambda\cos\lambda t. \qquad (2.8.2)$$

Here $x(t, \mu)$ denotes the strength of current in a tube generator with an oscillatory circuit, ω is the eigenfrequency of the oscillatory circuit of the tube generator, μ is a small parameter characterizing the connection of the circuit with the vacuum tube, λ is the frequency of the exciting current strength, and κ is a positive parameter.

$$\omega - \lambda = O(\mu^\gamma), \quad \gamma \geq 1, \qquad (2.8.3)$$

i.e. the eigenfrequency and the frequency of excitation are closely approximated, we have a dual-frequency resonance; the greater is γ, the "sharper" is the resonance. This will also be clear from the analytic constructions shown below.

A more general mathematical model is comprised of the systems of autonomous and nonautonomous Van der Pol oscillators, described by the systems of ordinary differential equations

$$\frac{d^2x_k}{dt^2} + \omega_k^2 x_k = \mu f_k(x, \dot{x}),$$
$$x = (x_1, \ldots, x_n), \quad \dot{x} = (\dot{x}_1, \ldots, \dot{x}_n), \quad k = 1, \ldots, n; \qquad (2.8.4)$$

$$\frac{d^2x_k}{dt^2} + \omega_k^2 x_k = \mu f_k(x, \dot{x}, t), \quad k = 1, \ldots, n, \qquad (2.8.5)$$

where $f_k(x, \dot{x}, t)$ are analytical functions of the variables x, $\dot{x}$ in the $2n$-dimensional open sphere K_{2n} with radius R

$$\sum_{k=1}^{n} (x_k^2 + \dot{x}_k^2) < R^2. \qquad (2.8.6)$$

The method of asymptotic series in trigonometric form was apparently first applied by Lindstedt (see Poincaré [3]) to the one-dimensional Van der Pol equation

$$\frac{d^2x}{dt^2} + \omega^2 x = \mu f(x, t), \qquad (2.8.7)$$

where $f(x, t)$ is a function analytic with respect to x and periodic with respect to t. His ideas and the analytic form he proposed were generalized and developed for Hamiltonian systems by Poincaré [2]. In the one-dimensional case, moreover when there are no external exciting forces, resonance phenomena do not occur, therefore for us the systems of Van der Pol oscillators (2.8.4) and (2.8.5) are of special interest. Unlike the one-dimensional equation, both resonance correlations of the form $(k, \omega) = 0$ and resonance between frequencies ω and excitation frequencies λ may appear.

With $\mu = 0$, generating systems for (2.8.4) and (2.8.5) consist of the second-order equations independent of each other

$$\frac{dx_k^{(0)}}{dt} + \omega_k^2 x_k^{(0)} = 0, \quad k = 1, \ldots, n, \qquad (2.8.8)$$

with their general solution expressed by the formulae

$$x_k^{(0)}(t) = A_k \cos\omega_k t + B_k \sin\omega_k t, \quad k = 1, \dots, n, \tag{2.8.9}$$

where A_k, B_k are arbitrary constants. Each function $x_k^{(0)}(t)$ has the period $2\pi\omega_k^{-1}$, but the general solution of (2.8.9) may be both a periodic and a nonperiodic vector function t, and its nature is determined by the arithmetic properties of the frequency vector ω. The geometric interpretation of paths on an n-dimensional torus given in Section 1.5 is fully applicable to the solution of (2.8.9).

Now consider the disturbed system (2.8.4). Instead of $2n$ unknown functions $x_k, \dot{x}_k$ we introduce new variables u_k, v_k by the formulae

$$\begin{aligned} x_k &= u_k \sin\omega_k t - v_k \cos\omega_k t, \\ \dot{x}_k &= \omega_k(u_k \cos\omega_k t + v_k \sin\omega_k t). \end{aligned} \tag{2.8.10}$$

The second equality of (2.8.10) means that the following conditions are imposed on the variables u_k, v_k:

$$\begin{gathered} \frac{du_k}{dt}\sin\omega_k t - \frac{dv_k}{dt}\cos\omega_k t = 0, \\ v_k(0) = -x_k(0), \quad u_k(0)\omega_k = \dot{x}_k(0). \end{gathered} \tag{2.8.11}$$

where $x_k(0)$, $\dot{x}_k(0)$ are the initial conditions for the initial equations. The conditions (2.8.11) exactly coincide with the similar conditions appearing in the classical method of variation of arbitrary constants (see Stepanov [1]).

Using the new functions, system (2.8.4) may be written as

$$\begin{aligned} \frac{du_k}{dt} &= \mu\bar{\varphi}_k(u,v) + \mu \sum_{\|s\|\geq 1} \varphi_k^{(s)}(u,v)e^{i(s,\omega)t}, \\ \frac{dv_k}{dt} &= \mu\bar{\psi}_k(u,v) + \mu \sum_{\|s\|\geq 1} \psi_k^{(s)}(u,v)e^{i(s,\omega)t}, \end{aligned} \tag{2.8.12}$$

where $\bar{\varphi}_k$, $\bar{\psi}_k$, $\varphi_k^{(s)}, \psi_k^{(s)}$ are analytic functions of variables u, v in an open $2n$-dimensional sphere K'_{2n}

$$\sum_{k=1}^{n} \left(u_k^2 + v_k^2\right) < R^2.$$

Now introduce the $2n$-dimensional vectors

$$\begin{gathered} z = (u_1, \dots, u_n, v_1, \dots, v_n), \quad \bar{Z} = (\bar{\varphi}_1, \dots, \bar{\varphi}_n, \bar{\psi}_1, \dots, \bar{\psi}_n), \\ Z = \left(\bar{\varphi}_1 + \sum \varphi_1^{(s)} e^{i(s,\omega)t}, \dots, \bar{\psi}_n + \sum \psi_n^{(s)} e^{i(s,\omega)t}\right). \end{gathered}$$

In this notation system (2.8.12) takes the form

$$\frac{dz}{dt} = \mu Z(z,t), \quad z(0) = z_0. \tag{2.8.13}$$

System (2.8.13) is a so-called standard system in Bogolyubov's sense (see Bogolyubov [1]). We will write a comparison system of the first approximation

$$\frac{d\bar{z}}{dt} = \mu \bar{Z}(\bar{z}), \quad \bar{z}(0) = z_0, \tag{2.8.14}$$

where the comparison function has the form

$$\bar{Z}(z) = \lim_{T \to \infty} \frac{1}{T} \int_0^T Z(z,t)\, dt.$$

Unfortunately, it is impossible to apply the remarkable Bogolyubov theorem (see Bogolyubov and Mitropolsky [1]) on ε-proximity of $z(t,\mu)$ and $\bar{z}(t,\mu)$ in an asymptotically large time interval $(T = O(\mu^{-1}))$ because the function $Z(z,t)$ does not have an average value over t, uniform in the whole domain of its existence. Therefore we will now consider one iteration method allowing us to construct the exact solution of system (2.8.13). This goes back to the work of Mandelstam and Papaleksi [1].

Theorem *Let:*

(1) *the vector-function $Z(z,t)$ be continuous together with its partial derivatives with respect to z in the domain $G_{2n+1} = \{(z,t): \ z \in K'_{2n}, \ t \in (-\infty,\infty))\}$ and its Euclidian norm satisfy, with respect to z, the Lipschitz condition with the constant L*

$$\|Z(z',t) - Z(z'',t)\| < L\|z' - z''\|, \quad z', z'' \in K'_{2n}; \tag{2.8.15}$$

(2) *$Z(z,t)$ be a periodic function t with frequencies $\omega_k = \lambda p_k q_k^{-1}$ (λ is any real positive number, p_k, q_k be integers, $k = 1, \dots, n$) or a quasi-periodic function with respect to t, consisting of a finite number of harmonics;*

(3) *the norms $\|Z\|$ and $\|\partial Z/\partial z\|$ be bounded in G_{2n+1}*

$$\|Z\| < K, \quad \|\partial Z/dz\| < K; \tag{2.8.16}$$

(4) *$\bar{z}(t,\mu) \in K'_{2n}$ as $t \in (-\infty,\infty)$ together with its ρ-neighborhood.*

Then for all $\varepsilon > 0$ and $A > 0$ there exists $\mu_0(\varepsilon, A, \rho) > 0$ such that with $\mu \in [0, \mu_0]$ and with $t \in [0, A\mu^{-1}]$ the following estimation holds:

$$\|z(t,\mu) - \bar{z}(t,\mu)\| < \varepsilon, \quad z(0) = \bar{z}(0) = z_0.$$

Proof In view of condition (2) of the theorem, the function $Z(z,t)$ can be represented as

$$Z(z,t) = \bar{Z}(t) + \sum_{\|s\| \geq 1} Z_s(z) e^{i(s,\omega)t}. \tag{2.8.17}$$

If it is periodic with respect to t, the norm of the summation index $||s||$ can also be unbounded, and if $Z(z,t)$ is quasi-periodic with respect to t, then in sum (2.8.17) $1 \leq ||s|| \leq N$.

Now we will construct the solution of system (2.8.13) by the method of iterations, supposing that the first approximation is determined by the relation

$$z_1(t,\mu) = z_0 + \mu \int_0^t \bar{Z}(\bar{z}(\tau,\mu))\, d\tau + \mu \sum_{||s|| \geq 1} \int_0^t Z_0(\bar{z}(\tau,\mu)) e^{i(s,\omega)\tau}\, d\tau. \tag{2.8.18}$$

In other words, as a zero approximation the solution of the averaged system (2.8.14) is taken. Integrating the last summand by parts, we find

$$\begin{aligned} z_1(t,\mu) - \bar{z}(t,\mu) &= \mu \sum_{||s|| \geq 1} Z_s(\bar{z}) \frac{e^{i(s,\omega)t}}{i(s.\omega)} \Bigg|_0^t \\ &- \mu^2 \sum_{||s|| \geq 1} \int_0^t \left(\frac{\partial Z_s}{\partial \bar{z}}, \bar{Z}(\bar{z}) \right) \frac{e^{i(s,\omega)t}}{i(s,\omega)}\, d\tau. \end{aligned} \tag{2.8.19}$$

In systems of Van der Pol equations the frequencies $\omega_1, \ldots, \omega_n$ are constant, therefore the existence or the absence of resonance relations between them do not depend on time. A system is either always or never a resonance one. From condition (2) it follows that for all integer vectors s for which $(s,\omega) \neq 0$ we have the estimation

$$|(s,\omega)| \geq \alpha > 0, \tag{2.8.20}$$

and it is important to note that $\alpha > 0$ as $||s|| \to \infty$. In addition, we constructed the function $Z(z,t)$ in such a way that those summands of Fourier representation, for which $(s,\omega) = 0$, were beforehand included into $\bar{Z}(z)$. Therefore in sum (2.8.17) the summation index-vector s only takes nonresonance values.

Then, taking into account the condition (3) of the theorem and estimation (2.8.20), from (2.8.19) we derive the estimation

$$||z_1(t,\mu) - \bar{z}(t,\mu)|| < \frac{2\mu K}{\alpha} + \frac{\mu^2 K^2 t}{\alpha} \tag{2.8.21}$$

and with $0 \leq \mu t \leq A$ we obtain

$$||z_1(t,\omega) - \bar{z}(t,\mu)|| < \mu C, \quad C = \frac{K(2 + AK)}{\alpha}. \tag{2.8.22}$$

Using the Liptschitz condition, it is easyto estimate the norm of the difference of the second and first approximations:

$$\|z_2(t,\mu) - z_1(t,\mu)\| < \mu \int_0^t \|Z(z,t) - Z(\bar{z},\tau)\| \, d\tau$$
$$< \mu L \int_0^t \|z_1(\tau,\mu) - \bar{z}(\tau,\mu)\| \, d\tau,$$

and with $0 \leq t \leq A\mu^{-1}$

$$\|z_2(t,\mu) - z_1(t,\mu)\| \leq \mu ACL. \tag{2.8.23}$$

For the m-th approximation we have

$$\|z_m(t,\mu) - z_{m-1}(t,\mu)\| < \mu C \frac{(AL)^{m-1}}{(m-1)!},$$

therefore

$$\|z_m(t,\mu) - \bar{z}(t,\mu)\| < \mu C e^{AL}.$$

Performing the limit transfer with $m \to \infty$, we obtain

$$\|z(t,\mu) - \bar{z}(t,\mu)\| \leq \mu C e^{AL}. \tag{2.8.24}$$

Now let there be given an arbitrary value $\varepsilon > 0$. Then if $\mu_0(\varepsilon, A, \rho)$ is chosen from the condition

$$\mu_0 C e^{AL} = \inf\{\varepsilon, \rho\}, \tag{2.8.25}$$

then with all $\mu \in [0,\, \mu_0]$ and with all $t \in [0,\, A\mu^{-1}]$ all approximations together with the exact solution $z(t,\mu)$ will belong to the ball K'_{2n}, and on the other hand, we will have

$$\|z(t,\mu) - \bar{z}(t,\mu)\| < \varepsilon. \tag{2.8.26}$$

So the theorem has been proved. One might say that it expresses the averaging principle for the Van der Pol system.

Now consider another aspect of the mathematical problem of Van der Pol asymptotic theory.

Let there be given an n-dimensional cube $P_n = \{0 \leq x_k \leq 1,\ k = 1, \ldots, n\}$ and an arbitrary point $\omega \in P_n$. We will study not a single system, but a set of Van der Pol equations systems of the form (2.8.4) considering that $\omega = (\omega_1, \ldots, \omega_n) \in P_n$, and the vector $f = (f_1, \ldots, f_n)$ is analytic with respect to x, $\dot{x}$ in the ball K_{2n}, and its expansion contains an infinite number of summands. Then for the arbitrary vector ω with irrational components from the cube P_n we see that it is always possible to choose a subset of vectors s such that on the one hand $\|s\| \to \infty$, and $|(s,\omega)| \to 0$ on the other. Here we come across a situation typical of problems with small denominators. In view of the fact that $|(s,\omega)| \to 0$, the quantity $C(\alpha)$ included into estimation (2.8.22) grows without limit, and for this reason the inequality (2.8.24) though not being violated, becomes senseless. Consequently, without additional conditions it is hardly possible to prove the averaging principles for such Van der Pol systems. The most natural additional conditions may be:

(a) the condition of analyticity of $Z(z,t)$ with respect to z in a $2n$-dimensional ball K'_{2n};

(b) introduction of a "metric concept", i.e. consideration not of all points of the cube P_n, but "almost all" (in the sense of Lebesgue measure, see Shilov [1]) its points, but such that for them $|(s,\omega)| \to 0$ would follow some regularity if $\|s\| \to \infty$.

We will study the above conditions in more detail.

Condition (a) Let the vector-function $Z(z)$ be analytic and 2π-periodic with respect to z in some domain $\|\operatorname{Im} z\| < \alpha$. Then in this domain we have the Fourier representation

$$Z(z) = \sum_{\|s\| \geq 0} Z_s e^{i(s,z)}, \tag{2.8.27}$$

and if $\|Z(z)\| \leq M$, then (see Arnol'd [1])

$$\|Z_s\| \leq M e^{-\alpha \|s\|}. \tag{2.8.28}$$

Inequality (2.8.28) shows a very fast exponential decrease of the Fourier coefficients of the analytic function.

Now let $Z(z,t)$ be obtained by means of the above-described transformations from the vector-function $f = (f_1, \dots, f_n)$ analytic with respect to x, $\dot{x}$ in a $2n$-dimensional ball K_{2n}. Then it is analytic with respect to z in the ball K', and periodic or quasi-periodic with respect to $t \in (-\infty, \infty)$. A question arises: what is the structure of the representation coefficients (2.8.17) and what is their decrease rate as $\|s\| \to \infty$?

Representation (2.8.17) was obtained from the expansion of vector f in terms of powers of $x, , \dot{x}$, their replacement by new variables u, v with consideration for (2.8.11) and as a result regrouping of an intermediate expansion for obtaining a standard Fourier representation. From these operations it follows that

$$Z_s(z) = \sum A_{k_1,\dots,k_{2n}} z_1^{k_1} \dots z_{2n}^{k_{2n}}, \tag{2.8.29}$$

or, in vector notation,

$$Z_s(z) = \sum_{\|k\| \geq \|s\|} A_k z^k. \tag{2.8.30}$$

If $\|f(x,\dot{x})\| < B$ when $(x,\dot{x}) \in K_{2n}$, then

$$|f_{k_1,\dots,k_{2n}}| < \frac{B}{k_1! \cdots k_{2n}!}. \tag{2.8.31}$$

Then

$$|A_{k_1,\dots,k_{2n}}| < \frac{B^*}{k_1! \cdots k_{2n}!}, \tag{2.8.32}$$

where $B^* = BN$, and N is an integer depending, generally speaking, on $k_1, \ldots, k_{2n}$. This implies

$$\|Z_s\| \leq \frac{B^*}{k_1^*! \cdots k_{2n}^*!}, \qquad \sum_{r=1}^{2n} |k_r^*| = \|s\|.$$

Condition (b) Almost all (in the sense of Lebesgue measure) points $\omega \in P_n$ satisfy (see Arnol'd [1])

$$|(s, \omega)| > \frac{R(\omega)}{\|s\|^{n+1}}, \tag{2.8.33}$$

where $R(\omega)$ depends on ω only. Now we can formulate the next theorem.

Theorem *Let:*

(1) *the vector-function $Z(z, t)$ be analytic with respect to z in a unit ball K'_{2n} and depend on $t \in (-\infty, \infty)$ in a periodic or conditionally periodic way;*
(2) *its norm be bounded in the domain $K'_{2n} \times (-\infty, \infty)$ by a constant K;*
(3) *the frequency vector ω belong to a unit n-dimensional cube P_n and satisfy (2.8.33);*
(4) *$\bar{z}(t, \mu) \in K'_{2n}$ with $t \in (-\infty, \infty)$ together with its ρ-neighborhood.*

Then for any $\varepsilon > 0$, $A > 0$ there exists $\mu_0(\varepsilon, A, \rho, \omega) > 0$ such that with $\mu \in [0, \mu_0]$ and $t \in [0, A\mu^{-1}]$

$$\|z(t, \mu) - \bar{z}(t, \mu)\| < \varepsilon, \quad z(0, \mu) = \bar{z}(0, \mu) = z_0. \tag{2.8.34}$$

The proof of this theorem is given in Arnol'd [1].

In conclusion we will show one more transformation $(x, \dot{x}) \to (u, v)$ instead of (2.8.10) and (2.8.11), which is often used by physicists and mathematicians. It is possible to seek the solution of equations (2.8.4) in the form

$$x_k(t) = a_k(t) \cos(\omega_k t + \varphi_k(t)), \quad k = 1, \ldots, n, \tag{2.8.35}$$

where $a_k(t)$, $\varphi_k(t)$ are slowly changing amplitudes and phases. In this case instead of (2.8.11) we will have the conditions

$$\frac{da_k}{dt} \cos(\omega_k t + \varphi_k) - a_k \frac{d\varphi_k}{dt} \sin(\omega_k t + \varphi_k) = 0, \quad k = 1, \ldots, n. \tag{2.8.36}$$

Taking account of these conditions, the initial functions x_k are determined from the differential equalities

$$\frac{dx_k}{dt} = -a_k \omega_k \sin(\omega_k t + \varphi_k), \quad k = 1, \ldots, n. \tag{2.8.37}$$

Naturally, substitution of variables (2.8.35), (2.8.37) can be effectively used for sufficiently small μ, because with $\mu = 0$ the amplitudes of oscillations a_k and their phases φ_k are constant (see formulae (2.8.9)).

We will show the aspects of the averaging principles on Van der Pol equations and, in particular, study the influence of resonance upon the qualitative behavior of solutions.

First consider the autonomous Van der Pol oscillator

$$\frac{d^2x}{dt^2} + \omega^2 x = \mu(1-x^2)\frac{dx}{dt}, \quad x(0) = x_0, \quad \left.\frac{dx}{dt}\right|_{t-0} = \dot{x}_0. \tag{2.8.38}$$

We will look for a transform $(x, \dot{x}) \to (a, \varphi)$,

$$x(t,\mu) = a\cos(\omega t + \varphi), \qquad \frac{dx(t,\mu)}{dt} = -a\omega\sin(\omega t + \varphi), \tag{2.8.39}$$

that will transform (2.8.38) into the system $\psi = \omega t + \varphi$

$$\begin{aligned} \frac{da}{dt} &= \mu a \sin^2\psi\,(1 - a^2\cos^2\psi), \\ \frac{d\psi}{dt} &= \omega + \mu\sin\psi\cos\psi\,(1 - a^2\cos^2\psi). \end{aligned} \tag{2.8.40}$$

The existence of the transform (2.8.39) implies that the additional conditions

$$\begin{gathered} \frac{da}{dt}\cos(\omega t + \varphi) - \frac{d\psi}{dt}a\sin(\omega t + \varphi) = 0, \\ a(0)\cos\varphi(0) = x_0, \quad a(0)\omega\sin\varphi(0) = -\dot{x}_0, \end{gathered}$$

typical for the Van der Pol method and for the method of variation of arbitrary constants are satisfied.

System (2.8.40) contains one slow variable a and one fast phase ψ ($\psi = \omega t +$ slow phase φ), therefore the method of the averaging principles for Van der Pol oscillator systems can be applied to it. Thus, we will look for a new substitution of variables

$$\begin{aligned} a &= \bar{a} + \mu u_1(\bar{a}, \bar{\psi}) + \mu^2 u_2(\bar{a}, \bar{\psi}) + \cdots, \\ \psi &= \bar{\psi} + \mu v_1(\bar{a}, \bar{\psi}) + \mu^2 v_2(\bar{a}, \bar{\psi}) + \cdots, \end{aligned} \tag{2.8.41}$$

that will transform the system (2.8.40) into the comparison system of any approximation

$$\begin{aligned} \frac{d\bar{a}}{dt} &= \mu A_1(\bar{a}) + \mu^2 A_2(\bar{a}) + \cdots, \\ \frac{d\bar{\psi}}{dt} &= \omega + \mu B_1(\bar{a}) + \mu^2 B_2(\bar{a}) + \cdots. \end{aligned} \tag{2.8.42}$$

Now if we perform all mathematical operations necessary for the averaging method, we will obtain an infinite system of partial differential equations for the calculation of $u_k(\bar{a}, \bar{\psi})$, $v_k(\bar{a}, \bar{\psi})$:

$$
\begin{aligned}
\omega \frac{\partial u_1}{\partial \bar{\psi}} &= \frac{\bar{a}(4-\bar{a}^2)}{8} - \frac{\bar{a}\cos 2\bar{\psi}}{2} + \frac{\bar{a}^3 \cos 4\bar{\psi}}{8} - A_1(\bar{a}), \\
\omega \frac{\partial v_1}{\partial \bar{\psi}} &= \frac{(2-\bar{a}^2)\sin 2\bar{\psi}}{4} - \frac{\bar{a}^2 \sin 4\bar{\psi}}{8} - B_1(\bar{a}), \\
\omega \frac{\partial u_2}{\partial \bar{\psi}} &= -A_1 \frac{\partial u_1}{\partial \bar{a}} - B_1 \frac{\partial u_1}{\partial \bar{\psi}} + \frac{u_1(1-\cos 2\bar{\psi})}{2} \\
&\quad - \frac{3\bar{a}^2 u_1(1-\cos 4\bar{\psi})}{8} - \frac{\bar{a}^3 v_1 \sin 4\bar{\psi}}{4} - A_2(\bar{a}), \\
\omega \frac{\partial v_2}{\partial \bar{\psi}} &= -A_1 \frac{\partial v_1}{\partial \bar{a}} - B_1 \frac{\partial v_1}{\partial \bar{\psi}} - \frac{\bar{a} u_1 \sin 2\bar{\psi}}{2} \\
&\quad - \frac{\bar{u}_1 \sin 4\bar{\psi}}{4} + \frac{(2-\bar{a}^2) v_1 \cos 2\bar{\psi}}{4} - \frac{\bar{a}^2 v_1(1+\cos 4\bar{\psi})}{8} - B_2(\bar{a}) \\
&\ldots\ldots\ldots\ldots\ldots\ldots\ldots\ldots\ldots\ldots
\end{aligned}
\tag{2.8.43}
$$

This infinite system can be integrated by the method of characteristics. From the whole set we will choose the simplest class of periodic solutions. Following the general rule of finding the unknown functions A_k, B_k, we find

$$
\begin{aligned}
A_1(\bar{a}) &= \frac{1}{2\pi} \int_0^{2\pi} \left[\frac{\bar{a}(4-\bar{a}^2)}{8} - \frac{\bar{a}\cos 2\bar{\psi}}{2} + \frac{\bar{a}^2 \cos 4\bar{\psi}}{8} \right] d\bar{\psi}, \\
B_1(\bar{a}) &= \frac{1}{2\pi} \int_0^{2\pi} \left[\frac{(2-\bar{a}^2)\sin 2\bar{\psi}}{4} - \frac{\bar{a}^2 \sin 4\bar{\psi}}{8} \right] d\bar{\psi},
\end{aligned}
\tag{2.8.44}
$$

or

$$
A_1(\bar{a}) = \frac{\bar{a}(4-\bar{a}^2)}{8}, \quad B_1(\bar{a}) = 0. \tag{2.8.45}
$$

Therefore

$$
\begin{aligned}
u_1(\bar{a}, \bar{\psi}) &= \frac{\bar{a}}{4\omega} \left[-\sin 2\bar{\psi} + \frac{\bar{a}^2 \sin 4\bar{\psi}}{8} \right] + \varphi_1(\bar{a}), \\
v_1(\bar{a}, \bar{\psi}) &= \frac{(\bar{a}^2-2)\cos 2\bar{\psi}}{8\omega} + \frac{\bar{a}^2 \cos 4\bar{\psi}}{32\omega} + \theta_1(\bar{a}),
\end{aligned}
\tag{2.8.46}
$$

where φ_1, θ_1 are arbitrary differentiable functions of the variable $\bar{a}$. The appearance of arbitrary functions in the structure of (2.8.46) is a corollary of the general theorem on the analytic structure of the general integral of partial differential equations (see Stepanov [1]). For the functions u_1, v_1 to have only trigonometric summands, it is necessary to assume

$$
\varphi_1(\bar{a}) \equiv \theta_1(\bar{a}) \equiv 0. \tag{2.8.47}
$$

In this case we obtain

$$\begin{aligned} u_1(\bar{a},\bar{\psi}) &= \frac{\bar{a}}{4\omega}\left[-\sin 2\bar{\psi} + \frac{\bar{a}^2 \sin 4\bar{\psi}}{2}\right], \\ v_1(\bar{a},\bar{\psi}) &= \frac{1}{8\omega}\left[(\bar{a}^2-2)\cos 2\bar{\psi} + \bar{a}^2\cos 4\bar{\psi}\right]. \end{aligned} \tag{2.8.48}$$

Then, substituting (2.8.48) and (2.8.45) into the third and fourth equations of system (2.8.43), we obtain

$$\begin{aligned} \omega\frac{\partial u_2}{\partial\bar{\psi}} &= \frac{\bar{a}(-32+29\bar{a}^2-4\bar{a}^4)\sin 2\bar{\psi}}{128\omega} + \frac{\bar{a}(16-5\bar{a}^2-24\bar{a}^4)\sin 4\bar{\psi}}{256\omega} \\ &\quad + \frac{\bar{a}^3(1-4\bar{a}^2)\sin 6\bar{\psi}}{128\omega} - \frac{\bar{a}^5\sin 8\bar{\psi}}{512\omega} - A_2(\bar{a}), \\ \omega\frac{\partial v_2}{\partial\bar{\psi}} &= \frac{-32+48\bar{a}^2-11\bar{a}^4}{256\omega} - \frac{(\bar{a}^2+\bar{a}^4)\cos 2\bar{\psi}}{64\omega} \\ &\quad + \frac{(-16+4\bar{a}^2-3\bar{a}^4)\cos 4\bar{\psi}}{128\omega} - \frac{\bar{a}^2\cos 6\bar{\psi}}{64\omega} + \frac{\bar{a}^4\cos 8\bar{\psi}}{256\omega} - B_2(\bar{a}). \end{aligned} \tag{2.8.49}$$

Using formulae similar to (2.8.44), we find

$$A_2(\bar{a}) = 0, \qquad B_2(\bar{a}) = \frac{-32+48\bar{a}^2-11\bar{a}^4}{256\omega}. \tag{2.8.50}$$

Then, integrating the equations for u_2, v_2, we obtain

$$\begin{aligned} u_2(\bar{a},\bar{\psi}) &= \frac{\bar{a}(32-29\bar{a}^2+4\bar{a}^4)\cos 2\bar{\psi}}{256\omega^2} - \frac{\bar{a}(16-5\bar{a}^2-24\bar{a}^4)\cos 4\bar{\psi}}{1024\omega^2} \\ &\quad - \frac{\bar{a}^3(1-4\bar{a}^2)\cos 6\bar{\psi}}{768\omega^2} + \frac{\bar{a}^5\cos 8\bar{\psi}}{4096\omega^2} + \varphi_2(\bar{a}), \\ v_2(\bar{a},\bar{\psi}) &= -\frac{\bar{a}^2(1+\bar{a}^2)\sin 2\bar{\psi}}{128\omega^2} + \frac{(-16+4\bar{a}^2-3\bar{a}^4)\sin 4\bar{\psi}}{512\omega^2} \\ &\quad - \frac{\bar{a}^2\sin 6\bar{\psi}}{384\omega^2} + \frac{\bar{a}^4\sin 8\bar{\psi}}{2048\omega^2} + \theta_2(\bar{a}), \end{aligned} \tag{2.8.51}$$

where by analogy φ_2, θ_2 are arbitrary differentiable functions of the variable $\bar{a}$. To keep the trigonometric form of the transform (2.8.41) it is necessary to assume

$$\varphi_2(\bar{a}) \equiv \theta_2(\bar{a}) \equiv 0. \tag{2.8.52}$$

So the above method makes it possible to integrate successively the infinite system (2.8.43) to any number s and consequently obtain in trigonometric form (with respect to $\bar{\psi}$) the transform (2.8.41)

$$a = \bar{a} + \sum_{k=1}^{s}\mu^k u_k(\bar{a},\bar{\psi}), \quad \psi = \bar{\psi} + \sum_{k=1}^{s}\mu^k v_k(\bar{a},\bar{\psi}) \tag{2.8.53}$$

which satisfies with error $O(\mu^{s+1})$ the system (2.8.40) equivalent to the initial equation (2.8.38). For the relations (2.8.53) to give explicit dependence of variables a and ψ on t, it is necessary to determine the functions $\bar{a}(t,\mu)$, $\bar{\psi}(t,\mu)$ from the comparative system of the s-th approximation

$$\frac{d\bar{a}}{dt} = \mu A_1(\bar{a}) + \cdots + \mu^s A_s(\bar{a}),$$
$$\frac{d\bar{\psi}}{dt} = \omega + \mu B_1(\bar{a}) + \cdots + \mu^s B_s(\bar{a}). \tag{2.8.54}$$

The function $\bar{a}(t,\mu)$ is found by way of direct integration:

$$\int\limits_{\bar{a}(0,\mu)}^{\bar{a}(t,\mu)} \frac{d\bar{a}}{A_1(\bar{a}) + \mu^2 A_3(\bar{a}) + \cdots + \mu^{s-1} A_s(\bar{a})} = \mu t, \tag{2.8.55}$$

where the initial conditions $\bar{a}(0,\mu)$, $\bar{\psi}(0,\mu)$ are determined from the system of functional equations

$$\bar{a}(0,\mu) + \sum_{k=1}^{s} \mu^k u_k(\bar{a}(0,\mu), \bar{\psi}(0,\mu)) = a(0),$$
$$\bar{\psi}(0,\mu) + \sum_{k=1}^{s} \mu^k v_k(\bar{a}(0,\mu), \bar{\psi}(0,\mu)) = \psi(0). \tag{2.8.56}$$

After finding the function $\bar{a}$ we easily find

$$\bar{\psi}(t,\mu) = \bar{\psi}(0,\mu) + \omega t + \sum_{k=1}^{s} \mu^k \int\limits_0^t B_k(\bar{a}(\tau,\mu))\, d\tau. \tag{2.8.57}$$

Now substituting the functions $\bar{a}$, $\bar{\psi}$ into relations (2.8.53) we find the functions $a(t,\mu)$, $\psi(t,\mu)$, and then, using formulae (2.8.39), we find the solution of the Van der Pol equation $x(t,\mu)$, satisfying x_0, $\dot{x}_0$.

Now we will perform a more detailed study of the functions $x(t,\mu)$ and $\dot{x}(t,\mu)$ (see 2.8.39). They constitute the solution of the Van der Pol equation in the form of slowly changing time functions. Though transform (2.8.41) is periodic with respect to $\bar{\psi}$, from this it does not follow that $x(t,\mu)$ is a periodic function of t.

Indeed, we will make sure of that, constructing first approximation theory. Then

$$\frac{d\bar{a}}{dt} = \mu A_1(\bar{a}), \qquad \frac{d\bar{\psi}}{dt} = \omega, \tag{2.8.58}$$

$$\int\limits_{\bar{a}(0,\mu)}^{\bar{a}(t,\mu)} \frac{dw}{A_1(w)} = \mu t, \qquad \bar{\psi} = \omega t + \bar{\psi}(0). \tag{2.8.59}$$

Considering (2.8.45) for the coefficient $A_1(\bar{a})$, from (2.8.59) one can find the explicit dependence of $\bar{a}$ on t:

$$\bar{a}(t,\mu) = 2\sqrt{\frac{Ce^{\mu t}}{Ce^{\mu t} - 1}}, \qquad C = \frac{\bar{a}_0^2}{\bar{a}_0^2 - 4}, \qquad \bar{a}_0 = \bar{a}(0). \tag{2.8.60}$$

Taking into account the transform of the first approximation

$$\begin{aligned} a(t,\mu) &= \bar{a}(t,\mu) + \mu u_1(\bar{a},\bar{\psi}), \\ \psi(t,\mu) &= \bar{\psi}(t,\mu) + \mu v_1(\bar{a},\bar{\psi}) \end{aligned} \tag{2.8.61}$$

we obtain a rather complicated dependence of the amplitude a and the fast phase ψ on time t:

$$\begin{aligned} a(t,\mu) &= 2\sqrt{\frac{Ce^{\mu t}}{Ce^{\mu t}-1}}\Bigg[1 - \frac{\mu}{4\omega}\sin(2\omega t + 2\bar{\psi}_0) \\ &\quad + \frac{\mu}{8\omega}\frac{Ce^{\mu t}}{Ce^{\mu t}-1}\sin(4\omega t + 4\bar{\psi}_0)\Bigg], \\ \psi(t,\mu) &= \omega t + \bar{\psi}_0 + \frac{\mu}{8\omega}\Bigg[-2\cos(2\omega t + 2\bar{\psi}_0) \\ &\quad + \frac{4Ce^{\mu t}}{Ce^{\mu t}-1}\left[\cos(2\omega t + 2\bar{\psi}_0) + \frac{1}{4\cos(4\omega t + 4\bar{\psi}_0)}\right]\Bigg\}. \end{aligned} \tag{2.8.62}$$

Formulae (2.8.62) show that the amplitude $a(t,\mu)$ is not a periodic function of t, therefore the initial function $x(t,\mu) = a(t,\mu)\cos\psi(t,\mu)$, generally speaking, is not a periodic time function, because upon respective substitutions we obtain

$$\begin{aligned} x(t,\mu) &= 2\sqrt{\frac{Ce^{\mu t}}{Ce^{\mu t}-1}}\Bigg[1 - \frac{\mu}{4\omega}\sin(2\omega t + 2\bar{\psi}_0) \\ &\quad + \frac{\mu}{8\omega}\frac{Ce^{\mu t}}{Ce^{\mu t}-1}\sin(4\omega t + 4\bar{\psi}_0)\Bigg]\cos\psi(t,\mu). \end{aligned} \tag{2.8.63}$$

The analysis of the formula (2.8.63) shows that if the initial value of the averaged amplitude $\bar{a}_0 \neq 2$, then for any $t \geq 0$ the function $x(t,\mu)$ is bounded, though not periodic. For any given but sufficiently small $\mu > 0$ the following asymptotic equality is true

$$\begin{gathered} x(t,\mu) \to 2\left[1 - \frac{\mu}{4\omega}\sin(2\omega t + 2\bar{\psi}_0) + \frac{\mu}{8\omega}\sin(4\omega t + 4\bar{\psi}_0)\right]\cos\psi(t,\mu) \\ \text{as t} \to \infty, \end{gathered} \tag{2.8.64}$$

and for any given $t \geq 0$ another asymptotic relation is true:

$$x(t,\mu) \to 2\sqrt{\frac{C}{C-1}}\cos(\omega t + \bar{\psi}_0) \quad \text{as} \quad \mu \to 0. \tag{2.8.65}$$

So, with arbitrary initial values x_0, $\dot{x}_0$ the approximate solution of the autonomous Van der Pol equation (2.8.38) is a bounded non-periodic time function (of course, the trivial case $x_0 = \dot{x}_0 = 0$ is excluded, i.e. the trivial solution $x(t,\mu) \equiv 0$).

Now we will use the system of s-th approximation to find the equilibrium, or stationary solutions. The necessary condition for the existence of equilibrium solutions is that the following equality shall hold true:

$$A_1(\bar{a}) + \mu A_2(\bar{a}) + \cdots + \mu^{s-1}A_s(\bar{a}) = 0, \tag{2.8.66}$$

but since $A_2(\bar{a}) = 0$, we have

$$A_1(\bar{a}) + \mu^2 A_3(\bar{a}) + \cdots + \mu^{s-1} A_s(\bar{a}) = 0. \tag{2.8.67}$$

The equation (2.8.67) has three real roots:

$$\bar{a}_1 = 0, \quad \bar{a}_2 = 2 + O(\mu^2), \quad \bar{a}_3 = -2 + O(\mu^2). \tag{2.8.68}$$

The first root corresponds to the trivial solution $x(t, \mu) \equiv 0$. The other two roots $\bar{a}_2$ and $\bar{a}_3$ generate two periodic solutions, because with a constant value of $\bar{a}$ we obtain

$$\begin{aligned} \bar{\psi}_2(t, \mu) &= \left[\omega + \mu^2 B_2(\bar{a}_2) + \cdots + \mu^s B_s(\bar{a}_2)\right] t + \bar{\psi}_{20}, \\ \bar{\psi}_3(t, \mu) &= \left[\omega + \mu^2 B_2(\bar{a}_3) + \cdots + \mu^s B_s(\bar{a}_3)\right] t + \bar{\psi}_{30}, \end{aligned} \tag{2.8.69}$$

and the final form of these periodic solutions will be:

$$\begin{aligned} x_2(t, \mu) &= \left[\bar{a}_2 + \mu u_1(\bar{a}_2, \bar{\psi}_2) + \cdots \right. \\ &\quad \left. + \mu^s u_s(\bar{a}_2, \bar{\psi}_2)\right] \cos\left[\bar{\psi}_2(t, \mu) + \mu v_1(\bar{a}_2, \bar{\psi}_2) + \cdots\right], \\ x_3(t, \mu) &= \left[\bar{a}_3 + \mu u_1(\bar{a}_3, \bar{\psi}_3) + \cdots \right. \\ &\quad \left. + \mu^s u_s(\bar{a}_3, \bar{\psi}_3)\right] \cos\left[\bar{\psi}_3(t, \mu) + \mu v_1(\bar{a}_3, \bar{\psi}_3) + \cdots\right]. \end{aligned} \tag{2.8.70}$$

The periods of these solutions are respectively equal to:

$$\begin{aligned} T_2 &= \frac{2\pi}{\omega + \mu^2 B_2(\bar{a}_2) + \cdots + \mu^s B_s(\bar{a}_2)}, \\ T_3 &= \frac{2\pi}{\omega + \mu^2 B_2(\bar{a}_3) + \cdots + \mu^s B_s(\bar{a}_3)}. \end{aligned} \tag{2.8.71}$$

The solution $\bar{a}_3 = -2 + O(\mu^2)$ has only theoretical meaning, because it is natural to consider $\bar{a}$ as a positive value.

Now apply the averaging principle described above to the non-autonomous Van der Pol oscillator

$$\begin{gathered} \frac{d^2x}{dt^2} + \omega^2 x = \mu(1 - x^2)\frac{dx}{dt} + \mu k\lambda \cos \lambda t, \\ x(0) = x_0, \qquad \left.\frac{dx}{dt}\right|_{t=0} = \dot{x}_0. \end{gathered} \tag{2.8.72}$$

Using the substitution of variables (2.8.39), it is easy to obtain a system of two equations (one for the amplitude, the other for the fast phase), equivalent to the equation (2.8.72):

$$\begin{aligned} \frac{da}{dt} &= \mu a \sin^2 \psi\, (1 - a^2 \cos^2 \psi) + \frac{\mu k \lambda}{\omega} \sin \psi \cos \lambda t. \\ \frac{d\psi}{dt} &= \omega + \mu \sin \psi \cos \psi\, (1 - a^2 \cos^2 \psi) - \frac{\mu k \lambda}{a\omega} \cos \psi \cos \lambda t. \end{aligned} \tag{2.8.73}$$

We will look for a substitution of variables

$$
\begin{aligned}
a &= \bar{a} + \mu u_1(\bar{a}, \bar{\psi}, t) + \mu^2 u_2(\bar{a}, \bar{\psi}, t) + \cdots , \\
\psi &= \bar{\psi} + \mu v_1(\bar{a}, \bar{\psi}, t) + \mu^2 v_2(\bar{a}, \bar{\psi}, t) + \cdots
\end{aligned}
\tag{2.8.74}
$$

that would transform system (2.8.73) into the comparison system of any approximation

$$
\begin{aligned}
\frac{d\bar{a}}{dt} &= \mu A_1(\bar{a}) + \mu^2 A_2(\bar{a}) + \cdots , \\
\frac{d\bar{\psi}}{dt} &= \omega + \mu B_1(\bar{a}) + \mu^2 B_2(\bar{a}) + \cdots .
\end{aligned}
\tag{2.8.75}
$$

For the realization of the transform (2.8.73) → (2.8.75), we obtain an infinite system of partial differential equations of the form

$$
\begin{aligned}
\frac{\partial u_1}{\partial t} + \omega \frac{\partial u_1}{\partial \bar{\psi}} &= \frac{\bar{a}(4 - \bar{a}^2)}{2} - \frac{\bar{a}\cos 2\bar{\psi}}{2} + \frac{\bar{a}^3 \cos 4\bar{\psi}}{8} + \frac{k\lambda \cos \lambda t \sin \bar{\psi}}{\omega} \\
&\quad - A_1(\bar{a}), \\
\frac{\partial v_1}{\partial t} + \omega \frac{\partial v_1}{\partial \bar{\psi}} &= \frac{(2 - \bar{a}^2)\sin 2\bar{\psi}}{4} - \frac{\bar{a}^2 \sin 4\bar{\psi}}{8} + \frac{k\lambda \cos \lambda t \cos \bar{\psi}}{\bar{a}\omega} - B_1(\bar{a}), \\
\frac{\partial u_2}{\partial t} + \omega \frac{\partial u_2}{\partial \bar{\psi}} &= -A_1 \frac{\partial u_1}{\partial \bar{a}} - B_1 \frac{\partial u_1}{\partial \bar{\psi}} + \frac{u_1(1 - \cos 2\bar{\psi})}{2} + \bar{a} v_1 \sin 2\bar{\psi} \\
&\quad - \frac{3\bar{a}^2 u_1 (1 - \cos 4\bar{\psi})}{8} - \frac{\bar{a}^3 v_1 \sin 4\bar{\psi}}{4} + \frac{k\lambda v_1 \cos \lambda t \cos \bar{\psi}}{2\omega} - A_2(\bar{a}), \\
\frac{\partial v_2}{\partial t} + \omega \frac{\partial v_2}{\partial \bar{\psi}} &= -A_1 \frac{\partial v_1}{\partial \bar{a}} - B_1 \frac{\partial v_1}{\partial \bar{\psi}} - \frac{\bar{a} u_1 \sin 2\bar{\psi}}{2} - \frac{\bar{a} u_1 \sin 4\bar{\psi}}{4} \\
&\quad + \frac{\bar{a}^2 v_1 (1 - \cos 4\bar{\psi})}{8} + \frac{(2 - \bar{a}^2) v_1 \cos 2\bar{\psi}}{4} - \frac{\bar{a} v_1 (1 + \cos 4\bar{\psi})}{6} \\
&\quad - \frac{k\lambda v_1 \cos \lambda t \sin \bar{\psi}}{2\bar{a}\omega} - \frac{k\lambda u_2 \cos \lambda t \cos \bar{\psi}}{\bar{a}\omega} - B_2(\bar{a}), \\
&\cdots\cdots\cdots\cdots\cdots\cdots\cdots\cdots\cdots\cdots
\end{aligned}
\tag{2.8.76}
$$

Calculating A_1, B_1 as average values of the right-hand members of the first equations of (2.8.76) with respect to $\bar{\psi}$ and t, we obtain

$$
\begin{aligned}
A_1(\bar{a}) &= \frac{\lambda}{4\pi^2} \int_0^{2\pi} \int_0^{2\pi\lambda^{-1}} \left[\frac{\bar{a}(4 - \bar{a}^2)}{8} - \frac{\bar{a}\cos 2\bar{\psi}}{2} \right. \\
&\quad \left. + \frac{\bar{a}^3 \cos 4\bar{\psi}}{8} + \frac{k\lambda \cos \lambda t \sin \bar{\psi}}{\omega} \right] d\bar{\psi}\, dt, \\
A_1(\bar{a}) &= \frac{\bar{a}(4 - \bar{a}^2)}{8}, \qquad B_1(\bar{a}) = 0.
\end{aligned}
\tag{2.8.77}
$$

Then the first two equations of system (2.8.76) are easily integrated by the method of characteristics, resulting in

$$\begin{aligned} u_1(\bar{a}, \bar{\psi}, t) &= -\frac{\bar{a}\sin 2\bar{\psi}}{4\omega} + \frac{\bar{a}^3 \sin 4\bar{\psi}}{32\omega} \\ &\quad - \frac{k\lambda}{2\omega}\left[\frac{\cos(\bar{\psi}+\lambda t)}{\omega+\lambda} + \frac{\cos(\bar{\psi}-\lambda t)}{\omega-\lambda}\right], \\ v_1(\bar{a}, \bar{\psi}, t) &= \frac{(\bar{a}^2-2)\cos 2\bar{\psi}}{8\omega} + \frac{\bar{a}^2\cos 4\bar{\psi}}{32\omega} \\ &\quad + \frac{k\lambda}{2\omega}\left[\frac{\sin(\bar{\psi}+\lambda t)}{\omega+\lambda} + \frac{\sin(\bar{\psi}-\lambda t)}{\omega-\lambda}\right]. \end{aligned} \tag{2.8.78}$$

For the next summands in transform (2.8.74) also to have a trigonometric form, we assume in the expressions for u_1, v_1 that the arbitrary functions $\varphi_1(\bar{a})$, $\theta_1(\bar{a})$ appearing during integration are equal to zero. Note that in this case, unlike the autonomous Van der Pol oscillator, the function u_1 may be very large if there is a resonance between the eigenfrequency of the oscillator ω and the frequency of the external excitation λ, i.e. if $\omega - \lambda = O(\alpha)$, where α is sufficiently small.

The algorithm for the calculation of A_1, B_1 in the first step of the iteration process and the method of integration of first-order partial differential equations can be used in the next steps of the iteration process, and in this way it is possible to determine all functions $u_s(\bar{a}, \bar{\psi}, t)$, $v_s(\bar{a}, \bar{\psi}, t)$ of transform (2.8.74). These expressions are very large, and we cannot write them here. However note that at any step of the iterations u_s, v_s are expressed by trigonometric functions of arguments $\bar{\psi}, t$. They can be periodic, quasi-periodic and generally more composite functions of time. This is determined by arithmetic properties of the frequencies ω and λ, as well as by the small functions μA_1, $\mu B_1, \ldots$ included in the right-hand members of equations (2.8.76) and determined at the previous iteration steps.

The above examples of autonomous and nonautonomous Van der Pol oscillators show that transforms of the form (2.8.41) and (2.8.74) allow us to effectively construct approximate solutions of Van der Pol equations in analytical form, with any prescribed accuracy. The transforms themselves exist in the class of periodic functions, and resonances and consequently splashes in the behavior of the deviation of the exact solution from the approximate one occur where the external excitation resounds with the eigenfrequency ω of the oscillator. Another situation is distinctive for Van der Pol oscillator systems even in the absence of external exciting forces. To illustrate this statement, consider a system of two weakly connected Van der Pol oscillators.

Consider the following system of two equations:

$$\begin{aligned} \frac{d^2x_1}{dt^2} + \omega_1^2 x_1 &= \mu(1-x_1^2)\frac{dx_2}{dt}, \\ \frac{d^2x_2}{dt^2} + \omega_2^2 x_2 &= \mu(1-x_2^2)\frac{dx_1}{dt}. \end{aligned} \tag{2.8.79}$$

Here $\omega = (\omega_1, \omega_2)$ is a two-dimensional vector of eigenfrequencies, and μ is a small positive parameter, as always.

Using the substitution of variables

$$\begin{aligned} x_k &= a_k(t)\cos(\omega_k t + \varphi_k(t)), \\ \frac{dx_k}{dt} &= -a_k(t)\,\omega_k \sin(\omega_k t + \varphi_k(t)), \qquad k = 1, 2, \end{aligned} \tag{2.8.80}$$

after some computations, quite intricate but obvious in essence, instead of (2.8.79) we obtain the equivalent system for slowly changing amplitudes $a_1(t)$, $a_2(t)$ and fast phases $\psi_1(t)$, $\psi_2(t)$:

$$\begin{aligned} \frac{da_1}{dt} &= \mu a_2 \frac{\omega_2}{\omega_1} \sin\psi_1 \sin\psi_2\,(1 - a_1^2\cos^2\psi_1), \\ \frac{da_2}{dt} &= \mu a_1 \frac{\omega_1}{\omega_2} \sin\psi_1 \sin\psi_2\,(1 - a_2^2\cos^2\psi_2), \\ \frac{d\psi_1}{dt} &= \omega_1 + \mu\frac{a_2\omega_2}{a_1\omega_1} \sin\psi_2 \cos\psi_1\,(1 - a_1^2\cos^2\psi_1), \\ \frac{d\psi_2}{dt} &= \omega_2 + \mu\frac{a_1\omega_1}{a_2\omega_2} \sin\psi_1 \cos\psi_2\,(1 - a_2^2\cos^2\psi_2). \end{aligned} \tag{2.8.81}$$

Now use the substitution of variables of the form (2.8.41), considering $\bar{a}$ as a two-dimensional vector with components $\bar{a}_1, \bar{a}_2$, and $\bar{\psi}$ with components $\bar{\psi}_1$, $\bar{\psi}_2$, to transform system (2.8.81) into a comparison system of any approximation of the form

$$\begin{aligned} \frac{d\bar{a}_1}{dt} &= \mu A_1^{(1)}(\bar{a}_1, \bar{a}_2) + \mu^2 A_2^{(1)}(\bar{a}_1, \bar{a}_2) + \cdots, \\ \frac{d\bar{a}_2}{dt} &= \mu A_1^{(2)}(\bar{a}_1, \bar{a}_2) + \mu^2 A_2^{(2)}(\bar{a}_1, \bar{a}_2) + \cdots, \\ \frac{d\bar{\psi}_1}{dt} &= \omega_1 + \mu B_1^{(1)}(\bar{a}_1, \bar{a}_2) + \mu^2 B_2^{(1)}(\bar{a}_1, \bar{a}_2) + \cdots, \\ \frac{d\bar{\psi}_2}{dt} &= \omega_2 + \mu B_1^{(2)}(\bar{a}_1, \bar{a}_2) + \mu^2 B_2^{(2)}(\bar{a}_1, \bar{a}_2) + \cdots. \end{aligned} \tag{2.8.82}$$

In "coordinate" form the substitution (2.8.41) for this case should be represented by the equalities

$$\begin{aligned} a_1 &= \bar{a}_1 + \mu u_1^{(1)}(\bar{a}_1, \bar{a}_2, \bar{\psi}_1, \bar{\psi}_2) + \mu^2 u_2^{(1)}(\bar{a}_1, \bar{a}_2, \bar{\psi}_1, \bar{\psi}_2) + \cdots, \\ a_2 &= \bar{a}_2 + \mu u_1^{(2)}(\bar{a}_1, \bar{a}_2, \bar{\psi}_1, \bar{\psi}_2) + \mu^2 u_2^{(2)}(\bar{a}_1, \bar{a}_2, \bar{\psi}_1, \bar{\psi}_2) + \cdots, \\ \psi_1 &= \bar{\psi}_1 + \mu v_1^{(1)}(\bar{a}_1, \bar{a}_2, \bar{\psi}_1, \bar{\psi}_2) + \mu^2 v_2^{(1)}(\bar{a}_1, \bar{a}_2, \bar{\psi}_1, \bar{\psi}_2) + \cdots, \\ \psi_2 &= \bar{\psi}_2 + \mu v_1^{(2)}(\bar{a}_1, \bar{a}_2, \bar{\psi}_1, \bar{\psi}_2) + \mu^2 v_2^{(2)}(\bar{a}_1, \bar{a}_2, \bar{\psi}_1, \bar{\psi}_2) + \cdots. \end{aligned} \tag{2.8.83}$$

Instead of the system of partial differential equations (2.8.43), for the wo-dimensional case we obtain the following system of equations:

$$
\begin{aligned}
\omega_1 \frac{\partial u^{(1)}}{\partial \bar{\psi}_1} + \omega_2 \frac{\partial u_1^{(1)}}{\partial \bar{\psi}_2} &= \frac{\bar{a}_2 \omega_2}{2\omega_1}\left(1 - \frac{\bar{a}_1^2}{4}\right)\cos(\bar{\psi}_1 - \bar{\psi}_2) \\
&- \frac{\bar{a}_2 \omega_2}{2\omega_1}\left(1 - \frac{\bar{a}_1^2}{4}\right)\cos(\bar{\psi}_1 + \bar{\psi}_2) - \frac{\bar{a}_1^2 \bar{a}_2 \omega_2}{8\omega_1}\cos(3\bar{\psi}_1 - \bar{\psi}_2) \\
&+ \frac{\bar{a}_1^2 \bar{a}_2 \omega_2}{8\omega_1}\cos(3\bar{\psi}_1 + \bar{\psi}_2) - A_1^{(1)}(\bar{a}_1, \bar{a}_2) \\
&\equiv F_1^{(1)}(\bar{a}_1, \bar{a}_2, \bar{\psi}_1, \bar{\psi}_2) - A_1^{(1)}(\bar{a}_1, \bar{a}_2), \\
\omega_1 \frac{\partial u_1^{(2)}}{\partial \bar{\psi}_1} + \omega_2 \frac{\partial u_1^{(2)}}{\partial \bar{\psi}_2} &= \frac{\bar{a}_1 \omega_1}{2\omega_2}\left(1 - \frac{\bar{a}_2^2}{4}\right)\cos(\bar{\psi}_1 - \bar{\psi}_2) \\
&- \frac{\bar{a}_1 \omega_1}{2\omega_2}\left(1 - \frac{\bar{a}_2^2}{4}\right)\cos(\bar{\psi}_1 + \bar{\psi}_2) - \frac{\bar{a}_1 \bar{a}_2^2 \omega_1}{8\omega_2}\cos(3\bar{\psi}_2 - \bar{\psi}_1) \\
&+ \frac{\bar{a}_1 \bar{a}_2^2 \omega_1}{8\omega_2}\cos(3\bar{\psi}_2 + \bar{\psi}_1) - A_1^{(2)}(\bar{a}_1, \bar{a}_2) \\
&\equiv F_1^{(2)}(\bar{a}_1, \bar{a}_2, \bar{\psi}_1, \bar{\psi}_2) - A_1^{(2)}(\bar{a}_1, \bar{a}_2), \\
\omega_1 \frac{\partial v_1^{(1)}}{\partial \bar{\psi}_1} + \omega_2 \frac{\partial v_1^{(1)}}{\partial \bar{\psi}_2} &= \frac{\bar{a}_2 \omega_2}{2\bar{a}_1\omega_1}\left(1 - \frac{\bar{a}_1^2}{4}\right)\cos(\bar{\psi}_1 - \bar{\psi}_2) \\
&- \frac{\bar{a}_2 \omega_2}{2\bar{a}_1\omega_1}\left(1 - \frac{\bar{a}_1^2}{4}\right)\cos(\bar{\psi}_1 + \bar{\psi}_2) - \frac{\bar{a}_1 \bar{a}_2 \omega_2}{8\omega_1}\cos(3\bar{\psi}_1 - \bar{\psi}_2) \\
&+ \frac{\bar{a}_1 \bar{a}_2 \omega_2}{8\omega_1}\cos(3\bar{\psi}_1 + \bar{\psi}_2) - B_1^{(1)}(\bar{a}_1, \bar{a}_2) \\
&\equiv \Phi_1^{(1)}(\bar{a}_1, \bar{a}_2, \bar{\psi}_1, \bar{\psi}_2) - B_1^{(1)}(\bar{a}_1, \bar{a}_2), \\
\omega_1 \frac{\partial v_1^{(2)}}{\partial \bar{\psi}_1} + \omega_2 \frac{\partial v_1^{(2)}}{\partial \bar{\psi}_2} &= \frac{\bar{a}_1 \omega_1}{2\bar{a}_2\omega_2}\left(1 - \frac{\bar{a}_2^2}{4}\right)\cos(\bar{\psi}_1 - \bar{\psi}_2) \\
&- \frac{\bar{a}_1 \omega_1}{2\bar{a}_2\omega_2}\left(1 - \frac{\bar{a}_2^2}{4}\right)\cos(\bar{\psi}_1 + \bar{\psi}_2) - \frac{\bar{a}_1 \bar{a}_2 \omega_1}{8\omega_2}\cos(3\bar{\psi}_2 - \bar{\psi}_1) \\
&+ \frac{\bar{a}_1 \bar{a}_2 \omega_1}{8\omega_2}\cos(3\bar{\psi}_2 + \bar{\psi}_1) - B_1^{(2)}(\bar{a}_1, \bar{a}_2) \\
&\equiv \Phi_2^{(2)}(\bar{a}_1, \bar{a}_2, \bar{\psi}_1, \bar{\psi}_2) - B_1^{(2)}(\bar{a}_1, \bar{a}_2), \\
\omega_1 \frac{\partial u_2^{(1)}}{\partial \bar{\psi}_1} + \omega_2 \frac{\partial u_2^{(1)}}{\partial \bar{\psi}_2} &= F_2^{(1)}(\bar{a}_1, \bar{a}_2, \bar{\psi}_1, \bar{\psi}_2) - A_2^{(1)}(\bar{a}_1, \bar{a}_2), \\
\omega_2 \frac{\partial u_2^{(2)}}{\partial \bar{\psi}_1} + \omega_2 \frac{\partial u_2^{(2)}}{\partial \bar{\psi}_2} &= F_2^{(2)}(\bar{a}_1, \bar{a}_2, \bar{\psi}_1, \bar{\psi}_2) - A_2^{(2)}(\bar{a}_1, \bar{a}_2), \\
\omega_1 \frac{\partial v_2^{(1)}}{\partial \bar{\psi}_1} + \omega_2 \frac{\partial v_2^{(1)}}{\partial \bar{\psi}_2} &= \Phi_2^{(1)}(\bar{a}_1, \bar{a}_2, \bar{\psi}_1, \bar{\psi}_2) - B_2^{(1)}(\bar{a}_1, \bar{a}_2), \\
\omega_1 \frac{\partial v_2^{(2)}}{\partial \bar{\psi}_1} + \omega_2 \frac{\partial v_2^{(2)}}{\partial \bar{\psi}_2} &= \Psi_2^{(2)}(\bar{a}_1, \bar{a}_2, \bar{\psi}_1, \bar{\psi}_2) - B_2^{(2)}(\bar{a}_1, \bar{a}_2), \\
&\cdots\cdots\cdots\cdots\cdots\cdots
\end{aligned}
\tag{2.8.84}
$$

In the infinite system (2.8.84) the first four equations are the result of identically equating the coefficients of μ in equations (2.8.81), the second four equations are the result of a similar operation at μ^2, etc.

Here we cannot write out the analytic expressions for the functions $F_2^{(1)}$, $F_2^{(2)}$, $\Phi_2^{(1)}$, $\Phi_2^{(2)}$ in full due to their super-intricacy, so we will describe them. Those functions are trigonometric polynomials of the form

$$\left(F_2^{(k)},\, \Phi_2^{(k)}\right) = \sum_{k_1=-3}^{3} \sum_{k_1=-3}^{3} \Big[b_{k_1,k_2}^{(k)} \cos(2k_1\bar{\psi}_1 + 2k_2\bar{\psi}_2) + c_{k_1,k_2}^{(k)} \sin(2k_1\bar{\psi}_1 + 2k_2\bar{\psi}_2) \Big], \tag{2.8.85}$$

where the coefficients $b_{k_1,k_2}^{(k)}$, $c_{k_1,k_2}^{(k)}$ are fractional rational functions with denominators including one of the following expressions:

$$\omega_1 \pm \omega_2, \quad 3\omega_1 \pm \omega_2, \quad 3\omega_2 \pm \omega_1. \tag{2.8.86}$$

It is essential that not all coefficients $b_{k_1,k_2}^{(k)}$, $c_{k_1,k_2}^{(k)}$ are nonzero, but $b_{0,0}^{(k)} \neq 0$, i.e. the trigonometric polynomials (2.8.85) contain "free" terms. Hence it follows that if the objective is to construct the substitution of variables (2.8.83) in trigonometric form (without "secular" terms), then the unknown functions $A_s^{(1)}$, $A_s^{(2)}$, $B_s^{(1)}$, $B_s^{(2)}$ included in the infinite system (2.8.84) should be determined as average values of the functions $F_s^{(1)}$, $F_s^{(2)}$, $\Phi_s^{(1)}$, $\Phi_s^{(2)}$ on the square of periods $[0,\, 2\pi] \times [0,\, 2\pi]$, i.e.

$$\left(A_s^{(j)},\, B_s^{(j)}\right) = \frac{1}{4\pi^2} \int_0^{2\pi}\!\!\int_0^{2\pi} \left(F_s^{(j)},\, \Phi_s^{(j)}\right) d\bar{\psi}_1 d\bar{\psi}_2. \tag{2.8.87}$$

It follows that

$$A_1^{(1)} = A_1^{(2)} = B_1^{(1)} = B_1^{(2)} = 0, \qquad A_2^{(j)},\, B_2^{(j)} = \{b_{0,0}^{(k)}\}. \tag{2.8.88}$$

As an example of $A_2^{(j)}, B_2^{(j)}$ take

$$A_2^{(1)}(\bar{a}_1, \bar{a}_2) = \frac{\bar{a}_1\omega_2}{4(\omega_1^2 - \omega_2^2)}\left(1 - \frac{\bar{a}_1^2}{4}\right)\left(1 - \frac{\bar{a}_2^2}{4}\right) + \frac{\bar{a}_2^2\omega_2^2(5\bar{a}_2^2 - 4)}{16\bar{a}_1\omega_1(\omega_1^2 - \omega_2^2)}\left(1 - \frac{\bar{a}_1^2}{4}\right) + \frac{3\bar{a}_1^3\bar{a}_2\omega_2^2}{64\omega_1(9\omega_1^2 - \omega_2^2)}. \tag{2.8.89}$$

Analyzing formulae (2.8.82), (2.8.84), and (2.8.88), two conclusions can be made that show the essential difference of the behavior of two weakly connected Van der Pol oscillators from the current strength fluctuations in one of them.

First, in the one-dimensional case $A_1 \neq 0$; therefore the averaged equation for the amplitude had the form $d\bar{a}/dt = \mu A_1 + \cdots$. But here $A_1^{(1)} = A_1^{(2)} = B_1^{(1)} = B_1^{(2)} = 0$, so $d\bar{a}/dt = \mu^2 A_2 + \cdots$. Yet for the fast phases in both cases the equations have the same form:

$$\frac{d\bar{\psi}}{dt} = \omega + O(\mu^2).$$

Second, integration of the first four partial differential equations of system (2.8.84) by the method of characteristics gives the following analytic expressions for the functions of the first approximation u_1, v_1:

$$\begin{aligned}
u_1^{(1)}(\bar{\psi}_1, \bar{\psi}_2) &= \frac{\bar{a}_2\omega_2}{2\omega_1}\left(1 - \frac{\bar{a}_1^2}{4}\right)\frac{\sin(\bar{\psi}_1 - \bar{\psi}_2)}{\omega_1 - \omega_2} \\
&\quad - \frac{\bar{a}_2\omega_2}{2\omega_1}\left(1 - \frac{\bar{a}_1^2}{4}\right)\frac{\sin(\bar{\psi}_1 + \bar{\psi}_2)}{\omega_1 + \omega_2} - \frac{\bar{a}_1^2\bar{a}_2\omega_2}{8\omega_1}\frac{\sin(3\bar{\psi}_1 - \bar{\psi}_2)}{3\omega_1 - \omega_2} \\
&\quad + \frac{\bar{a}_1^2\bar{a}_2\omega_2}{8\omega_1}\frac{\sin(3\bar{\psi}_1 + \bar{\psi}_2)}{3\omega_1 + \omega_2}, \\
u_1^{(2)}(\bar{\psi}_1, \bar{\psi}_2) &= \frac{\bar{a}_1\omega_1}{2\omega_2}\left(1 - \frac{\bar{a}_2^2}{2\omega_2}\right)\frac{\sin(\bar{\psi}_2 - \bar{\psi}_1)}{\omega_2 - \omega_1} \\
&\quad - \frac{\bar{a}_1\omega_1}{2\omega_2}\left(1 - \frac{\bar{a}_2^2}{4}\right)\frac{\sin(\bar{\psi}_2 + \bar{\psi}_1)}{\omega_2 + \omega_1} - \frac{\bar{a}_1\bar{a}_2\omega_1}{8\omega_2}\frac{\sin(3\bar{\psi}_2 - \bar{\psi}_1)}{3\omega_2 - \omega_1} \\
&\quad + \frac{\bar{a}_1\bar{a}_2\omega_1}{8\omega_2}\frac{\sin(3\bar{\psi}_2 + \bar{\psi}_1)}{3\omega_2 + \omega_1}, \\
v_1^{(1)}(\bar{\psi}_1, \bar{\psi}_2) &= \frac{\bar{a}_2\omega_2}{2\bar{a}_1\omega_1}\left(1 - \frac{\bar{a}_1^2}{4}\right)\frac{\sin(\bar{\psi}_1 - \bar{\psi}_2)}{\omega_1 - \omega_2} \\
&\quad - \frac{\bar{a}_2\omega_2}{2\bar{a}_1\omega_1}\left(1 - \frac{\bar{a}_1^2}{4}\right)\frac{\sin(\bar{\psi}_1 + \bar{\psi}_2)}{\omega_1 + \omega_2} - \frac{\bar{a}_1\bar{a}_2\omega_2}{8\omega_1}\frac{\sin(3\bar{\psi}_1 - \bar{\psi}_2)}{3\omega_1 - \omega_2} \\
&\quad + \frac{\bar{a}_1\bar{a}_2\omega_2}{8\omega_1}\frac{\sin(3\bar{\psi}_1 + \bar{\psi}_2)}{3\omega_1 + \omega_2}, \\
v_1^{(2)}(\bar{\psi}_1, \bar{\psi}_2) &= \frac{\bar{a}_1\omega_1}{2\bar{a}_2\omega_2}\left(1 - \frac{\bar{a}_2^2}{4}\right)\frac{\sin(\bar{\psi}_2 - \bar{\psi}_1)}{\omega_2 - \omega_1} \\
&\quad - \frac{\bar{a}_1\omega_1}{2\bar{a}_2\omega_2}\left(1 - \frac{\bar{a}_2^2}{4}\right)\frac{\sin(\bar{\psi}_2 + \bar{\psi}_1)}{\omega_2 + \omega_1} - \frac{\bar{a}_1\bar{a}_2\omega_1}{8\omega_2}\frac{\sin(3\bar{\psi}_2 - \bar{\psi}_1)}{3\omega_2 - \omega_1} \\
&\quad + \frac{\bar{a}_1\bar{a}_2\omega_1}{8\omega_2}\frac{\sin(3\bar{\psi}_2 + \bar{\psi}_1)}{3\omega_2 + \omega_1}.
\end{aligned} \tag{2.8.90}$$

Formulae (2.8.90) show that already in the first approximation, in the oscillator system the resonance terms may appear due to the presence of a denominator of the form $\omega_1 - \omega_2$, $3\omega_1 - \omega_2$, if these values are close to zero.

This phenomenon never occurs in the one-dimensional case. During construction of the second approximation, i.e. under integration of the second four equations of the system (2.8.84), resonances may increase, because upon integration of the functions u_1, v_1, the functions u_2, v_2 will contain denominators of the form $(\omega_1 - \omega_2)^2$, $(3\omega_1 - \omega_2)^2$. However, the functions u_2, v_2 are multiplied by μ^2, which, generally speaking, may decrease the influence of small denominators in transform (2.8.83).

Now write the equation of the second approximation for the slow amplitudes $\bar{a}_1$ and $\bar{a}_2$, which should be studied in order to probe the interaction of two oscillators:

$$\begin{aligned}
\frac{d\bar{a}_1}{dt} = \mu^2 A_2^{(1)}(\bar{a}_1, \bar{a}_2) \equiv \mu^2 \Bigg[& \frac{\bar{a}_1\omega_2}{4(\omega_1^2 - \omega_2^2)} \left(1 - \frac{\bar{a}_1^2}{4}\right)\left(1 - \frac{\bar{a}_2^2}{4}\right) \\
& + \frac{\bar{a}_2^2\omega_2^2(5\bar{a}_1^2 - 4)}{16\bar{a}_1\omega_1(\omega_1^2 - \omega_2^2)}\left(1 - \frac{\bar{a}_1^2}{4}\right) + \frac{3\bar{a}_1^3\bar{a}_2\omega_2^2}{64\omega_1(9\omega_1^2 - \omega_2^2)}\Bigg], \\
\frac{d\bar{a}_2}{dt} = \mu^2 A_2^{(2)}(\bar{a}_1, \bar{a}_2) \equiv \mu^2 \Bigg[& \frac{\bar{a}_2\omega_1}{4(\omega_2^2 - \omega_1^2)} \left(1 - \frac{\bar{a}_1^2}{4}\right) \\
& + \frac{\bar{a}_1^2\omega_1^2(5\bar{a}_2^2 - 4)}{16\bar{a}_2\omega_2(\omega_2^2 - \omega_1^2)}\left(1 - \frac{\bar{a}_2^2}{4}\right) + \frac{3\bar{a}_2^3\bar{a}_1\omega_1^2}{64\omega_2(9\omega_2^2 - \omega_1^2)}\Bigg].
\end{aligned} \tag{2.8.91}$$

System (2.8.91) is nonlinear with respect to $\bar{a}_1$ and $\bar{a}_2$, therefore it is impossible to obtain an exact solution in compact form. Such systems should be solved by methods of numerical integration. Even a study of equilibrium solutions of the second approximation, i.e. finding the solution of the algebraic system

$$A_2^{(1)} = 0, \quad A_2^{(2)} = 0$$

with respect to $\bar{a}_1$ and $\bar{a}_2$ entails extensive work.

In conclusion we point out once more that, using the average principle and the asymptotic theory of ordinary differential equations, it is certainly possible to study systems of Van der Pol oscillators. The features of resonance systems would be relevant to them, if the predefined frequency parameters $(\omega_1, \dots, \omega_n)$ are rationally commensurable. This may cause a great increase of oscillation amplitudes in electric circuits, and a considerable change of phases. Obtaining qualitative estimations for those changes adds up to solving partial differential equations of the form (2.8.84), systems of ordinary differential equations of the form (2.8.91), and to the construction of substitutions of variables of the form (2.8.74). As the reader may have noticed, the solution of equations (2.8.84) can be obtained constructively as exact formulae, and the solution of the equations (2.8.91), by the use of numerical methods. Obtaining the initial functions, i.e. the construction of substitution of variables (2.8.74), upon solving the first two problems, is quite obvious.

2.9 Study of Multifrequency Systems with Their Solutions Not Remaining Close to Resonance Points

It is possible to choose three most interesting aspects of the comparison of solutions of exact and averaged equations

$$\frac{dz}{dt} = Z(z, t, \mu), \qquad z(0) = z_0, \tag{2.9.1}$$

$$\frac{d\bar{z}}{dt} = \overline{Z}(\bar{z}, t, \mu), \qquad \bar{z}(0) = \bar{z}_0. \tag{2.9.2}$$

Aspect 1 Let there be given an arbitrary positive number $\varepsilon > 0$ and a time interval $[0, T]$. It is required to find the conditions to be satisfied by the function $Z(z, t, \mu)$, so that the following inequality should hold:

$$||z(t, \mu) - \bar{z}(t, \mu)|| < \varepsilon, \quad t \in [0, T]. \tag{2.9.3}$$

If T is an arbitrarily small number, then the ε-estimation for the norm $||z - \bar{z}||$ can be directly deduced from the classical theorem on the continuous dependence of the solution on small changes of the right-hand members. If T is a finite quantity, then it is very difficult to obtain such estimations for the norm $||z - \bar{z}||$. For some classes of equations regular with respect to a small parameter μ, time intervals $T = O(\mu^{-1})$ or $T = O(\mu^{-\alpha})$ $(\alpha > 0)$ are considered, which obviously makes the problem even more intricate. This can be illustrated by the averaging principle for Van der Pol systems, described in Section 2.8.

Aspect 2 Let there be given a time interval $[0, T]$. It is necessary to determine a maximum $K > 0$ that bounds the norm

$$||z(t, \mu) - \bar{z}(t, \mu)|| < K, \quad \text{if} \quad t \in [0, T].$$

Knowledge of the upper bound of the norm of the deviation of the approximate solution from the exact solution plays an important part in applied problems.

Aspect 3 Let there be given a positive number $K > 0$. It is necessary to find those time intervals in which

$$||z(t, \mu) - \bar{z}(t, \mu)|| < K.$$

Actually these aspects are the essence of ordinary differential equation theory from nonlinear analysis; therefore analytical, qualitative and numerical methods have been developed for their study. Among these, due to the constructive nature of asymptotic methods they are especially effective for the study of the first and the third aspects.

Revert to the $(m + n)$-dimensional multifrequency rotary system of differential equations

$$\begin{aligned} \frac{dx}{dt} &= \mu X(x, y), \\ \frac{dy}{dt} &= \omega(x) + \mu Y(x, y), \end{aligned} \tag{2.9.4}$$

where x, X are m-dimensional vectors, and y, ω, Y are n-dimensional vectors. Vector-functions $X(x,y)$, $Y(x,y)$ are defined and 2π-periodic in the domain

$$G_{m+n} = \{(x,y)\colon\ x \in P_m, \quad \|\operatorname{Im} y\| < \kappa < 1\}. \tag{2.9.5}$$

Along with (2.9.4) consider the corresponding averaged system of the first approximation:

$$\begin{aligned} \frac{d\bar{x}}{dt} &= \mu\overline{X}(\bar{x}), \\ \frac{d\bar{y}}{dt} &= \omega(\bar{x}) + \mu\overline{Y}(\bar{x}). \end{aligned} \tag{2.9.6}$$

Average values of $\overline{X}$ and $\overline{Y}$ are calculated by the formula

$$\left(\overline{X}(\bar{x}),\, \overline{Y}(\bar{x})\right) = \frac{1}{(2\pi)^n}\int_0^{2\pi}\cdots\int_0^{2\pi}(X(x,y),\, Y(x.y))\, dy_1\cdots dy_n. \tag{2.9.7}$$

It is most difficult to obtain ε-estimations for the norm $\|x(t,\mu) - \bar{x}(t,\mu)\|$ with $T = O(\mu^{-1})$ due to the possible occurrence of resonances between the frequencies $\omega_1(x)$, $\omega_2(x)$, $\ldots$, $\omega_n(x)$. Unlike Van der Pol oscillator systems where the frequencies $\omega_1, \ldots, \omega_n$ are constant, here we do not know beforehand what will happen with the frequencies if t changes, because we do not know the solution $x(t,\mu)$ itself. The complexity of this question is illustrated by the study performed by Arnol'd [3] who was the first to obtain an effective estimation for the deviation norm $\|x(t,\mu) - \bar{x}(t,\mu)\|$ in the two-frequency case. With $n \geq 3$ a similar result can be found in Neishtadt [1].

Now we will describe the results of proving the averaging method for systems of the form (2.9.4), obtained by Popova [1, 2].

First we will give some auxiliary statements and estimations.

(1) Let $0 < K < 1 < N < \infty$. Let $P_{m,N}$ denote the set of all points $x \in P_m$, in which $(k,\, \omega(x)) \neq 0$ with all integer vectors k having the norm $0 < \|k\| \leq N$. In other words, $P_{m,N}$ is the set P_m minus zero-resonances (exact resonances).

(2) Let $P_{m,N,K}$ denote the set of those points $x \in P_m$ that belong to $P_{m,N}$ together with their K-neighborhood. It is evident that for $x \in P_{m,N,K}$ the following estimation is true:

$$|(k,\, \omega(x))| > K, \quad 0 < \|k\| \leq N. \tag{2.9.8}$$

(3) The complement of the set $P_{m,N,K}$ with respect to the set P_m will be denoted by $R_{m,N,K}$:

$$R_{m,N,K} = P_m \setminus P_{m,N,K}. \tag{2.9.9}$$

(4) Let $(x(t,\mu),\, y(t,\mu))$ be the solution of the system (2.9.4) provided $t \in [0,\, \mu^{-1}]$, $x \in P_m$. Divide the time interval $[0,\, \mu^{-1}]$ into two sets: the nonresonance set $p_{m,N,K}$ for which $x \in P_{m,N,K}$, and the resonance set $r_{m,N,K}$ for which $x \in R_{m,N,K}$.

(5) Suppose that at least one component of the function $X(x,y)$, e.g. the s-th one, at any point $(x,y) \in G_{m+n}$ satisfies the condition

$$|X_s(x,y)| > C_1^{-1}, \tag{2.9.10}$$

where $C_1 > 0$. This inequality does not allow the return of the path (or more exactly, of the m-dimensional projection of the phase path) into the small neighborhood of the point after the time interval $\Delta t = |t - t_1| \neq 0$.

(6) Suppose that the frequency vector $\omega(x)$ on the solutions of the system (2.9.4) satisfies the inclusion

$$\omega(x(t,\mu)) \in C_t^{n_1}[0,\, \mu^{-1}], \tag{2.9.11}$$

where $1 \leq n_1 < \infty$ and $|\omega_n(x)| > \gamma > 0$ when $x \in P_m$. Here we demand not the analyticity of the vector $\omega(x)$, but only the n_1 times differentiability with respect to $t \in [0,\, \mu^{-1}]$. Consider the $(n-1)$-dimensional vector-function

$$\lambda(x) = \left(\frac{\omega_1(x)}{\omega_n(x)}, \ldots, \frac{\omega_{n-1}(x)}{\omega_n(x)}\right) \tag{2.9.12}$$

and let $I_{N,\mathrm{res}}$ denote the set of integer resonance vectors k, $0 \leq ||k|| \leq N$, such that $(k, \omega(x)) = 0$ for some $x \in P_m$. Vectors of dimension $(n-1)$, obtained from the vectors $k \in I_{N,\mathrm{res}}$ by truncation of the n-th component, will be denoted by k^-. It is easy to verify that $||k^-|| \geq 1$.

(7) Let $x_{r_s} = x(t_{r_s}, \mu)$ be the resonance points of system (2.9.4), i.e. $(k,\, \omega(x_{r_s})) = 0$. Suppose that at time t all derivatives of some resonance heterodyne frequency $(k^-, \lambda(x(t,\mu)))$ with respect to t to the order $n_{k_s} - 1$ (where $n_{k_s} \leq n_1$, $1 \leq n_1$) inclusive become zero:

$$\left|\left(k^-,\, \frac{d^j \lambda\left(x\left(t_{r_s}, \mu\right)\right)}{dt^j}\right)\right| = 0 \tag{2.9.13}$$

when $j = 1, \ldots, n_{k_s} - 1$. Regarding the n_{k_s}-th order derivative we will assume that with $\mu \leq \mu^*$ for any $t \in [0,\, \mu^{-1}]$ the following inequality is true:

$$\left|\left(k^-,\, \frac{d^{n_{k_s}} \lambda}{dt^{n_{k_s}}}\right)\right| \geq C_2^{-1} \mu^{n_{k_s}} ||k||^{-\nu}, \tag{2.9.14}$$

where the constant C_2 is positive and does not depend on $\mu, K, ||k||$, $\nu = n+1$.

Conditions (2.9.13) and (2.9.14) guarantee that the system (2.9.4) cannot remain on any resonance of order $l \leq N$ within the time $[0,\, \mu^{-1}]$.

Lemma 1 *Let $A_{N,n}$ denote the number of n-dimensional integer vectors with their norm exactly equal to N. The following estimations are true:*

$$\begin{aligned} &A_{N,n} \le 2nN^{n-1} && \text{for } n \ge 1, \text{ if } N \ge 1, \\ &A_{N,n} \le 2^{\frac{n^2}{n-2}} N^{n-1} && \text{for } n \ge 1, \text{ if } N \ge 2, \\ &A_{N,n} \le \frac{3}{\sqrt{\pi(4n-3)}} \left(\frac{6N}{n-1}\right)^{n-1} && \text{for } n \ge 2, \text{ if } N \ge n-1. \end{aligned} \tag{2.9.15}$$

To obtain a ε-estimation for the norm $\|x(t,\mu) - \bar{x}(t,\mu)\|$, it is necessary to impose the following additional conditions on the right-hand members of system (2.9.4).

(8) The norms satisfy

$$\|X(x,y)\| < C, \quad \|\omega(x)\| < C \tag{2.9.16}$$

in their definitional domains.

(9) For any integer vector $\rho = (\rho_1, \dots, \rho_n)$ and any $s = 1, \dots, m$ the following inequalities should hold:

$$\max_{0 \le \|\rho\| \le l} \sup_{(x,y) \in G_{m+n}} \left| \frac{\partial^{\|\rho\|} X_s(x,y)}{\partial y_1^{\rho_1} \cdots \partial y_n^{\rho_n}} \right| < C_l^{(s)}. \tag{2.9.17}$$

(10) The Fourier coefficients of the function $X(x,y)$ satisfy the estimation

$$\sup_{x \in P_m} \sum_{\|k\| \ge 1} \left\| \frac{\partial X_k(x)}{\partial x} \right\| < C'. \tag{2.9.18}$$

(11) The norm of the matrix $\partial\omega/\partial x$ is bounded:

$$\left\| \frac{\partial \omega}{\partial x} \right\| < C''. \tag{2.9.19}$$

Using all these conditions it is possible to formulate a number of lemmas necessary for the further computations; their proofs are given in Popova [1, 2].

Lemma 2 *With the above assumptions, the path $x(t,\mu)$ of the system (2.9.4) within the time $[0, \mu^{-1}]$ cannot pass more than $2nN^{n-2} - 2$ resonance points. The time of stay in the neighborhood of the resonance point of radius K does not exceed*

$$|\Delta t_{\text{res}}| \le C_3 \mu^{-1} \left(K \|k\|^{\nu+1}\right)^{1/n_1} \tag{2.9.20}$$

where

$$C_3 = \left(2C'' C_2 n_1! \gamma^{-1}\right)^{1/n_1}. \tag{2.9.21}$$

Lemma 3 *Let*

$$\left\| \frac{dz}{dt} \right\| \le a\|z\| + b(t), \quad \|z(0)\| < C$$

where a, b, C are not negative. Then

$$\|z(t)\| \leq m\left[C + \int_0^t b(\tau)\, d\tau\right] e^{amt}. \tag{2.9.22}$$

Here m is the length of the vector z.

Now let the scalar function f of the vector variables x, y belong to the class C_y^l:

$$f(x,y) \in C_y^l[G_{m+n}], \tag{2.9.23}$$

i.e. be l times continuously differentiable with respect to y, and let it be representable by the n-fold Fourier series

$$f(x,y) = \sum_{\|k\|\geq 0} f_k(x) e^{i(k,y)}. \tag{2.9.24}$$

As above (see Section 1.1) let,

$$S_N(x,y) = \sum_{0\leq\|k\|\leq N} f_k(x) e^{i(k,y)}. \tag{2.9.25}$$

Introduce one more item of notation:

$$\begin{aligned} |f(x,y)|_{\|\rho\|} &= \max_{0\leq\|\alpha\|\leq\|\rho\|} \sup_{(x,y)\in G_{m+n}} \left| \frac{\partial^{\|\alpha\|} f(x,y)}{\partial y_1^{\alpha_1} \cdots \partial y_n^{\alpha_n}} \right|, \\ |f(x,y)|_{\|\rho\|} &\leq C^*, \quad \text{if} \quad \|\rho\| \leq l. \end{aligned} \tag{2.9.26}$$

Lemma 4 *With the above listed conditions satisfied for any vector $\rho = (\rho_1, \dots, \rho_n)$ such that $0 \leq \|\rho\| < l - n - 1$, the following estimate is true*

$$|f(x,y) - S_N(x,y)| < \frac{C(n, \|\rho\|, N)}{l - n - \|\rho\|} C^* \left(N + \frac{1}{2}\right)^{n+\|\rho\|-l}, \tag{2.9.27}$$

where

$$C(n, \|\rho\|, N) = \begin{cases} 2n^{l+1-\|\rho\|}, & \text{if} \quad N \geq 1, \\ \dfrac{n^{l+2-\|\rho\|}}{2^{n-2}}, & \text{if} \quad N \geq 2, \\ \dfrac{3n^{l-\|\rho\|}}{\sqrt{\pi(4n-3)}} \left(\dfrac{6N}{n-1}\right)^{n-1}, & \text{if} \quad n \geq 2, \ N \geq n - 1 \geq 1. \end{cases}$$

This lemma allows us to estimate the residual Fourier series and its derivatives depending on the degree of smoothness of the function expanded in the Fourier series. From (2.9.27) it is easy to deduce an estimation for the vector-function

$$\left\| X(x,y) - \sum_{0\leq\|k\|\leq N} X_k(x) e^{i(k,y)} \right\| \leq \frac{C(n,N)}{l-n} \left(N + \frac{1}{2}\right)^{n-1} \sum_{s=1}^m C_s^*, \tag{2.9.28}$$

where

$$C(n,N) = \begin{cases} 2n^{l+1}, & \text{if } \ N \geq 1, \\ \dfrac{n^{l+2}}{2^{2n-2}}, & \text{if } \ N \geq 2, \\ \dfrac{3n^l}{\sqrt{\pi(4n-3)}} \left(\dfrac{6N}{n-1}\right)^{n-1}, & \text{if } \ n \geq 2, \ N \geq n-1 \geq 1. \end{cases}$$

Lemma 5 *Let condition (2.9.18) be satisfied for $l > 2\nu + n + 1$. Then in the domain $G_{N,K} = P_{m,N,K} \times Q_n$ there exists an m-dimensional vector-function*

$$z = x + Z_N(x, y) \tag{2.9.29}$$

such that

$$\begin{aligned} &\|z - x\| < C_4 \mu K^{-n_1}, \\ &\left\| \frac{dz}{dt} - \mu \overline{X}(z) \right\| < C_8 \mu^2 K^{-n_1} + C_9 \mu^2 K^{-n_1} N^{\nu} + C_{10} \mu^2 K^{-2n_1} + \mu \delta, \end{aligned} \tag{2.9.30}$$

for $N \approx C' \delta^{1/(l-n)}$ and $\mu \leq C_4^{-1} K^{n_1+1}$. The function $Z_N(x, y)$ is 2π-periodic with respect to y in $G_{N,K}$, and the constants C do not depend on μ, N, K.

The above-described conditions guarantee the non-remaining of the solutions of the rotary system (2.9.4) (or more exactly, the projection of $x(t, \mu)$) in the neighborhood of any resonance point with all $t \in [0, \mu^{-1}]$. Together with the above lemmas they allow us to formulate and prove the theorems proving the averaging principle for rotary multifrequency systems when the averaging operator M_y is applied to them.

One of those theorems is given below.

Theorem (see Grebenikov and Popova [1]) *Let:*

1. *the vector-function $X(x, y)$ in the domain G_{m+n} satisfy the conditions (2.9.10), (2.9.16) - (2.9.19);*
2. *the vector-function $\omega(x)$ satisfy the conditions (2.9.13), (2.9.14);*
3. *the solution $x(t, \mu)$ belong to P_m for all $t \in [0, \infty)$ and all $\mu \in [0, \mu_0]$;*
4. *the order of resonances l be higher than* $\max(2n, 2\nu + n + 1)$.

Then for any $\mu \in [0, \mu_0]$ there exists a number $a > 0$, such that for all $t \in [0, \mu^{-1}]$ the following estimation is true

$$\|x(t, \mu) - \bar{x}(t, \mu)\| < C_{19} \mu^{a(1-n/(l-n))}, \tag{2.9.31}$$

where $\bar{x}(t, \mu)$ is the solution of the equation of the first approximation

$$\frac{d\bar{x}}{dt} = \mu \overline{X}(\bar{x}). \tag{2.9.32}$$

The parameters included in (2.9.31) are expressed by the following relations:

if $n_1 = 1$, then

$$
\begin{aligned}
a &= \left(2 + 2\,\frac{\nu+1}{l-n}\right)^{-1}, \\
C_{19} &\le 2C_{11} n\,(C_8')^n \Big\{ 2C_{12}\sqrt{C_4}\,(C_8')^{\nu+1} \\
&\quad + C_{16}\left[C_8 \mu^{a\nu/(l-n)} + C_9\,(C_8')^{\nu}\right]\mu^{\theta}\ln\left(1 + C_1^{-1} C_4^{-1/2}\mu^{-b}\right) \qquad (2.9.33) \\
&\quad + C_{15} + \left[\sqrt{C_4}(1 + C_{15}) + \frac{C_{10}C_{16}}{\sqrt{C_4}(2n_1 - 1)}\right]\mu^{\kappa_1}\Big\}, \\
\theta &= a\left(1 + \frac{\nu+2}{l-n}\right), \quad b = a\left(1 + \frac{\nu+1}{l-n}\right), \quad \kappa_1 = \frac{\nu+1}{2(l-n+\nu+1)};
\end{aligned}
$$

if $n_1 \ge 2$, then

$$
\begin{aligned}
a &= \left[1 + (2n_1 - 1)\left(n_1 + \frac{\nu+1}{l-n}\right)\right]^{-1}, \\
C_{19} &\le 2C_{11} n\,(C_8')^n \Big\{ 2C_{12}\,(C_8')^{(\nu+1)/n_1}\, C_4^{1/n_1(n_1+1)} \\
&\quad + C_4^{1/(n_1+1)}\,(1 + C_{15})\,\mu^{\kappa_2} + C_{16} C_4^{(1-n_1)/(n_1+1)}\Big[C_8 \mu^{a\nu/(l-n)} \qquad (2.9.34) \\
&\quad + C_9\,(C_8')^{\nu}\,(n_1 - 1)^{-1}\mu^{\kappa_3} + \frac{1}{2n_1 - 1} C_{10} C_4^{-n_1/(n_1+1)}\Big] + C_{15}\Big\}, \\
\kappa_2 &= a(n_1 - 1)\left(n_1 + \frac{\nu+1}{l-n}\right), \quad \kappa_3 = a\left[n_1^2 + \frac{(n_1 - 1)\nu + 1}{l-n}\right].
\end{aligned}
$$

Proof Let the successive time intervals composing the nonresonance set $p_{N,K}$ be denoted by $\left[t^{(j)}_{\mathrm{l,\,nonres}},\, t^{(j)}_{\mathrm{r,\,nonres}}\right]$ $(j = 1, \ldots)$. Note that $\left[t^{(j)}_{\mathrm{r,\,nonres}},\, t^{(j)}_{\mathrm{l,\,nonres}}\right] \in r_{N,K}$. For definiteness, assume that $t^{(1)}_{\mathrm{l,\,res}} = 0 \in r_{N,K}$. Denote

$$
\bar{x}_j = \bar{x}\left(x^{(j)}_{\mathrm{l,\,nonres}},\, t^{(j)}_{\mathrm{l,\,nonres}},\, t\right), \quad z(t,\mu) = z(x(t,\mu),\, y(t,\mu)),
$$
$$
x^{(j)}_{\alpha} = x\left(t^{(j)}_{\alpha,\,\mathrm{nonres}}\right), \quad \alpha = (\mathrm{l},\, \mathrm{r}).
$$

Then

$$
\|x(t,\mu) - \bar{x}(t,\mu)\| \le \|x(t,\mu) - \bar{x}_j(t,\mu)\| + \sum_{j\ge1}\|\bar{x}_{j+1}(t,\mu) - \bar{x}_j(t,\mu)\|. \qquad (2.9.35)
$$

Using Lemma 3 from Grebenikov and Ryabov [1, p.40] that allows us to estimate the deviation of smoothed solutions, conditioned by the deviation of the initial points, we obtain

$$
\begin{aligned}
\|x(t,\mu) - \bar{x}(t,\mu)\| &\le \|x(t,\mu) - \bar{x}_j(t,\mu)\| \\
&+ C_{11}\sum_{j\ge 1}\left\{\left\|x_{\mathrm{l}}^{(j+1)} - x_{\mathrm{r}}^{(j)}\right\| + \left\|\bar{x}^{(j)}\left(t_{\mathrm{r,\,nonres}}^{(j)}\right) - \bar{x}^{(j)}\left(t_{\mathrm{l,\,nonres}}^{(j+1)}\right)\right\|\right\},
\end{aligned}
\tag{2.9.36}
$$

where $C_{11} = e^L$, and L is a Lipschitz constant for $\overline{X}(x)$ in the domain $x \in P_m$.

The interval $\left[t_{\mathrm{r,\,nonres}}^{(j)},\, t_{\mathrm{l,\,nonres}}^{(j+1)}\right]$ belongs to the resonance set $r_{N,K}$, and its length can be estimated by means of Lemma 2. Therefore

$$
\begin{aligned}
\left\|x_{\mathrm{l}}^{(j+1)} - x_{\mathrm{r}}^{(j)}\right\| + \left\|x_{\mathrm{r}}^{(j)} - \bar{x}^{(j)}\left(t_{\mathrm{r,\,nonres}}^{(j)}\right)\right\| &< 2C_{12}(KN^{\nu+1})^{1/n_1}, \\
C_{12} &= CC_3.
\end{aligned}
\tag{2.9.37}
$$

In the resonance domain $G_{N,K} = P_{m,N,K} \times Q_n$ construct the vector-function $z(x,y)$ with the properties specified in Lemma 5. On each interval

$$
\left[t_{\mathrm{l,\,nonres}}^{(j)},\, t_{\mathrm{r,\,nonres}}^{(j)}\right] \subset p_{N,K}
$$

the following inequality is true:

$$
\begin{aligned}
\left\|\frac{dz}{dt} - \frac{d\bar{x}^{(j)}}{dt}\right\| &\le \left\|\frac{dz}{dt} - \mu\overline{X}(z)\right\| + \mu\left\|\overline{X}(z) - \overline{X}\left(\bar{x}^{(j)}\right)\right\| \\
&\le C_{13}\mu\left\|z - \bar{x}^{(j)}\right\| + \left\|\frac{dz}{dt} - \mu\overline{X}(z)\right\|, \\
C_{13} &= L.
\end{aligned}
\tag{2.9.38}
$$

Taking into account (2.9.10), we obtain

$$
\left\|x\left(t_{\mathrm{l,\,nonres}}^{(j)} + \tau\right) - x\left(t_{\mathrm{l,\,nonres}}^{(j)}\right)\right\| > C_1^{-1}\mu\tau.
$$

Hence, $x(t,\mu) \in P_{m,N,K+C_1^{-1}\mu\tau}$ for

$$
t \in \left[t_{\mathrm{l,\,nonres}}^{(j)} + \tau,\; t_{\mathrm{r,\,nonres}}^{(j)} - \tau\right], \qquad \tau < \frac{1}{2}\left[t_{\mathrm{r,\,nonres}}^{(j)} - t_{\mathrm{l,\,nonres}}^{(j)}\right].
$$

Then from Lemma 5 the following estimation is true on the specified interval:

$$
\begin{aligned}
\left\|\frac{dz}{dt} - \mu\overline{X}(z)\right\| &< \mu^2(C_8 + C_9N^\nu)(N + C_1^{-1}\mu\tau)^{-n_1} \\
&\quad + \mu^2 C_{10}(K + C_1^{-1}\mu\tau)^{-2n_1} + \mu\varepsilon.
\end{aligned}
$$

Since $\left|t_{\mathrm{r,\,nonres}}^{(j)} - t_{\mathrm{l,\,nonres}}^{(j)}\right| < \mu^{-1}$, then

$$
\begin{aligned}
\int\limits_{t_{\mathrm{l,\,nonres}}^{(j)}}^{t_{\mathrm{r,\,nonres}}^{(j)}} \left\|\frac{dz}{dt} - \mu\overline{X}(z)\right\| dt &< \mu C_1(C_8 + C_9N^\nu)C_{14}(K) \\
&\quad + \frac{\mu C_1 C_{10}}{2n_1 - 1}K^{-2n_1+1} + \varepsilon
\end{aligned}
\tag{2.9.39}
$$

where

$$C_{14}(K) = \begin{cases} \ln\left(1 - \dfrac{1}{C_1 K}\right), & \text{if} \quad n_1 = 1, \\ (n_1 - 1)^{-1} K^{-n_1+1}, & \text{if} \quad n_1 > 1. \end{cases}$$

From Lemma 5 it follows that

$$\left\| z\left(t^{(j)}_{1,\,\text{nonres}}\right) - \bar{x}\left(t^{(j)}_{1,\,\text{nonres}}\right) \right\| < \mu C_4 K^{-n_1}. \tag{2.9.40}$$

Now, applying Lemma 3 to inequality (2.9.38) and taking account of (2.9.39) and (2.9.40), we obtain for $t^{(j)}_{\text{r, nonres}} \in [0,\, \mu^{-1}]$

$$\left\| z\left(t^{(j)}_{\text{r, nonres}}\right) - \bar{x}^{(j)}\left(t^{(j)}_{\text{r, nonres}}\right) \right\| < m\big[\mu C_4 K^{-n_1} + \varepsilon + \mu C_1 (C_8 + C_9 N^{\nu}) C_{14} + \mu C_1 C_{10} (2n_1 - 1)^{-1} K^{-2n_1+1}\big] e^{m C_{13}}.$$

Then the second summand under the summation sign in (2.9.36) can be estimated by means of the relation

$$\begin{aligned} &\left\| x\left(t^{(j)}_{\text{r, nonres}}\right) - \bar{x}^{(j)}\left(t^{(j)}_{\text{r, nonres}}\right) \right\| \\ &\quad < \left\| x\left(t^{(j)}_{\text{r, nonres}}\right) - z\left(t^{(j)}_{\text{r, nonres}}\right) \right\| + \left\| z\left(t^{(j)}_{\text{r, nonres}}\right) - \bar{x}^{(j)}\left(t^{(j)}_{\text{r, nonres}}\right) \right\| \\ &\quad < \mu C_4 (1 + C_{15}) K^{-n_1} + C_{16} \varepsilon \\ &\qquad + \mu C_{16} \big[(C_8 + C_9 N^{\nu}) C_{14} + (2n_1 - 1)^{-1} C_{10} K^{-2n_1+1}\big], \end{aligned} \tag{2.9.41}$$

where $C_{15} = m e^{m C_{13}}$, $C_{16} = C_1 C_{15}$.

Note that relations (2.9.38) – (2.9.41) are true for any $t \in p_{N,K}$. Consequently for any $t \in [0,\, \mu^{-1}]$ estimation (2.9.37) or (2.9.41) is true. Summing up on all resonance and nonresonance time intervals, we obtain that for $t \in [0,\, \mu^{-1}]$

$$\begin{aligned} &\| x(t, \mu) - \bar{x}(t, \mu) \| \\ &\qquad < 2C_{11} \left(n N^n - 1\right) \Big\{ 2C_{12} \left(K N^{\nu+1}\right)^{1/n_1} + C_4 (1 + C_{15}) \mu K^{-n_1} \\ &\qquad + C_{10} \big[\mu (C_8 + C_9 N^{\nu}) C_{14} + C_{10} (2n_1 - 1)^{-1} \mu K^{-2n_1+1}\big] \Big\} + C_{16} \varepsilon. \end{aligned} \tag{2.9.42}$$

For the complete justification of the averaging method it is necessary to show that the right-hand member of equation (2.9.42) with $\mu \to 0$ also approaches zero with the appropriately selected $\mu_0 > 0$ and $K > 0$, when $\varepsilon > 0$ is defined. To obtain a classical ε-estimation it is necessary that the following conditions are simultaneously satisfied as $\mu \to 0$:

$$\begin{aligned} N^n \left(K N^{\nu+1}\right)^{1/n_1} &\to 0, \quad N^n \mu K^{-n_1} \to 0, \\ \mu N^{\nu+n} K^{-n_1+1} &\to 0, \quad n_1 \geq 2, \\ \mu N^{\nu+n} \ln\left(1 + C_1^{-1} K^{-1}\right) &\to 0, \quad n_1 = 1, \\ \mu N^n K^{-2n_1+1} &\to 0, \quad N^n \varepsilon \to 0. \end{aligned} \tag{2.9.43}$$

Select $\mu_0^a = \varepsilon$ and $K = B\mu^b$, where $a > 0$, $b > 0$. From Lemma 4 it follows that $N \sim \varepsilon^{-1/(l-n)}$ or $N \sim \mu^{-a/(l-n)}$. Relations (2.9.43) hold for a and b such that the following system of inequalities is satisfied:

$$\begin{aligned}
f_1 &\equiv -\frac{an}{l-n} + \frac{1}{n_1}\left(b - \frac{a(\nu+1)}{l-n}\right) > 0,\\
f_2 &\equiv -\frac{an}{l-n} + 1 - bn_1 > 0,\\
f_3 &\equiv 1 - \frac{a(\nu+n)}{l-n} - b(n_1 - 1) > 0, n_1 \geq 2,\\
f_4 &\equiv 1 - \frac{a(\nu+n)}{l-n} - \theta > 0, \qquad 0 < \theta < 1, \quad n_1 = 1,\\
f_5 &\equiv 1 - \frac{an}{l-n} - b(2n_1 - 1) > 0,\\
f_6 &\equiv a - \frac{an}{l-n} > 0.
\end{aligned} \tag{2.9.44}$$

From the condition $f_6 > 0$ it follows that the condition $l > 2n$ should be satisfied. Take separately the cases $n_1 \geq 2$ and $n_1 = 1$.

Case $n_1 \geq 2$. For the conditions $f_1 = f_6$, $f_5 = f_6$ to be satisfied, a and b should satisfy the system

$$a = 1 - b(2n_1 - 1), \qquad n_1 a = b - \frac{a(\nu+1)}{l-n}.$$

Hence we obtain

$$a = \left[1 + (2n_1 + 1)\left(n_1 + \frac{\nu+1}{l-n}\right)\right]^{-1}, \qquad b = a\left[n_1 + \frac{\nu+1}{l-n}\right].$$

With these a and b we get

$$f_2 - f_6 = a(n_1 - 1)\left(n_1 + \frac{\nu+1}{l-n}\right), \qquad f_3 - f_5 = a\left[n_1^2 + \frac{(n_1-1)\nu + 1}{l-n}\right].$$

Hence follows that

$$f_2 > f_6, \qquad f_3 > f_5 = f_1 = f_6.$$

For the conditions of Lemma 5 to be satisfied it is necessary that μ should satisfy an additional inequality from which at $B = C_4^{1/(n_1+1)}$ follows the inequality

$$1 - b(n_1 + 1) \geq 0. \tag{2.9.45}$$

Substituting the expression for b into (2.9.45), we obtain

$$1 + (2n_1 - 1)\left(n_1 + \frac{\nu+1}{l-n}\right) - \left(n_1 + \frac{\nu+1}{l-n}\right)(n_1 + 1) \geq 0.$$

The last inequality is always true for $n_1 \geq 2$. Reverting to (2.9.42) with $t \in [0, \mu^{-1}]$, $\mu \in [0, \mu_0]$, $n_1 \geq 2$, we obtain

$$
\begin{aligned}
&\|x(t,\mu) - \bar{x}(t,\mu)\| < C_{17}\mu^{a(1-n/(l-n))}, \\
&C_{17} = 2nC_{11}(C_8')^n\Big\{2C_{12}(C_8')^{(\nu+1)/n_1}C_4^{1/(n_1+1)} \\
&\quad + C_4^{1/(n_1+1)}(1+C_{15})\mu^{a(n_1-1)[n_1+(\nu+1)/(l-n)]} \\
&\quad + C_{16}C_4^{(1-n_1)/(n_1+1)}\Big[\big(C_8\mu^{a\nu/(l-n)} + C_9(C_8')^\nu\big)(n_1-1)^{-1} \\
&\quad \times \mu^{a[n_1^2+((n_1-1)\nu+1)/(l-n)]} + (2n_1-1)^{-1}C_4^{-n_1/(n_1+1)}C_{10}\Big] + C_{15}\Big\}.
\end{aligned}
\tag{2.9.46}
$$

For the inequality (2.9.46) to be nonformal, it is necessary that the smoothness degree of the function $X(x,y)$ (number l) be higher than $\sup\{2n, 2\nu+n+1\}$.

Case $n_1 = 1$. Then $f_2 = f_5$. If the parameter b is calculated by the above formula, then

$$b = \left(1 + \frac{\nu+1}{l-n}\right)\left(2 + \frac{\nu+1}{l-n}\right)^{-1} > \frac{1}{2},$$

and inequality (2.9.45) does not hold. So we will introduce the bound $b \leq 1/2$ and demand that the equality $f_1 = f_6$ should hold. Then

$$b = a\left(1 + \frac{\nu+1}{l-n}\right), \tag{2.9.47}$$

where

$$a = \left(2 + 2\frac{\nu+1}{l-n}\right)^{-1}. \tag{2.9.48}$$

With $b \leq 1/2$ it is always true that $f_2 > f_1$. Let

$$\theta = 1 + \frac{a}{l-n} - b. \tag{2.9.49}$$

Then $f_4 = f_1$ and consequently, with the conditions (2.9.46) and (2.9.48) satisfied, we will have $f_5 = f_2 > f_1 = f_4 = f_6$.

Taking into account all these relations, (2.9.42) can be given the following analytic form:

$$
\begin{aligned}
&\|x(t,\mu) - \bar{x}(t,\mu)\| \leq C_{18}\mu^{a[1-n/(l-n)]}, \\
&C_{18} = 2nC_{11}(C_8')^n\Big\{2C_{12}\sqrt{C_4'}(C_8')^{\nu+1} \\
&\quad + C_{16}\big[C_8\mu^{a\nu/(l-n)} + C_9(C_8')^\nu\big]\mu^\theta \ln\Big(1 + C_1^{-1}\sqrt{C_4^{-1}\mu^b}\Big) \\
&\quad + \mu^{\kappa_1}\big[\sqrt{C_4}(1+C_{14}) + C_{10}C_{16}\sqrt{C_4}\big] + C_{15}\Big\}, \\
&\kappa_1 = \frac{1}{2}(\nu+1)(l-n+\nu+1)^{-1} < \frac{1}{2}.
\end{aligned}
\tag{2.9.50}
$$

The parameters a, b, θ are chosen in accordance with formulae (2.9.47) – (2.9.49).

In ε-terminology the statement of the theorem is as follows:

For any $\varepsilon > 0$ *there exists* $\mu_0(\varepsilon) \ll 1$, *such that for any* $\mu \in [0, \mu_0(\varepsilon)]$ *on the interval* $t \in [0, \mu^{-1}]$, $\|x(t,\mu) - \bar{x}(t,\mu)\| < \varepsilon$ *is true.*

2.10 Study of Multifrequency Systems Belonging to Class II

We have more than once pointed out that integral average values of the function $f(y)$ depending on n variables $y_1, \dots, y_n$ (and 2π-periodic with respect to each of them), constructed by means of different smoothing operators, may differ from each other. The equality

$$M_y[f(y)] = M_t[f(\omega t)] = f_0$$

is only satisfied when $(k,\omega) \neq 0$ for any integer vector k with a nonzero norm. If for some vectors k (with $\|k\| \neq 0$) the resonance condition $(k,\omega) = 0$ is satisfied, then, as is well known,

$$M_y[f(y)] \neq M_t[f(\omega t)],$$

because here $M_y[f] = f_0$, and

$$M_t[f(\omega t)] = \sum_{\|k\| \geq 0}' f_k, \tag{2.10.1}$$

where the prime at the sum, as before, means that k only takes those values that satisfy $(k,\omega) = 0$.

This circumstance makes us separately consider the question of proving the average method of multifrequency systems with no coinciding average values over time and fast variables.

Again write the rotary multifrequency system:

$$\begin{aligned} \frac{dx}{dt} &= \mu X(x,y), \\ \frac{dy}{dt} &= \omega(x) + \mu Y(x,y), \end{aligned}$$

$$X(x,y) = \sum_{\|k\| \geq 0} X_k(x) e^{i(k,y)}, \quad Y(x,y) = \sum_{\|k\| \geq 0} Y_k(x) e^{i(k,y)}, \tag{2.10.2}$$

$$(k,y) = \sum_{s=1}^{n} k_s y_s, \quad \|k\| = \sum_{s=1}^{n} |k_s|, \quad k_s = 0, \pm 1, \pm 2, \dots$$

We know the sufficient conditions ensuring the departure of the solution $x(t,\mu)$ from the neighborhood of radius $O(\sqrt{\mu})$ of any resonance point within the time of order $O(\mu^{-1/2})$. For a dual frequency system this condition was found by Arnold [3]:

$$\left| \left(\frac{\partial \omega_1}{\partial x}, X \right) \omega_2 - \left(\frac{\partial \omega_2}{\partial x}, X \right) \omega_1 \right| > 0 \tag{2.10.3}$$

for all $y = (y_1, y_2) \in Q_2$ if $x \in P_m$. If the number of frequencies is more than two $(n > 2)$, one of the sufficient conditions for not remaining close to the resonance of radius $O(\sqrt{\mu})$ can be written by means of the inequality (see Grebenikov and Ryabov [1]):

$$0 < c^{-1} < \left| \left(k, \left(\frac{\partial \omega}{\partial x}, X \right) \right) \right| < c, \tag{2.10.4}$$

if $(x, y) \in G_{m+n}$. It is the left-hand member of the inequality (2.10.4) that ensures the non-remaining of the solution $x(t, \mu)$ in the resonance neighborhood of the form $|(k, \omega(x))| < c_2\sqrt{\mu}$.

Now we will study the behavior of multifrequency rotary systems that have the property of remaining close to a resonance point of the form $(k, \omega(x)) = 0$. We related such systems to class II.

We will use the operator of smoothing for continuous disturbances, first introducing phase coordinate disturbances by the formulae

$$p = x - x_0, \quad q = y - \omega_0 t - y_0, \tag{2.10.5}$$

where $x = x_0$, $y = \omega_0 t + y_0$ is the solution of the corresponding generating (undisturbed) system determined by the initial point $(x_0, y_0) \in G_{m+n}$. In the new variables system (2.10.2) takes the form

$$\begin{aligned} \frac{dp}{dt} &= \mu X(p + x_0,\, q + \omega_0 t + y_0), \\ \frac{dq}{dt} &= \omega(p + x_0) - \omega_0 + \mu Y(p + x_0,\, q + \omega_0 t + y_0), \end{aligned} \tag{2.10.6}$$

$$\begin{aligned} X &= \sum_{\|k\| \geq 0} X_k(p + x_0) \exp[i(k,\, q + \omega_0 t + y_0)], \\ Y &= \sum_{\|k\| \geq 0} Y_k(p + x_0) \exp[i(k,\, q + \omega_0 t + y_0)]. \end{aligned} \tag{2.10.7}$$

For simplicity we assume that $y_0 = 0$. System (2.10.6) is convenient for the performance of smoothing with respect to t, because it contains time in an explicit form. Let the frequency vector $\omega(x_0) = \omega_0$ at the initial time satisfy the resonance condition $(k, \omega_0) = 0$, $\|k\| \neq 0$. Then

$$\begin{aligned} \overline{X}(p, q) &= M_t[X] = {\sum_{\|k\| \geq 0}}' X_k(p + x_0) e^{i(k,q)}, \\ \overline{Y}(p, q) &= M_t[Y] = {\sum_{\|k\| \geq 0}}' Y_k(p + x_0) e^{i(k,q)}. \end{aligned} \tag{2.10.8}$$

The respective system of averaged equations of the first approximation will have the form

$$\begin{aligned} \frac{d\bar{p}}{dt} &= \mu \overline{X}(\bar{p}, \bar{q}), \\ \frac{d\bar{q}}{dt} &= \omega(\bar{p} + x_0) - \omega_0 + \mu \overline{Y}(\bar{p}, \bar{q}), \end{aligned} \tag{2.10.9}$$

and reverting to the initial variables, we obtain

$$\begin{aligned}\frac{d\bar{x}}{dt} &= \mu\overline{X}(\bar{x},\bar{y}),\\ \frac{d\bar{y}}{dt} &= \omega(\bar{x}) + \mu\overline{Y}(\bar{x},\bar{y}).\end{aligned} \tag{2.10.10}$$

First we will assume that the Fourier series of functions X, Y contain a finite number of summands:

$$\begin{aligned}X(x,y) &= \sum_{0\le\|k\|\le N} X_k(x)e^{i(k,y)},\\ Y(x,y) &= \sum_{0\le\|k\|\le N} Y_k(x)e^{i(k,y)},\end{aligned} \tag{2.10.11}$$

and try to find some sufficient conditions guaranteeing the remaining solution $x(t,\mu)$ in the vicinity of a resonance point, e.g. the initial point $(x_0, y_0) \in G_{m+n}$.

So, let the initial vector of frequencies $\omega_0 = (\omega_1(x_0), \ldots, \omega_n(x_0))$ be such that

$$(k, \omega_0) = 0, \quad k \in I_{N,\,\mathrm{res}}, \tag{2.10.12}$$

where $I_{N,\mathrm{res}}$ denotes a set of resonance vectors with the norm $\|k\| \le N$. If $I_{N,\mathrm{nonres}}$ denotes a set of nonresonance vectors k with the norm $\|k\| \le N$, then, obviously, there exists a positive number $\alpha_N > 0$ such that

$$0 < \alpha_N = \inf_{k\in I_{N,\,\mathrm{nonres}}} |(k,\omega_0)|. \tag{2.10.13}$$

With $N \to \infty$ the value α_N, generally speaking, may become indefinitely small. However sometimes $\alpha_N \not\to 0$ with $N \to \infty$. For example, this is the case described in Section 2.8.

Lemma 1 *Let*

(1) *the norm of the matrix* $\partial\omega/\partial x$ *satisfy*

$$\left\|\frac{\partial\omega}{\partial x}\right\| = \sum_{s=1}^{n}\sum_{r=1}^{m}\left|\frac{\partial\omega_s}{\partial x_r}\right| < \frac{\alpha_N}{N} \tag{2.10.14}$$

in the domain G_{m+n}*;*

(2) *the norm* $\|X\|_{(x,y)\in G_{m+n}}$ *be less than* $C_2 < 1$.

Then for all $t \in [0,\, \mu^{-1}]$ *the following inequalities hold:*

$$|(k,\, \omega(x(t,\mu)))| < \alpha_N, \quad k \in I_{N,\,\mathrm{res}}, \tag{2.10.15}$$

$$|(k,\, \omega(x(t,\mu)))| \ge \alpha_N, \quad k \in I_{N,\,\mathrm{nonres}}. \tag{2.10.16}$$

Condition (2.10.14) expresses the sufficient condition for the solution of initial system (2.10.2) to remain in the neighborhood of the resonance point $(x_0, y_0) \in G_{m+n}$ of radius α_N within the time interval $t \in [0,\, \mu^{-1}]$. But it should be noted that this neighborhood may contain a great (maybe, infinite) number of resonance neighborhoods of radius $\sqrt{\mu}$, because as $\mu \to 0$ these neighborhoods decrease without limit, and the value α_N independent of μ may not approach zero. So the condition (2.10.14) gives quite a crude division of the phase space into resonance and nonresonance domains, and consequently the class of multifrequency systems satisfying inequality (2.10.14) may also contain systems with solutions passing through resonances in the process of evolution.

Now consider multifrequency rotary systems for which the following condition holds true at any point of G_{m+n}:

$$C_1^{-1}\mu^\alpha < \left|\left(k,\left(\frac{\partial\omega}{\partial x}, X(x,y)\right)\right)\right| < C_1\mu^\alpha, \tag{2.10.17}$$

where $\alpha > 0,\ 1 \leq \|k\| \leq N$.

Lemma 2 *If in G_{m+n} condition (2.10.17) is satisfied, then the solution of system (2.10.2), starting at the resonance point (x_0, y_0), remains in its neighborhood of radius $O(\mu^\alpha)$ provided $t \in [0,\ \mu^{-1}]$.*

The lemma points out that the solution $x(t,\mu)$ of system (2.10.2) behaves so that $(k,\omega(x))$ with $k \in I_{N,\,\mathrm{res}}$ may change by $O(\mu^\alpha)$, if $t \in [0,\ \mu^{-1}]$. Hence the condition (2.10.17) guarantees that the solution remains in the neighborhood of radius $O(\mu^\alpha)$, but unfortunately from this condition it does not follow that the solution of the averaged system $\bar{x}(t,\mu)$ also has the property of remaining in the vicinity of a resonance point of radius $O(\mu^\alpha)$. Now compare the solutions $x(t,\mu)$ and $\bar{x}(t,\mu)$ for (2.10.2) and (2.10.10).

Theorem *Let:*

(1) *the functions (2.10.11) be twice differentiable in G_{m+n};*
(2) *the frequency vector $\omega(x)$ be also twice differentiable in $x \in P_m$;*
(3) $\|X\| < C_2,\ \|Y\| < C_2,\ \|\omega\| < C_2$ *in the domain G_{m+n};*
(4) *the initial point (x_0, y_0) be a resonance point, i.e. $(k,\ \omega(x_0)) = 0,\ k \in I_{N,\,\mathrm{res}}$;*
(5) *the solution of the averaged system (2.10.10) for $t \in [0,\ \infty)$ satisfy the inclusion $(\bar{x}(t,\mu),\ \bar{y}(t,\mu)) \in G_{m+n}$;*
(6) *in G_{m+n} (2.10.17) be true.*

Then for any $\varepsilon > 0$ there exists $\mu_0(\varepsilon, C_1, C_2) > 0$ such that for all $\mu \in [0,\ \mu_0]$ and for all $t \in [0,\ \mu^{-1}]$ the following inequality is satisfied

$$\|x(t,\mu) - \bar{x}(t,\mu)\| < \varepsilon.$$

Proof Find a transform $(x,y) \to (\bar{x},\bar{y})$

$$x = \bar{x} + \sum_{s\geq 1}\mu^s u_s(\bar{x},\bar{y}), \qquad y = \bar{y} + \sum_{s\geq 1}\mu^s v_s(\bar{x},\bar{y}), \tag{2.10.18}$$

that transforms the multifrequency system (2.10.2) into the comparison system of any approximation

$$\begin{aligned}\frac{d\bar{x}}{dt} &= \mu\overline{X}(\bar{x},\bar{y}) + \sum_{s\geq 2}\mu^s A_s(\bar{x}),\\ \frac{d\bar{y}}{dt} &= \omega(\bar{x}) + \mu\overline{Y}(\bar{x},\bar{y}) + \sum_{s\geq 2}\mu^s B_s(\bar{x}).\end{aligned} \tag{2.10.19}$$

If we substitute (2.10.18) into the initial system (2.10.2) and demand that system (2.10.19) should be finally obtained, then we obtain an infinite system of first-order partial differential equations for the successive calculation of the functions $u_1, v_1, \ldots,\ A_2, B_2, \ldots$:

$$\begin{aligned}\left(\frac{\partial u_1}{\partial \bar{y}}, \omega(\bar{x})\right) &= X(\bar{x},\bar{y}) - \overline{X}(\bar{x},\bar{y}) = \sum_{1\leq\|k\|\leq N}'' X_k(\bar{x})e^{i(k,\bar{y})},\\ \left(\frac{\partial v_1}{\partial \bar{y}}, \omega(\bar{x})\right) &= \left(\frac{\partial \omega(\bar{x})}{\partial \bar{x}}, u_1\right) + Y(\bar{x},\bar{y}) - \overline{Y}(\bar{x},\bar{y})\\ &= \left(\frac{\partial \omega(\bar{x})}{\partial \bar{x}}, u_1\right) + \sum_{1\leq\|k\|\leq N}'' Y_k(\bar{x})e^{i(k,\bar{y})},\\ &\cdots\cdots\cdots\cdots\cdots\cdots\cdots\cdots\\ \left(\frac{\partial u_s}{\partial \bar{y}}, \omega(\bar{x})\right) &= U_s(\bar{x},\bar{y},u_1,v_1,\ldots,u_{s-1},v_{s-1};\ A_2,B_2,\ldots,B_{s-1},A_s),\\ \left(\frac{\partial v_s}{\partial \bar{y}}, \omega(\bar{x})\right) &= V_s(\bar{x},\bar{y},u_1,v_1,\ldots,v_{s-1},u_s;\ A_2,B_2,\ldots,A_s,B_s),\\ &\cdots\cdots\cdots\cdots\cdots\cdots\cdots\cdots\end{aligned} \tag{2.10.20}$$

In equations (2.10.20) the symbol $\sum''$ means that the summation is only performed with respect to nonresonance values of $k \in I_{N,\,\text{nonres}}$, for which $(k,\ \omega_0) \neq 0$. One of the solutions of system (2.10.20) can be easily found by the method of characteristics, and it has the form

$$u_1(\bar{x},\bar{y}) = \sum_{1\leq\|k\|\leq N}'' \frac{X_k(\bar{x})e^{i(k,\bar{y})}}{i(k,\ \omega(\bar{x}))} + \varphi_1(\bar{x},x_0,y_0), \tag{2.10.21}$$

where $\varphi_1(\bar{x},x_0,y_0)$ is an arbitrary vector-function differentiable with respect to $\bar{x}$.

If the initial conditions $(\bar{x}_0,\bar{y}_0)$ for an averaged system coincide with the initial conditions for the initial system (x_0,y_0), then it is evident that

$$u_1(\bar{x}(0,\mu),\ \bar{y}(0,\mu)) = 0.$$

Hence at the initial time

$$\varphi_1(\bar{x}(0,\mu),x_0,y_0) = -\sum_{1\leq\|k\|\leq N}'' \frac{X_k(x_0)e^{i(k,y_0)}}{i(k,\ \omega_0)}, \tag{2.10.22}$$

If the initial conditions $(\bar{x}_0, \bar{y}_0)$ for an averaged system are defined from $\varphi_1 = 0$, then for the function u_1 we will have the following expression:

$$u_1(\bar{x}, \bar{y}) = \sum_{1 \le \|k\| \le N}'' \frac{X_k(\bar{x}) e^{i(k, \bar{y})}}{i(k, \omega(\bar{x}))}. \tag{2.10.23}$$

In both cases in the domain G_{m+n} the following estimation holds:

$$\|u_1(\bar{x}, \bar{y})\| < \frac{2}{\bar{\alpha}_N} C_2, \quad \bar{\alpha}_N = \alpha_N(1 - C_3(\mu)), \tag{2.10.24}$$

where $C_3(\mu) \to 0$ as $\mu \to 0$. Direct differentiation of the function u_1 and the condition of the theorem allow us to write the estimations

$$\left\| \frac{\partial u_1}{\partial \bar{x}} \right\| < \frac{1}{\bar{\alpha}_N^2} C_2 (C_2 N + \alpha_N), \tag{2.10.25}$$

$$\left\| \frac{\partial u_1}{\partial \bar{y}} \right\| < \frac{N}{\bar{\alpha}_N} C_2. \tag{2.10.26}$$

Now we will introduce a new function

$$\delta x(t, \mu) = x(t, \mu) - \bar{x}(t, \mu) - \mu u_1(\bar{x}, \bar{y}) \tag{2.10.27}$$

and for its norm $\|\delta x\|$ construct an estimation at $t \in [0, \mu^{-1}]$. This function satisfies the m-dimensional system of ordinary differential equations

$$\frac{d\delta x}{dt} = \mu X(x, y) - \mu X(\bar{x}, \bar{y}) + O(\mu^2), \quad \delta x(0, \mu) = 0. \tag{2.10.28}$$

Indeed, differentiating (2.10.27) with respect to t and using equations (2.10.2), (2.10.10), (2.10.20), and estimations (2.10.25) and (2.10.26), we obtain

$$\begin{aligned} \frac{d\delta x}{dt} &= \frac{dx}{ds} - \frac{d\bar{x}}{dt} - \mu \left(\frac{\partial u_1}{\partial \bar{x}}, \frac{d\bar{x}}{dt} \right) - \mu \left(\frac{\partial u_1}{\partial y}, \frac{d\bar{y}}{dt} \right) \\ &= \mu X(x, y) - \overline{X}(\bar{x}, \bar{y}) - \mu^2 \left(\frac{\partial u_1}{\partial \bar{x}}, \overline{X} \right) - \mu \left(\frac{\partial u_1}{\partial \bar{y}}, \omega(\bar{x}) \right) - \mu^2 \left(\frac{\partial u_1}{\partial \bar{y}}, \overline{Y} \right) \\ &= \mu X(x, y) - \mu \overline{X}(\bar{x}, \bar{y}) - \mu \sum_{1 \le \|k\| \le N}'' X_k(\bar{x}) e^{i(k, \bar{y})} + O(\mu^2). \end{aligned}$$

Now if we consider that

$$\overline{X}(\bar{x}, \bar{y}) + \sum_{1 \le \|k\| \le N}'' X_k(\bar{x}) e^{i(k, \bar{y})} = X(\bar{x}, \bar{y}),$$

then equation (2.10.28) is obtained. In coordinate form this will be written as

$$\begin{aligned} \frac{d(\delta x_s)}{dt} &= \mu [X^{(s)}(x, y) - X^{(s)}(\bar{x}, \bar{y})] + O(\mu^2), \\ \delta x_s(0, \mu) &= 0, \quad s = 1, \dots, m. \end{aligned} \tag{2.10.29}$$

Then, applying the Lipschitz inequality to the difference $X^{(s)}(x,y)-X^{(s)}(\bar{x},\bar{y})$, we obtain

$$\begin{aligned} -\mu L_x^{(s)}\|\delta x\| + \mu\big[X^{(s)}(x,y) - X^{(s)}(\bar{x},\bar{y})\big] + O(\mu^2) \le \frac{d(\delta x_s)}{dt} \\ \le \mu L_x^{(s)}\|\delta x\| + \mu\big[X^{(s)}(x,y) - X^{(s)}(\bar{x},\bar{y})\big] + O(\mu^2), \end{aligned} \tag{2.10.30}$$

where $\|\delta x\| = \sum\limits_{s=1}^{m} |\delta x_s|$, and $L_x^{(s)}$ is a Lipschitz constant for the function $X^{(s)}(x,y)$ by $x \in P_m$. Write the differential equations

$$\begin{aligned} \frac{dw_s}{dt} &= -\mu L_x^{(s)} w_s + \mu\big[X^{(s)}(x,y) - X^{(s)}(\bar{x},\bar{y})\big] + O(\mu^2), \\ & w_s(0,\mu) = 0, \quad s = 1,\dots,m; \end{aligned} \tag{2.10.31}$$

$$\begin{aligned} \frac{dz_s}{dt} &= \mu L_x^{(s)} z_s + \mu\big[X^{(s)}(x,y) - X^{(s)}(\bar{x},\bar{y})\big] + O(\mu^2), \\ & z_s(0,\mu) = 0, \quad s = 1,\dots,m. \end{aligned} \tag{2.10.32}$$

Since $w_s(0,\mu) = \delta x_s(0,\mu) = z_s(0,\mu) = 0$, then in view of (2.10.30)

$$w_s(t,\mu) \le \delta x_s(t,\mu) \le z_s(t,\mu), \quad s = 1,\dots,m. \tag{2.10.33}$$

For the function w_s and z_s we have the following integral equalities:

$$\begin{aligned} w_s(t,\mu) = \frac{O(\mu^2)}{\mu L_x^{(s)}}\big(1 - \exp\{-\mu L_x^{(s)} t\}\big) \\ + \mu \exp\{-\mu L_x^{(s)} t\} \int_0^t \exp\{\mu L_x^{(s)} \tau\} K(\tau,\mu)\, d\tau, \end{aligned} \tag{2.10.34}$$

$$\begin{aligned} z_s(t,\mu) = \frac{O(\mu^2)}{\mu L_x^{(s)}}\big(\exp\{\mu L_x^{(s)} t\} - 1\big) \\ + \mu \exp\{\mu L_x^{(s)} t\} \int_0^t \exp\{-\mu L_x^{(s)} \tau\} K(\tau,\mu)\, d\tau, \end{aligned} \tag{2.10.35}$$

$$K(\tau,\mu) = \sum_{1\le\|k\|\le N} X_k(x(\tau,\mu))[\exp\{i(k,y(\tau,\mu))\} - \exp\{i(k,\bar{y}(\tau,\mu))\}]. \tag{2.10.36}$$

Since the condition (2.10.17) is an integral estimation for the whole domain G_{m+n}, and the condition (5) of the theorem is true, then

$$0 \le \left\|\left(k, \left(\frac{\partial\omega}{\partial x}, X(x,y) - X(\bar{x},\bar{y})\right)\right)\right\| < 2C_1\mu^\alpha \tag{2.10.37}$$

at all k satisfying $1 \le \|k\| \le N$.

For multifrequency systems (2.10.2) the norm $\|\partial\omega/\partial x\|$ has order $O(1)$ in the domain $x \in P_m$, and therefore condition (2.10.37) can be satisfied everywhere in G_{m+n} when and only when $\|X(x,y) - X(\bar{x},\bar{y})\| < C_1\mu^\alpha$ for any couple of points (x,y) and $(\bar{x},\bar{y})$ from G_{m+n}. From this it follows that $\|K(\tau,\mu)\| < \widetilde{C}_1\mu^\alpha$, $\alpha > 0$.

Then, using (2.10.34) and (2.10.35), we obtain

$$\|w(t,\mu)\| < C_4\mu^\alpha, \quad \|z(t,\mu)\| < C_4\mu^\alpha, \tag{2.10.38}$$

if $0 < \alpha < 1$, or

$$\|w(t,\mu)\| < C_4\mu, \quad \|z(t,\mu)\| < C_4\mu, \tag{2.10.39}$$

if $\alpha \geq 1$ for all $t \in [0,\ \mu^{-1}]$. Then from (2.10.33) we have the estimations

$$\|\delta x(t,\mu)\| < C_4\mu^\alpha, \tag{2.10.40}$$

if $0 < \alpha < 1$, or

$$\|\delta x(t,\mu)\| < C_4\mu, \tag{2.10.41}$$

if $\alpha \geq 1$ for all $t \in [0,\ \mu^{-1}]$. However, since in the first approximation $\|x(t,\mu) - \bar{x}(t,\mu)\| \leq \|\delta x(t,\mu)\| + \mu\|u_1(\bar{x},\bar{y})\|$, from this it follows that

$$\|x(t,\mu) - \bar{x}(t,\mu)\| < C_5\mu$$

with $t \in [0,\ \mu^{-1}]$. The transfer to "ε–terminology" is trivial. The theorem is proved.

2.11 Multifrequency Systems with their Solutions Not Leaving the Neighborhood of a Resonance Point

Earlier we discussed multifrequency rotary systems, the solutions of which may remain near the resonance point, in a neighborhood of radius μ^α $(\alpha > 0)$. If as $\mu \to 0$ the neighborhoods of resonance points to not contract to a point, then the issue of ε-proximity of exact and averaged solutions requires separate study. Popova [1] was the first to pay attention to this, and the results described below belong to her for the most part.

So consider a $(m+n)$-th order multifrequency system

$$\begin{aligned} \frac{dx}{dt} &= \mu X_1(x) + \mu^2 X_2(x,y), \\ \frac{dy}{dt} &= \omega(x) + \mu Y(x,y) \end{aligned} \tag{2.11.1}$$

with initial conditions $x(0,\mu) = x_0$, $y(0,\mu) = y_0$. In (2.11.1) $X_s = (X_{s_1},\dots,X_{s_m})$ $(s = 1,2)$, $x = (x_1,\dots,x_m)$, $y = (y_1,\dots,y_n)$, $Y = (Y_1,\dots,Y_n)$, $\omega = (\omega_1,\dots,\omega_n)$ is a frequency vector, $0 \leq \mu \leq \mu_0 < 1$. Note that the vector-function $X_1(x)$ depends only on the slow variables x. Vector-functions X_2, Y are determined in the open domain $G_{m+n} = \{P_m \times Q_n\}$ as usual and are 2π-periodic with respect to y in Q_n, and $X_1(x)$ and $\omega(x)$ are determined in P_m.

In the system (2.11.1) perform the usual substitution

$$p = x - x_0, \quad q = y - \omega_0 t - y_0.$$

With the help of this substitution the initial system will be written as

$$\begin{gathered}\frac{dp}{dt} = \mu X_1(p + x_0) + \mu^2 X_2(p + x_0,\, q + \omega_0 t + y_0), \\ \frac{dq}{dt} = \omega(p + x_0) - \omega_0 + \mu Y(p + x_0,\, q + \omega_0 t + y_0), \\ p(0, \mu) = q(0, \mu) = 0. \end{gathered} \tag{2.11.2}$$

Apply the operator of averaging for continuous disturbances to the right-hand members of the above system. As a result, we obtain the averaged system of the first approximation

$$\begin{aligned}\frac{d\bar{p}}{dt} &= \mu\overline{X}_1(\bar{p} + x_0) + \mu^2\overline{X}_2(\bar{P} + x_0,\, \bar{q} + y_0), \\ \frac{d\bar{q}}{dt} &= \omega(\bar{p} + x_0) - \omega_0 + \mu\overline{Y}(\bar{p} + x_0,\, \bar{q} + y_0).\end{aligned} \tag{2.11.3}$$

Below we will prove theorems on ε-proximity of the solutions of systems (2.11.2) and (2.11.3) in the time interval $t \in [0, \mu^{-1}]$ with the different properties of smoothness of the functions X_2, Y in the case of initial resonance. In addition, we will show that the proximity of slow and fast variables of the initial and averaged systems is guaranteed within $t \in [0, \mu^{\gamma-1}]$, where $0 < \gamma < 1$, under weaker conditions imposed on the right-hand member of the system.

First we write some additional estimations. If the following two systems are given

$$\begin{aligned}\frac{dp}{dt} &= \mu X(p.q), \\ \frac{dq}{dt} &= \omega(p) - \omega_0 + \mu Y(p, q);\end{aligned} \tag{2.11.4}$$

$$\begin{aligned}\frac{d\tilde{p}}{dt} &= \mu X_N(\tilde{p}, \tilde{q}), \\ \frac{d\tilde{q}}{dt} &= \omega(\tilde{p}) - \omega_0 + \mu Y_N(\tilde{p}, \tilde{q}), \\ p(0, \mu) &= \tilde{p}(0, \mu), \quad q(0, \mu) = \tilde{q}(0, \mu),\end{aligned} \tag{2.11.5}$$

then, as shown in Section 2.3 above,

$$\begin{aligned}\|q(t, \mu) - \tilde{q}(t, \mu)\| &\le C_1^{(p)} e^{\lambda_1 t} + C_2^{(p)} e^{\lambda_2 t} + a, \\ \|p(t, \mu) - \tilde{p}(t, \mu)\| &\le (L_\omega + \mu l_p)^{-1}\big[C_1^{(p)}\lambda_1 e^{\lambda_1 t} + C_2^{(p)}\lambda_2 e^{\lambda_2 t} \\ &\qquad - \mu l_q \|q(t, \mu) - \tilde{q}(t, \mu)\| - \mu\varepsilon_2\big].\end{aligned} \tag{2.11.6}$$

In these inequalities

$$a = \frac{\varepsilon_1(L_\omega + \mu l_p) - \mu L_p \varepsilon_2}{\mu L_p l_q - L_q L_\omega - \mu L_q l_p}, \quad C_1^{(p)} = \frac{a\lambda_2 + \mu\varepsilon_1}{\lambda_1 - \lambda_2}, \quad C_2^{(p)} = -a - C_1^{(p)},$$

$$\lambda_{1,2} = \frac{1}{2}\mu(L_p + l_q) \pm \frac{1}{2}\sqrt{\mu^2(L_p + l_q)^2 - 4(\mu^2 L_q l_p - \mu^2 L_p l_q - \mu L_q L_\omega)}. \tag{2.11.7}$$

Suppose that

$$L_p l_q \geq L_q l_p, \quad \varepsilon_1 = \varepsilon_2\sqrt{\mu} = \varepsilon. \tag{2.11.8}$$

If the Lipchitz constants L_p, L_q, l_p, l_q, L_ω do not depend on the small parameter μ, the following estimations can be obtained:

$$\begin{aligned}
&0 < \lambda_1 < K_1\sqrt{\mu} \\
&\quad \equiv \frac{1}{2}\Big[\sqrt{\mu_0}(L_p + l_q) + \sqrt{\mu_0(L_p + l_q)^2 + 8L_qL_\omega}\Big]\sqrt{\mu}, \\
&\lambda_2 < 0, \quad |\lambda_2| < K_2\sqrt{\mu} = \frac{1}{2}\sqrt{\mu}\sqrt{\mu_0(L_p + l_q)^2 + 8L_qL_\omega}, \\
&|a| < K_3\varepsilon \equiv \frac{L_\omega + \mu_0 l_p + L_p\sqrt{\mu_0}}{L_qL_\omega - \mu_0(L_pl_q - L_ql_p)}\,\varepsilon, \\
&\big|C_1^{(p)}\big| < K_4\varepsilon \equiv \frac{1}{2}|a|\varepsilon + \frac{1}{2}(2L_qL_\omega)^{-1/2}, \quad |C_2^{(p)}| < K_5\varepsilon.
\end{aligned} \tag{2.11.9}$$

Relations (2.11.9) are true provided that

$$\mu_0 < \frac{L_qL_\omega}{L_pl_q - L_ql_p},$$

which does not in the least limit the commonality of the considered class of equations. From (2.11.6) and (2.11.9) we conclude that for any point of time from the domain $[0,\, 1/\sqrt{\mu}]$ the following estimations are true:

$$\|q - \tilde{q}\| < K_6\varepsilon, \quad \|p - \tilde{p}\| < K_7\varepsilon\sqrt{\mu}, \tag{2.11.10}$$

where

$$\begin{aligned}
&K_6 = K_4e^{K_1} + K_3 + K_5, \\
&K_7 = L_\omega^{-1}\big[K_4K_1e^{K_1} + K_2K_5 + l_qK_4\sqrt{\mu_0} + 1\big].
\end{aligned}$$

Inequalities (2.11.10) show that the proximity of the solutions of systems (2.11.4) and (2.11.5) of order $O(\mu)$ for slow variables p, $\tilde{p}$ and of order $O(\sqrt{\mu})$ for fast variables q, $\tilde{q}$ will take place even if the value ε_2 is not small together with μ, but in this case the proximity of functions X and X_N should be at least of order $O(\sqrt{\mu})$.

In case $\varepsilon_1 = \varepsilon_2 = \varepsilon$ the estimations (2.11.10) also hold, and the constants K_3, K_4, K_7 can be replaced by smaller ones:

$$\overline{K}_3 = \frac{L_\omega + \mu_0|L_p - l_q|}{L_qL_\omega - \mu_0|L_pl_q - L_ql_p|}, \quad \overline{K}_4 = \frac{1}{2}\overline{K}_3 + \frac{\sqrt{\mu_0}}{2\sqrt{2L_qL_\omega}},$$

$$\overline{K}_7 = L_\omega^{-1}\big[\overline{K}_4K_1e^{K_1} + K_2K_5 + l_qK_6\sqrt{\mu_0} + \sqrt{\mu_0}\big].$$

Now let the Fourier series of functions X_2 and Y be trigonometric N-fold polynomials:

$$\begin{aligned}
X_2(x,y) &= \sum_{0\leq\|k\|\leq N} X_{2k}(x)e^{i(k,y)}, \\
Y(x,y) &= \sum_{0\leq\|k\|\leq N} Y_k(x)e^{i(k,y)}.
\end{aligned} \tag{2.11.11}$$

In addition, let the point x_0 be a resonance point at the initial time $(t = 0)$, i.e. $(k, \omega(x_0)) = 0$ for $k \in I_{N,\mathrm{res}}$, and within $t \in [0, \mu^{-1}]$ this resonance is isolated. This means that for any integer vector k such that $\|k\| \leq N$ and $(k, \omega(x_0)) \neq 0$, the following inequality is satisfied:

$$|(k, \omega(x(t, \mu)))| \geq \beta > 0 \tag{2.11.12}$$

for any $t \in [0, \mu^{-1}]$, where $\beta \not\to 0$ as $\mu \to 0$.

Theorem *Let:*

(1) *conditions (2.11.11) and (2.11.12) be satisfied and the coefficients X_{2k} and Y_k be differentiable in the domain P_m;*
(2) *the following inequalities be satisfied in G_{m+n}*

$$\|X_1\| < C, \quad \|X_2\| < C, \quad \|Y\| < C,$$
$$\left\| \frac{\partial X_s}{\partial x} \right\| = \sum_{r=1}^{m} \sum_{j=1}^{m} \left| \frac{\partial X_{sr}}{\partial x_j} \right| < C_1, \quad s = 1, 2,$$
$$\left\| \frac{\partial Y}{\partial x} \right\| < C_1;$$

(3) *in the domain P_m the function $\omega(x)$ be continuously differentiable with respect to x and the following inequalities be satisfied*

$$\|\omega\| < C, \quad \left| \frac{\partial \omega_s}{\partial x_j} \right| < C, \quad j, s = 1, \ldots, m;$$

(4) *the solution of the averaged system (2.11.3) within the time $t \in [0, \mu^{-1}]$ remain in the domain G_{m+n}.*

Then there exists $\mu^* > 0$ such that for any $0 \leq \mu \leq \mu^*$ and for any point of time $t \in [0, \mu^{-1}]$ the following inequalities are satisfied

$$\|p(t, \mu) - \bar{p}(t, \mu)\| < A\mu^2, \quad \|q(t, \mu) - \bar{q}(t, \mu)\| < B\mu, \tag{2.11.13}$$

where A, B do not depend on μ.

Proof Construct the first approximation for the solution of system (2.11.2) by the method of Krylov–Bogolyubov, i.e. find the solution of (2.11.2) in the form

$$\begin{aligned} p(t, \mu) &= \bar{p}(t, \mu) + \sum_{s \geq 1} \mu^s u_s(\bar{p}, \bar{q}, t), \\ q(t, \mu) &= \bar{q}(t, \mu) + \sum_{s \geq 1} \mu^s v_s(\bar{p}, \bar{q}, t). \end{aligned} \tag{2.11.14}$$

Substituting (2.11.14) into (2.11.2), expanding its right-hand members at the point $(\bar{p}, \bar{q})$ into series in terms of μ, and equating the coefficients at μ and μ^2, we will obtain the following systems of partial differential equations for the calculation of the vector-functions $u_1(\bar{p}, \bar{q}, t)$, $v_1(\bar{p}, \bar{q}, t)$ and $u_2(\bar{p}, \bar{q}, t)$:

$$\begin{aligned}
&\left(\frac{\partial u_1}{\partial \bar{q}},\, \omega(\bar{p}+x_0)-\omega_0\right)+\frac{\partial u_1}{\partial t}=0,\\
&\left(\frac{\partial v_1}{\partial \bar{q}},\, \omega(\bar{p}+x_0)-\omega_0\right)+\frac{\partial v_1}{\partial t}=\left(\overline{\frac{\partial \omega(\bar{p}+x_0)}{\partial \bar{p}}},\, u_1\right)\\
&\qquad\qquad +Y(\bar{p}+x_0,\,\bar{q}+\omega_0 t+y_0)-\overline{Y}(\bar{p}+x_0,\,\bar{q}+y_0),\\
&\left(\frac{\partial u_2}{\partial \bar{q}},\, \omega(\bar{p}+x_0)-\omega_0\right)+\frac{\partial u_2}{\partial t}=X_2(\bar{p}+x_0,\,\bar{q}+\omega_0 t+y_0)\\
&\qquad\qquad -\overline{X}_2(\bar{p}+x_0,\,\bar{q}+y_0)+\left(\overline{\frac{\partial X_1(\bar{p}+x_0)}{\partial \bar{p}}},\, u_1\right)\\
&\qquad\qquad -\left(\frac{\partial u_1}{\partial \bar{p}},\, X_1(\bar{p}+x_0)\right)-\left(\frac{\partial u_1}{\partial \bar{q}},\, Y\right),
\end{aligned} \tag{2.11.15}$$

where $\overline{\partial X_1/\partial \bar{p}}$, $\overline{\partial \omega/\partial \bar{p}}$ denote some average values of the functions $\partial X_1/\partial \bar{p}$ and $\partial \omega/\partial \bar{p}$ in the corresponding domains.

From the first m equations it follows that it is possible to choose a vector-function $u_1 = 0$. Since when averaging for continuous disturbances the functions $\overline{X}_2(\bar{p}+x_0,\, \bar{q}+y_0)$ and $\overline{Y}(\bar{p}+x_0,\, \bar{q}+y_0)$ include all harmonics for which $(k, \omega_0) = 0$, the second members of equations for v_1 and u_2 contain only the terms for which $(k, \omega_0) \neq 0$, i.e. only nonresonance terms. As before, denote the sum for $k \in I_{N,\text{nonres}}$ by

$$\sum_{k \in I_{N,\text{nonres}}} a_k = \sum_{1 \le \|k\| \le N}'' a_k.$$

The general solution for v_1 and u_2 has the form

$$\begin{aligned}
v_1(\bar{p}, \bar{q}, t) &= \sum_{1 \le \|k\| \le N}'' \frac{Y_k(\bar{p}+x_0)\exp\{i(k,\, \bar{q}+\omega_0 t+y_0)\}}{i(k,\, \omega(\bar{p}+x_0))}\\
&\quad + F_1[\bar{q}-t(\omega(\bar{p}+x_0)-\omega_0),\, \varphi_1(\bar{p})],\\
u_2(\bar{p}, \bar{q}, t) &= \sum_{1 \le \|k\| \le N}'' \frac{X_{2k}(\bar{p}+x_0)\exp\{i(k,\, \bar{q}+\omega_0 t+y_0)\}}{i(k,\, \omega(\bar{p}+x_0))}\\
&\quad + F_2[\bar{q}-t(\omega(\bar{p}+x_0)-\omega_0),\, \varphi_2(\bar{p})],
\end{aligned} \tag{2.11.16}$$

where F_1, F_2 are any differentiable vector-functions of $(m+n)$ variables (their dimensions are n and m respectively) and φ_1, φ_2 are arbitrary differentiable vector-functions of m variables $\bar{p}$. Assume $f_1 \equiv f_2 \equiv 0$. Then the functions v_{1s} and u_{2j} $(s = 1, \dots, n;\ j = 1, \dots, m)$ are trigonometric polynomials with respect to $\bar{q}$ of degree not higher than n.

It is easy to verify that the vector-functions v_1 and u_2 are bounded on norm together with their partial derivatives with respect to $\bar{p}$ and $\bar{q}$:

$$\begin{aligned} &\|v_1\| < \frac{1}{\beta} C, \quad \left\| \frac{\partial v_1}{\partial \bar{p}} \right\| \leq \frac{1}{\beta} C_1 + \frac{1}{\beta^2} mNCC_1, \\ &\left\| \frac{\partial v_1}{\partial \bar{q}} \right\| \leq \frac{1}{\beta} NC, \quad \|u_2\| \leq \frac{1}{\beta} C, \\ &\left\| \frac{\partial u_2}{\partial p} \right\| \leq \frac{1}{\beta} C_1 + \frac{1}{\beta^2} mNCC_1, \quad \left\| \frac{\partial u_2}{\partial \bar{q}} \right\| \leq \frac{1}{\beta} NC. \end{aligned} \tag{2.11.17}$$

Consider the functions

$$\begin{aligned} p_2(t, \mu) &= \bar{p}(t, \mu) + \mu^2 u_2(\bar{p}, \bar{q}, t), \\ q_1(t, \mu) &= \bar{q}(t, \mu) + \mu v_1(\bar{p}, \bar{q}, t), \end{aligned} \tag{2.11.18}$$

that are an approximation of system (2.11.2) to the exact solution (2.11.14). Differentiating expressions for p_2, q_1, we obtain

$$\begin{aligned} \frac{dp_2}{dt} &= \frac{d\bar{p}}{dt} + \mu^2 \Bigg[\left(\frac{\partial u_2}{\partial \bar{p}},\ \mu X_1(\bar{p} + x_0) + \mu^2 \overline{X}_2(\bar{p} + x_0, \bar{q} + y_0) \right) \\ &\quad + \left(\frac{\partial u_2}{\partial \bar{q}},\ \omega(\bar{p} + x_0) - \omega_0 + \mu \overline{Y}(\bar{p} + x_0,\ \bar{q} + y_0) \right) + \frac{\partial u_2}{\partial t} \Bigg] \\ &= \mu X_1(\bar{p} + x_0) - \mu^3 \left(\overline{\frac{\partial X_1}{\partial p}},\ u_2 \right) + \mu^2 X_2(p_2 + x_0,\ q_1 + \omega_0 t + y_0) \\ &\quad - \mu^4 \left(\overline{\frac{\partial X_2}{\partial p}},\ u_2 \right) - \mu^3 \left(\overline{\frac{\partial X_2}{\partial q}},\ v_1 \right) + \mu^3 \left[\left(\frac{\partial u_2}{\partial \bar{p}},\ X_1 + \mu \overline{X}_2 \right) + \left(\frac{\partial u_2}{\partial \bar{q}},\ \overline{Y} \right) \right] \\ &= \mu X_1(p_2 + x_0) + \mu^2 X_2(p_2 + x_0,\ q_1 + \omega_0 t + y_0) + \mu^3 D_1, \end{aligned}$$

$$\begin{aligned} \frac{dq_1}{dt} &= \frac{d\bar{q}}{dt} + \mu \Bigg[\left(\frac{\partial v_1}{\partial \bar{p}},\ \mu X_1(\bar{p} + x_0) + \mu^2 X_2(\bar{p} + x_0,\ \bar{q} + \omega_0 t + y_0) \right) \\ &\quad + \left(\frac{\partial v_1}{\partial \bar{q}},\ \omega(\bar{p} + x_0) - \omega_0 + \mu \overline{Y}(\bar{p} + x_0,\ \bar{q} + y_0) \right) + \frac{\partial v_1}{\partial t} \Bigg] \\ &= \omega(p_2 + x_0) - \omega_0 + \mu Y(p_2 + x_0,\ q_1 + \omega_0 t + y_0) + \mu^2 D_2, \end{aligned}$$

where $\overline{\partial X_1/\partial p},\ \overline{\partial X_2/\partial p},\ \overline{\partial X_2/\partial q}$ are average values of the corresponding functions in their definitional domains. In G_{m+n} with $t \in [0,\ \mu^{-1}]$ the following estimations hold:

$$\|D_1\| < \frac{2}{\beta}C\Big(2C_1 + NC + \frac{mN}{\beta}CC_1\Big),$$
$$\|D_2\| < \frac{2}{\beta}C\Big[\Big(\frac{m}{2} + \frac{3}{2}\Big)C_1 + NC + \frac{mN}{\beta}CC_1\Big].$$

The system of equations for the functions $p_2(t,\mu)$ and $q_1(t,\mu)$

$$\frac{dp_2}{dt} = \mu X_1(p_2 + x_0) + \mu^2 X_2(p_2 + x_0, q_1 + \omega_0 t + y_0) + \mu^3 D_1,$$
$$\frac{dq_1}{dt} = \omega(p_2 + x_0) - \omega_0 + \mu Y(p_2 + x_0, q_1 + \omega_0 t + y_0) + \mu^2 D_2$$

and the system (2.11.2) satisfy the conditions described in Section 2.3, with constants

$$L_p = C_1(1+\mu), \quad L_q = \mu NC, \quad l_p = C_1, \quad l_q = NC, \quad L_\omega = mnC.$$

Then estimations (2.11.6) are true, and

$$|\lambda_1| < K'_{1\mu}, \quad |\lambda_2| < K'_{2\mu}, \quad |a| < K'_{3\mu}, \quad |C_1^{(p)}| < K'_{4\mu}, \quad |C_2^{(p)}| < K'_{5\mu}.$$

Consequently for all $t \in [0,\ \mu^{-1}]$

$$\|p - p_2\| < K'_7\mu^2, \quad \|q - q_1\| < K'_6\mu,$$

where the constants K'_6, K'_7 do not depend on μ. But on the other hand, with $t \in [0,\ \mu^{-1}]$ we have

$$\|p_2 - \bar{p}\| = \|\mu^2 u_2\| < \frac{1}{\beta}\mu^2 C, \quad \|q_1 - \bar{q}\| = \|\mu v_1\| \le \frac{1}{\beta}\mu C.$$

So

$$\|p - \bar{p}\| \le \|p - p_2\| + \|p_2 - \bar{p}\| < A\mu^2, \quad \|q - \bar{q}\| < B\mu,$$

where the constants $A = K'_7 + C\beta^{-1}$, $B = K'_6 + C\beta^{-1}$ do not depend on μ. The transfer to ε-terminology is trivial. The theorem is proved.

In conclusion we note that equations (2.11.1) may relate to any of the classes II, III or IV, and this fully depends on the width of the resonance zone $\|(k, \omega(x))\| \le \alpha$.

2.12 Comments and References

In this chapter we described the main presently known mathematical results of proving the applicability of asymptotic theory to multifrequency systems of ordinary differential equations typical for modern resonance dynamics. These are equations determined on tori or toroidal manifolds, with their right-hand members being multiperiodic functions with respect to fast angular variables, and the frequencies being functions of slow phase variables. Sufficiently well-investigated equations include those of Van der Pol, Mathieu, Duffing, and Hill, and one might say, simple cases of multifrequency systems defined on tori. For these equations the phase space has dimension 2, and though the subspace of fast variables (torus) has dimension 1, here also occurs the problem of small denominators, like the Poincaré problem of reflection of a circumference on itself. Generally speaking, this is sufficient to understand the difficulty of the mathematical problems of the construction of exact, but not asymptotic solutions of equations defined on *many-dimensional tori.*

For the interested reader we formulate some mathematical problems of the asymptotic theory of multifrequency systems.

Problem 1. We do not know a single proof of the averaging principle for multifrequency systems for which it is sufficient to impose conditions not on the right-hand members of the initial equations, but on their averaged values. Consequently the matter concerns the separation of a subclass of multifrequency systems with the above-mentioned properties from the class of multifrequency systems defined on many-dimensional tori.

Problem 2. For dual frequency systems the resonance of frequencies $k_1\omega_1 + k_2\omega_2 = 0$ (except the obvious case $k_1 = k_2 = 0$) may take place only when both components k_1 and k_2 of the integer vector k are nonzero. Another situation occurs in the case of multifrequency systems (or more exactly, n-frequency systems with $n > 2$). Here the existence of s-resonance manifolds is possible, with $2 < s < n$, in the space of frequencies $\omega_1, \omega_2, \dots, \omega_n$, for which the mathematical justification of the averaging principles has not been investigated by mathematicians. As regards the construction of asymptotic approximations of such multifrequency system solutions, it should be expected that their analytic form will be simpler in comparison with the general case.

Problem 3. Finally, we do not know any proof of the averaging principle for multifrequency systems in the case when the basic frequencies ω_k depend not only on slow variables x, but also on fast variables y, i.e. $\omega = \omega(x, y)$. In this case the undisturbed system

$$\frac{dx}{dt} = 0,$$
$$\frac{dy}{dt} = \omega(x, y),$$

is, generally speaking, nonintegrable, and therefore even the geometric interpretation of an undisturbed solution on tori is an intricate problem.

Section 2.5 Strict definitions and properties of the averaging operator for continuous disturbances are given in Volosov [1], and Grebenikov [1, 3].

Section 2.6 A new analytic form for disturbances of phase coordinates in the case of 0-resonances was first derived in Grebenikov and Ryabov [2], and detailed comments were given in Grebenikov [3].

Section 2.8 The proof of the convergence of the iteration method for the Van der Pol scalar equation was first obtained by Mandelstam and Papaleksi [1]. The extension for the many-dimensional case is given in Grebenikov and Ryabov [1]).

For several results and references on the comparison principle in the theory of nonlinear systems with slow and fast variables see Martynyuk [1].

3 Some Resonance Problems of Nonlinear Mechanics

3.0 Introduction

The previous chapter described the mathematical aspects of nonlinear differential equation theory, based on the averaging principles. We have more than once pointed out that the main mathematical difficulties of this theory arise due to the possible occurrence of small denominators in asymptotic formulae. Problems with small denominators (or problems with frequency resonances, which is the same) are not particularly abstract, but reflect the real picture of dynamic processes going on in micro- and macrocosms. Therefore here we will discuss some resonance problems of analytical dynamics that have certain theoretical and practical importance.

3.1 Newtonian Three-Body Problem

In the Preface we said that historically the problem of small denominators arose during the investigation of the movement of Jupiter and Saturn around the Sun. Ignoring the existence of other planets, it turns out that the dynamic model consisting of three homogeneous material balls mutually attracting each other in accordance with Newton's law is an extremely complex resonance system investigated by practically all outstanding naturalists of the past. In the mathematical model of the three-body problem the matter concerns the study of the motion of each of the three bodies P_0, P_1, P_2 with arbitrary masses m_0, m_1, m_2 for arbitrary initial conditions and in an infinite time interval. Assuming that m_0 is the mass of the Sun, m_1 is the mass of Jupiter, and m_2 is the mass of Saturn, the quantitative relations

$m_1 \approx 10^{-3} m_0$, $m_2 \approx 3 \times 10^{-4} m_0$ are true, and consequently in the three body problem Sun–Jupiter–Saturn one mass is much greater than the other two masses. This model is called *the planetary variation of the three-body problem.* From the mathematical point of view this means that in the problem a small parameter is naturally present. If the masses of all the bodies are approximately the same, it is called *the star variant of the three-body problem.*

If a rectangular Cartesian coordinate system with origin in the center of the Sun (the so-called rectangular heliocentric system) is introduced, the differential equations of planetary motion around the Sun are written as (see Duboshin [1])

$$\begin{aligned}
\frac{d^2 x_1}{dt^2} + \frac{(m_0 + m_1) x_1}{r_1^3} &= m_2 \left[\frac{x_2 - x_1}{\Delta_{12}^3} - \frac{x_2}{r_2^3} \right], \\
\frac{d^2 y_1}{dt^2} + \frac{(m_0 + m_1) y_1}{r_1^3} &= m_2 \left[\frac{y_2 - y_1}{\Delta_{12}^3} - \frac{y_2}{r_2^3} \right], \\
\frac{d^2 z_1}{dt^2} + \frac{(m_0 + m_1) z_1}{r_1^3} &= m_2 \left[\frac{z_2 - z_1}{\Delta_{12}^3} - \frac{z_2}{r_2^3} \right], \\
\frac{d^2 x_2}{dt^2} + \frac{(m_0 + m_2) x_2}{r_2^3} &= m_1 \left[\frac{x_1 - x_2}{\Delta_{12}^3} - \frac{x_1}{r_1^3} \right], \\
\frac{d^2 y_2}{dt^2} + \frac{(m_0 + m_2) y_2}{r_2^3} &= m_1 \left[\frac{y_1 - y_2}{\Delta_{12}^3} - \frac{y_1}{r_1^3} \right], \\
\frac{d^2 z_2}{dt^2} + \frac{(m_0 + m_2) z_2}{r_2^3} &= m_1 \left[\frac{z_1 - z_2}{\Delta_{12}^3} - \frac{z_1}{r_1^3} \right],
\end{aligned} \tag{3.1.1}$$

where x_s, y_s, z_s are the rectangular heliocentric coordinates of the planet P_s $(s = 1, 2)$,

$$\begin{aligned}
\Delta_{12}^2 &= (x_2 - x_1)^2 + (y_2 - y_1)^2 + (z_2 - z_1)^2, \\
r_s^2 &= x_s^2 + y_s^2 + z_s^2, \quad s = 1, 2.
\end{aligned} \tag{3.1.2}$$

The system of fundamental units was chosen so that the gravitational constant is equal to 1. The order of system (3.1.1) is 12, and though the latter is quite compact in form, it is impossible to obtain its exact solution in an analytic form because it is essentially nonlinear. In the past the search for its first integrals was, as one might say, a point of honor for every outstanding mathematician. And though they did not manage to do very much, it would not be an exaggeration to say that the three-body problem was the main generator of ideas in the fields of the analytic, qualitative and numerical theory of ordinary differential equations. This is easily explicable. In the 18th and 19th centuries the majority of mathematical models were drawn by scientists from their study of the visible cosmos, they knew practically nothing about the microcosm. And their idea of the cosmos was in the first place the solar system where movements of certain periodicity had been observed from ancient times. Therefore many things created in mathematics and mechanics were first of all verified on the model of the solar system.

But equations (3.1.1) have one essential drawback. Though they are compact (there are no infinite expansions in them), the basic frequencies of the problem (the functions $\omega(x)$ in multifrequency rotary systems described in the previous paragraphs) are not perceived in explicit form, so the above asymptotic theory cannot be directly applied to them. On the other hand, in the planetary system the basic frequencies of the three-body problem are visible, as the saying goes, "on the surface." Indeed, as far back as the epoch of the outstanding astronomer of antiquity Hipparchos (2nd century BC) the average angular velocities of the then known planets of the solar system were calculated, those planets included the Sun and the Moon which, as the ancients thought, were moving around the Earth. The angular velocities of the planets around the Earth (in the Earth-centered system of the world) are just the angular velocities of the planets around the Sun (in the heliocentric system of the world proposed by the great Polish astronomer Nikolai Copernik in 1543). It is very important to note that though the ancient astronomers were not armed with telescopes, their visual observations of planetary motions relative to the stationary stars made it possible to calculate the angular velocities of the planets (astronomers call them, as we have already mentioned, *average motions of the planets*) with a relatively high accuracy.

So, to investigate the three-body problem by asymptotic methods, it is necessary to turn from configuration space P_0xyz to phase space where some of the variables personify the slow movements, and the other, fast movements. The term that is most frequently used for planetary problems is Keplerian phase space. We will not overburden the text by a great number of formulae for the transition from one space to the other, but we will only give a qualitative description of this geometrical operation (for the details see Duboshin [1], Grebenikov [1], etc.).

The three-dimensional space P_0xyz, as mentioned above, used to be called configuration space, and this name seems to be the most obvious, because in this three-dimensional rectangular Euclidean space our daily life is enacted.

Each equation of system (3.1.1) is a second-order equation; therefore from the mathematical point of view the most usual form is the notation of equations of motion with the use of a system in normal form, where each equation is of first order. And this means that instead of the three-dimensional space P_0xyz the six-dimensional phase space $P_0xyz\dot{x}\dot{y}\dot{z}$ of coordinates and velocities is introduced, but this also does not solve our problem. It is only the transition to the six-dimensional Keplerian phase space that makes it possible to describe the motions in the three-body problem by means of a multifrequency rotary system. Keplerian space also has a simple geometric interpretation. (It is enough to recall Kepler's three laws of planetary motion around the Sun.)

According to Kepler's laws, each planet (if it was the only one) moves around the Sun along an elliptical orbit with the Sun located at one of the foci, and the orbital revolution itself is performed so that the squared periods of revolution of two planets around the Sun (e.g. expressed in terrestrial years) relate to each other as the cubes of the major semiaxes of their elliptical orbits. From this naturally follows the expediency of the following operation: since on the one hand the position of a planet in the six-dimensional space $P_0xyz\dot{x}\dot{y}\dot{z}$ at each point of time is determined by the six quantities $(x(t),\ y(t),\ z(t),\ \dot{x}(t),\ \dot{y}(t),\ \dot{z}(t))$, and on the other hand by the dimensions of the ellipse, by its position in space, as well as by the position of the planet itself on the elliptical orbit at the same point of time, exact formulae are established between the quantitative characteristics of these two phase spaces, as described e.g. in Duboshin [1], Grebenikov [1], etc. The characteristics or coordinates of the (six-dimensional) Keplerian space are the following quantities:

a) the major semiaxis a and the eccentricity e of the elliptical orbit;
b) two angles: the inclination i and the ascending node longitude Ω, determining the orientation of the elliptical orbit's plane relatively to some basic coordinate system;
c) one angle: the perihelion argument ω determining the orientation of the ellipses in its plane;
d) an angle: the average (or true) anomaly M (or v) characterizing the position of a planet on the orbit itself at every point of time. This angle is an analogue of the azimuth in the polar coordinate system.

The following characteristics of planetary orbits are also widely used in celestial mechanics: π is the longitude of the orbit perihelion $(\pi = \omega + \Omega)$, n the average angular motion of the planet, l the average longitude of the planet in the orbit $(l = \pi + M)$, $M = n(t - t_0) + M_0$, and M_0 is the value of the average anomaly at the initial time t_0, so in undisturbed motion the average anomaly is a linear function of time t.

From the meaning of these characteristics and from Kepler's laws it follows that the characteristics $(a,\ e,\ i,\ ,\Omega,\ \omega)$ in the problem of two bodies (the Sun and the planet) do not depend on time, i.e. they are constant, and only one characteristic M varies with time. If each planet is influenced by the attraction of other planets, the above phase picture is ruined, but due to the smallness of planetary masses this evolution goes on very slowly. The first five parameters of planetary orbits (the major semiaxis, the eccentricity, the inclinations, the longitudes of ascending nodes and the arguments of perihelia), are constant in the absence of any extraneous disturbances, but when disturbed they start to change slowly, i.e. they are slow phase variables (variables of type x in multifrequency rotary systems). The sixth variable M is a fast phase variable in the same terms (a variable of type y in multifrequency rotary systems). Hence it follows that the description of dynamics in the problem of three bodies (Sun–Jupiter–Saturn) with the help of Keplerian phase variables adds up to the study of a multifrequency rotary system of ordinary differential equations with slow and fast variables; the vector of slow variables is ten-dimensional, and the vector of fast variables is two-dimensional.

In these Keplerian variables the planetary three-body problem is described by the following system of 12th order differential equations (see Grebenikov [3], and Subbotin [1]) (the symbols most common in astronomy are used here):

$$
\begin{aligned}
\frac{da_s}{dt} &= \frac{2}{n_s a_s}\frac{\partial R_s}{\partial l_s},\\
\frac{de_s}{dt} &= -\frac{\sqrt{1-e_s^2}}{n_s a_s^2 e_s}\frac{\partial R_s}{\partial \pi_s} - \frac{e_s\sqrt{1-e_s^2}}{n_s a_s^2\left(1+\sqrt{1-e_s^2}\right)}\frac{\partial R_s}{\partial l_s},\\
\frac{di_s}{dt} &= -\frac{1}{n_s a_s^2 \sin i_s\sqrt{1-e_s^2}}\frac{\partial R_s}{\partial \Omega_s} - \frac{1}{n_s a_s^2\sqrt{1-e_s^2}}\operatorname{tg}\frac{i_s}{2}\left(\frac{\partial R_s}{\partial \pi_s}+\frac{\partial R_s}{\partial l_s}\right),\\
\frac{d\Omega_s}{dt} &= \frac{1}{n_s a_s^2 \sin i_s\sqrt{1-e_s^2}}\frac{\partial R_s}{\partial i_s},\\
\frac{d\pi_s}{dt} &= \frac{1}{n_s a_s^2\sqrt{1-e_s^2}}\operatorname{tg}\frac{i_s}{2}\frac{\partial R_s}{\partial i_s} + \frac{\sqrt{1-e_s^2}}{n_s a_s^2 e_s}\frac{\partial R_s}{\partial e_s},\\
\frac{dl_s}{dt} &= n_s - \frac{2}{n_s a_s}\frac{\partial R_s}{\partial a_s} + \frac{1}{n_s a_s^2\sqrt{1-e_s^2}}\operatorname{tg}\frac{i_s}{2}\frac{\partial R_s}{\partial i_s} +\\
&\quad + \frac{e_s\sqrt{1-e_s^2}}{n_s a_s^2\left(1+\sqrt{1-e_s^2}\right)}\frac{\partial R_s}{\partial e_s}, \qquad s = 1, 2.
\end{aligned}
\tag{3.1.3}
$$

In equations (3.1.3) the following symbols are used: π_s are the longitudes of perihelia of the orbits of the planets P_s ($\pi_s = \omega_s + \Omega_s$) where ω_s are the arguments of perihelia of the orbits of planets P_s, and they should not be confused with frequencies in multifrequency systems; the average motions of planets n_1, n_2 are the basic frequencies of the problem (earlier they were denoted by $\omega_j(x)$). The average anomalies of planets are calculated by the formula

$$M_s = n_s(t - t_0) + M_{s,0}, \tag{3.1.4}$$

from which it is clear that in undisturbed motion they are linear functions of time, and in disturbed motion n_s they become slowly changing functions of time, and consequently the average anomalies M_s (and therefore l_s as well) in their turn become quasi-linear functions of time. Kepler's third law can be given the mathematical notation refined by Newton:

$$n_s = \sqrt{\frac{f(m_0 + m_s)}{a_s^3}}. \tag{3.1.5}$$

Here f is the gravitational constant (we assumed it to be equal to one). It is from formula (3.1.5) that it follows that even if the masses m_s of all planets are supposed to be zero (certainly, except the mass of the Sun m_0), the average angular velocity of a planet n_s (i.e. the basic frequency in terms of slow and fast variables) does not become zero.

The discourse on system (3.1.3) would not be sufficiently complete while the explicit dependence of the right-hand members on the small parameter μ is not shown. The planet P_1 is influenced by the attraction from the direction of the planet P_2, therefore the disturbing (perturbation) function R_1 is proportional to the mass m_2, and the disturbing function R_2 is proportional to the mass m_1. If the masses are considered small quantities (of order μ), i.e. if we consider that $m_s = L_s\mu$ (where L_s is some constant of proportionality), then all disturbing functions R_s are proportional to the small parameter μ.

Finally, for equations (3.1.3) to have the form of a multifrequency rotary system it is necessary that the right-hand members should be Fourier series with respect to the fast variables y. Construction of concrete Fourier series for the disturbing functions R_s and their partial derivatives is one of the most intricate problems of analytical celestial mechanics: great mathematicians spent decades trying to solve it. First of all it is necessary to mention the titanic work performed by Le Verrier at the Paris Observatory after the great triumph that resulted in the discovery of Neptune. Le Verrier obtained a series for a disturbing function (see Le Verrier [1]) in which all the terms of the Fourier series with coefficients up to the seventh power inclusive with respect to the small quantities were determined: the eccentricities of the orbits of the two planets e_1, e_2, the ratio of the major semiaxes of their orbits a_1/a_2 and $\sin^2(J/2)$, where J is the angle between the planes of their orbits.

Now we introduce the following symbols:

$$
\begin{gathered}
a_s = x_s, \quad e_s = x_{s+2}, \quad i_s = x_{s+4}, \quad \Omega_s = x_{s+6}, \quad \pi_s = x_{s+8}, \\
l_s = y_s, \quad \omega_s = n_s, \quad x = (x_1, \dots, x_{10}), \quad y = (y_1, y_2), \quad \omega = (\omega_1, \omega_2),
\end{gathered}
$$

$$
\begin{aligned}
\mu X_s &= \frac{2}{n_s a_s} \frac{\partial R_s}{\partial l_s}, \\
\mu X_{s+2} &= -\frac{\sqrt{1-e_s^2}}{n_s a_s e_s} \frac{\partial R_s}{\partial \pi_s} - \frac{e_s\sqrt{1-e_s^2}}{n_s a_s^2\left(1+\sqrt{1-e_s^2}\right)} \frac{\partial R_s}{\partial l_s}, \\
\mu X_{s+4} &= -\frac{1}{n_s a_s^2 \sin i_s \sqrt{1-e_s^2}} \frac{\partial R_s}{\partial \Omega_s} - \frac{1}{n_s a_s^2 \sqrt{1-e_s^2}} \operatorname{tg}\frac{i_s}{2}\left(\frac{\partial R_s}{\partial \pi_s} + \frac{\partial R_s}{\partial l_s}\right), \\
\mu X_{s+6} &= \frac{1}{n_s a_s^2 \sin i_s \sqrt{1-e_s^2}} \frac{\partial R_s}{\partial i_s}, \\
\mu X_{s+8} &= \frac{1}{n_s a_s^2\sqrt{1-e_s^2}} \operatorname{tg}\frac{i_s}{2}\frac{\partial R_s}{\partial i_s} + \frac{\sqrt{1-e_s^2}}{n_s a_s^2 e_s}\frac{\partial R_s}{\partial e_s}, \\
\mu Y_s &= -\frac{2}{n_s a_s}\frac{\partial R_s}{\partial a_s} + \frac{1}{n_s a_s^2\sqrt{1-e_s^2}} \operatorname{tg}\frac{i_s}{2}\frac{\partial R_s}{\partial i_s} + \frac{e_s\sqrt{1-e_s^2}}{n_s a_s^2\left(1+\sqrt{1-e_s^2}\right)}\frac{\partial R_s}{\partial e_s}, \\
& s = 1, 2, \\
X &= (X_1, \dots, X_{10}), \quad Y = (Y_1, Y_2).
\end{aligned}
\tag{3.1.6}
$$

Then system (3.1.3) can be represented as a multifrequency rotary system

$$\frac{dx}{dt} = \mu X(x, y), \qquad \frac{dy}{dt} = \omega(x) + \mu Y(x, y), \tag{3.1.7}$$

where the vector-functions X and Y are 2π-periodic with respect to the average longitudes y_1, y_2.

As the two-planet three-body problem is now described not by equations (3.1.1) but by equations (3.1.7), it is possible to apply to the latter all the mathematical results stated in the previous sections. First of all we will give a general averaging scheme embracing practically all the averaging schemes applied by the classical work of celestial mechanics. For that we will introduce a $(m+n)$-dimensional column vector $R(x, y)$ with its components being the right-hand members of a multifrequency system:

$$R(x,y) = \begin{bmatrix} \mu X_1(x,y) \\ \cdots\cdots\cdots \\ \mu X_m(x,y) \\ \omega_1(x) + \mu Y_1(x,y) \\ \cdots\cdots\cdots \\ \omega_n(x) + \mu Y_n(x,y) \end{bmatrix}. \tag{3.1.8}$$

For this study we will also introduce the rectangular numerical matrix

$$A_{sn} = (a_{ij}), \quad i = 1, \ldots, s; \quad j = 1, \ldots, n; \quad s < n, \tag{3.1.9}$$

and generate an s-dimensional vector D with the help of the equality

$$D = (A_{sn}, y). \tag{3.1.10}$$

Let $\operatorname{rank} A_{sn} = s$, and for definiteness suppose that

$$\Delta_{ss} \equiv \det(a_{ij}) \neq 0, \quad i, j = 1, \ldots, s. \tag{3.1.11}$$

Now solving the system of algebraic equations (3.1.10) with respect to $y_1, \ldots, y_s$, we obtain

$$y_i = f_i(D, y_{s+1}, \ldots, y_n), \quad i = 1, \ldots, s, \tag{3.1.12}$$

where f_i are linear functions of their arguments. Then change the fast variables $y_1, \ldots, y_s$ in $R(x, y)$ by using (3.1.12):

$$R(x,y) = \widetilde{R}(x, D, y_{s+1}, \ldots, y_n). \tag{3.1.13}$$

The function $\widetilde{R}$ is periodic with respect to D, $y_{s+1}, \ldots, y_n$, but the period with respect to each of these arguments can differ from 2π. The values of the new periods are determined by elements of the numerical matrix (3.1.9). Denote the new periods of the function $\widetilde{R}$ with respect to $y_{s+1}, \dots, y_n$ by $T_{s+1}, \dots, T_n$ respectively and calculate the average value of $\widetilde{R}$ over the variables $y_{s+1}, \dots, y_n$ within the corresponding new periods, in the process of integration considering x and D as constants:

$$\overline{\widetilde{R}}(x, D) = \frac{1}{T_{s+1} \cdots T_n} \int_0^{T_{s+1}} \cdots \int_0^{T_n} \widetilde{R}(x, D, y_{s+1}, \dots, y_n)\, dy_{s+1} \dots dy_n. \qquad (3.1.14)$$

In Grebenikov [1] it is shown that this formula contains as particular cases the averaging operators described in Section 1.2. However for the three-body problem concrete averaging schemes were used that are called by the name of the author who either was the first to use them or obtained essential results via using them. For instance, Gauss's averaging scheme is the averaging of a disturbing function over all fast variables (averaging operator M_y).

The Fatou averaging scheme (see Fatou [1]) is expressed by the equality

$$\overline{R}_1(a_1, a_2, \dots, \pi_1, \pi_2, l_1) = \frac{1}{2\pi} \int_0^{2\pi} R(a_1, a_2, \dots, l_1, l_2)\, dl_2, \qquad (3.1.15)$$

where l_2 is the average longitude of the planet regarded as the disturbed (but not disturbing) body.

On the contrary, in N.D. Moiseev's scheme the averaging operation is performed over the average longitude of the disturbed planet (see Moiseev [1]):

$$\overline{R}_1(a_1, a_2, \dots, \pi_1, \pi_2, l_2) = \frac{1}{2\pi} \int_0^{2\pi} R(a_1, a_2, \dots, l_1, l_2)\, dl_1, \qquad (3.1.16)$$

The schemes of Moiseev and Fatou are typical for averaging over some of the fast variables (operator $M_{y^{(s)}}$ from Section 1.2).

Now consider two schemes applied in the resonance three-body problem. These are the so-called first and second schemes of Delaunay–Hill (see Moiseev [1], and Subbotin [1]).

Using the general averaging scheme for the problems of celestial mechanics, instead of the rapid variable l_2 introduce the function

$$D = k_1^* l_1 + k_2^* l_2. \qquad (3.1.17)$$

This is called the Delaunay anomaly. If the integers k_1^*, k_2^* are arbitrary, the Delaunay anomaly will be the same fast angular variable as l_1 and l_2, but if they are such that $k_1^* n_1^{(0)} + k_2^* n_2^{(0)} = \alpha$ is a small quantity (α-resonance of the frequencies of undisturbed motion of the two planets P_1 and P_2), then the Delaunay anomaly becomes a slow angular variable. At an exact resonance of average angular velocities of the planets ($\alpha = 0$) the undisturbed Delaunay anomaly becomes constant. If we now express from (3.1.17) the longitude l_2 by D, and substitute this expression into the Fourier series for the disturbing function, we have

$$R_1(a_1, a_2, \dots, l_1, l_2) = \widetilde{R}(a_1, a_2, \dots, D, l_1). \tag{3.1.18}$$

Then upon averaging over l_1 we obtain

$$\overline{\widetilde{R}}_1(a_1, a_2, \dots, D) = \frac{1}{T}\int_0^T \widetilde{R}(a_1, a_2, \dots, D, l_1)\, dl_1. \tag{3.1.19}$$

Here T is the least common multiple for the numbers

$$\frac{2\pi\|k^*\|}{(k_1k_2^* - k_2k_1^*)}$$

with $k_1, k_2 = 1, 2, \dots$. It is easy to see that in the averaged function $\overline{\widetilde{R}}_1$ all resonances are preserved, i.e. all the harmonics with the small denominator $k_1^*n_1^{(0)} + k_2^*n_2^{(0)}$. From this it uniquely follows that the scheme of Delaunay–Hill is a particular case of the averaging operator for continuous disturbances.

In the two-planet problem the second averaging scheme of Delaunay–Hill is also used. It is constructed as follows.

Introduce the so-called "generalized Delaunay–Hill anomaly" by the formula

$$\widetilde{D} = \frac{n_2^{(0)}M_1}{n_1} + \Omega_1 - M_2, \tag{3.1.20}$$

where M_1 and M_2 are the average anomalies of the planets P_1 and P_2, and $n_2^{(0)}$ is an undisturbed average angular velocity of the disturbed planet. Then by means of (3.1.20) exclude the difference $\Omega - M_2$ from the Fourier series of the disturbing force working upon the planet P_1. After that average the disturbing function R_1 over the average anomaly M_1 by means of the operator

$$\overline{\widetilde{R}}_1(a_1, a_2, \dots, \pi_2, \widetilde{D}) = \lim_{T\to\infty} \frac{1}{T}\int_0^T \widetilde{R}_1(a_1, \dots, \widetilde{D}, M_1)\, dM_1. \tag{3.1.21}$$

Now applying any of the averaging schemes of celestial mechanics and the Krylov–Bogolyubov transform to system (3.1.7), one can obtain averaged equations of any approximation (including the first one), and then the principal procedure of application of the whole arsenal of the asymptotic theory of differential equations can be followed.

For instance, applying the first Delaunay–Hill scheme to equations (3.1.7) and assuming that $m_1 = 0$ (in this case it is said that we are dealing with a limited three-body problem, where the motion of a planet with zero mass under the influence of a second planet with mass $\mu = m_2$ is studied), then in the notation usual for astronomers, the averaged equations of the first approximation will have the form

$$\begin{aligned}
&\frac{d\bar{a}}{dt} = 2k_1^*\sqrt{\frac{\bar{a}}{m_0}}\frac{\partial\overline{\overline{R}}}{\partial D}, \quad \frac{d\bar{p}}{dt} = 2k_2^*\sqrt{\frac{\bar{p}}{m_0}}\frac{\partial\overline{\overline{R}}}{\partial D},\\
&\frac{d\bar{i}}{dt} = \frac{\operatorname{ctg}\bar{i}}{\sqrt{m_0\bar{p}}}\frac{\partial\overline{\overline{R}}}{\partial\bar{\pi}} - \frac{k_2^*}{\sin\bar{i}\sqrt{m_0\bar{p}}}\frac{\partial\overline{\overline{R}}}{\partial D},\\
&\frac{d\overline{\Omega}}{dt} = \frac{1}{\bar{n}\bar{a}^2\sin\bar{i}\sqrt{1-\bar{e}^2}}\frac{\partial\overline{\overline{R}}}{\partial\bar{i}},\\
&\frac{d\bar{\pi}}{dt} = \frac{1}{\bar{n}\bar{a}^2\sqrt{1-\bar{e}^2}}\operatorname{tg}\frac{\bar{i}}{2}\frac{\partial\overline{\overline{R}}}{\partial\bar{i}} + \frac{\sqrt{1-\bar{e}^2}}{\bar{n}\bar{a}^2\bar{e}}\frac{\partial\overline{\overline{R}}}{\partial\bar{e}},\\
&\frac{dD}{dt} = \left(k_1^*\bar{n} - k_2^*n_2^{(0)}\right) - 2k_1^*\sqrt{\frac{\bar{a}}{m_0}}\frac{\partial\overline{\overline{R}}}{\partial\bar{a}} + \frac{k_2^*}{\sin\bar{i}\sqrt{m_0\bar{p}}}\frac{\partial\overline{\overline{R}}}{\partial\bar{i}}.
\end{aligned} \tag{3.1.22}$$

Unlike equations (3.1.3) here instead of the eccentricity e an elliptical parameter is introduced by the formula $p = a(1-e^2)$.

Since Delaunay–Hill averaging preserves resonance terms in the Fourier series of the disturbing function, then while solving the partial differential equations for the functions $u_1, v_1, u_2, v_2, \ldots$ constituting the Krylov–Bogolyubov transform that transfers system (3.1.3) into system (3.1.22) we obtain that the above listed functions will contain small terms periodic with respect to D. The harmful influence of resonance harmonics (or small denominators, which is the same) is localized in the averaged solutions of system (3.1.22), i.e. in the averaged Keplerian elements $\bar{a}$, $\bar{p}$, $\bar{i}$, $\overline{\Omega}$, $\bar{\pi}$, D, and it will not occur in the Krylov–Bogolyubov transform. Naturally, this reasoning is true provided that the partial differential equations for u_s, v_s are solved by the method described in Section 2.6, or by a similar one.

To be able to use the Krylov–Bogolyubov transform for the asymptotic representation of the solution of the initial equations (3.1.3), it is necessary, as has been emphasized more than once, to obtain a solution of the averaged equations (3.1.22) in explicit form. Unfortunately it is impossible to integrate exactly equations (3.1.22), but in the plane version of the problem (when all three bodies move perpetually in one fixed plane, e.g. in the plane $\bar{i} \equiv \overline{\Omega} \equiv 0$) the system is reduced to a fourth-order system (without equations for $\bar{i}$ and $\overline{\Omega}$); its four first integrals are known (see Grebenikov [1], Moiseev [1], Subbotin [1]):

$$\begin{aligned}
&\sqrt{\bar{a}} - \frac{k_1^*}{k_2^*}\sqrt{\bar{p}} = C_1,\\
&\frac{m_0}{2\bar{a}} + \frac{k_2^*n_2^{(0)}}{k_1^*}\sqrt{m_0\bar{a}} + \overline{\overline{R}}\,(\bar{a},\bar{e},\bar{\omega},D) = C_2,\\
&\bar{\omega} + 2\int\sqrt{\frac{\bar{p}}{m_0}}\frac{\partial\overline{\overline{R}}}{\partial\bar{p}}\,dt = C_3,\\
&t - \int\left[k_1^*\sqrt{\frac{m_0}{\bar{a}^3}} - k_2^*n_2^{(0)} - 2k_1^*\sqrt{\frac{\bar{a}}{m_0}}\frac{\partial\widetilde{R}}{\partial\bar{a}} - 2k_2^*\sqrt{\frac{\bar{p}}{m_0}}\frac{\partial\overline{\overline{R}}}{\partial\bar{p}}\right]^{-1} dD = C_4.
\end{aligned} \tag{3.1.23}$$

However to obtain an explicit dependence of the unknown functions $\bar{a}$, $\bar{p}$, $\bar{\omega}$, D on time t it is necessary to perform the operation of inversion of these first integrals, which in itself is a very difficult analytical problem.

This problem for different averaging schemes was considered in the papers of Gomtsjan [1], Grebenikov [1], Grebenikov and Isaeva [1], Karaganchu [1], Prikhod'ko [1], etc. Here we will only say that if the problem of inversion of the first integrals of a first-order averaged equation is constructively solved, then obtaining the initial Keplerian coordinates is the problem that we have already solved, and as a result we obtain

$$\begin{aligned} a(t,\mu) &= \bar{a}(t,\mu) + \mu u_1\left(\bar{a},\dots,\bar{l}\right) + \dots, \\ &\dots\dots\dots\dots\dots\dots \\ l(t,\mu) &= \bar{l}(t,\mu) + \mu v_1\left(\bar{a},\dots,\bar{l}\right) + \dots \end{aligned} \tag{3.1.24}$$

Explicit expressions for the functions u_1, v_1 in the case of a three-body problem were also found in the paper referenced above.

3.2 The Problem of Justification of the Averaging Principle in the Bounded Newtonian Three-Body Problem

In the monograph by Grebenikov and Ryabov [1] justification of the averaging schemes applied in the bounded circular three-body problem were described. In particular it was shown that for Gauss's scheme (the operator of averaging over all fast variables — in this problem the fast variables are the average longitudes of the disturbed and the disturbing planet) the following estimations necessarily hold with $t \in [0, \mu^{-1}]$

$$\begin{aligned} |a(t,\mu) - \bar{a}(t,\mu)| &\sim \sqrt{\mu}\ln^2\mu^{-1}, \\ |e(t,\mu) - \bar{e}(t,\mu)| &\sim \sqrt{\mu}\ln^2\mu^{-1}, \\ |i(t,\mu) - \bar{i}(t,\mu)| &\sim \sqrt{\mu}\ln^2\mu^{-1}, \\ |\omega(t,\mu)\bar{\omega}(t,\mu)| &\sim \sqrt{\mu}\ln^2\mu^{-1} \end{aligned} \tag{3.2.1}$$

in that part of (six-dimensional) Keplerian phase space where the following inequality is satisfied:

$$\left|\frac{\partial R}{\partial M}\right| > 0; \tag{3.2.2}$$

here R is the disturbing function of the bounded circular three-body problem, and M is the average anomaly of the disturbed planet.

So we have to find in the phase space (a, e, i, Ω, ω, M) connected domains where inequality (3.2.2) is satisfied. If a Le Verrier series (se Le Verrier [1]) for the disturbing function is used, then inequality (3.2.2) takes the form

$$\left|\sum_{k_1=0}^{\infty}\sum_{k_2=-\infty}^{\infty}\sum_{k_3=-\infty}^{\infty} k_1 C_{k_1,k_2,k_3}(a,e,i) \times \sin(k_1 M + k_2(\Omega - M_2) + k_3\omega)\right| > 0. \tag{3.2.3}$$

Here the coefficients C_{k_1,k_2,k_3} are infinite power series of the eccentricity of the orbit of the disturbed planet e, the ratio of major semiaxes $\alpha = a/a_2$ and $\sin^2(i/2)$ (i is the angle between the planets' orbital planes).

Solving inequality (3.2.3) without additional bounds is extremely difficult; therefore in the expansion of the disturbing function we only keep terms containing $e, e^2, i = 0$ (the plane variant of the bounded three-body problem) and $a/a_2 = \alpha < 1$. The latter means that within the time interval $[0,\ \mu^{-1}]$ the disturbed planet remains in the area bounded by the circular orbit of the disturbing planet, i.e. there is no crossing of paths.

Under these conditions inequality (3.2.3) becomes substantially simpler, because in this case

$$\begin{aligned}\frac{\partial R}{\partial M} = \frac{\mu}{a_2}\sum_{k_2=-\infty}^{\infty}\Bigg\{&\left[\frac{C_1^{(k)}}{2} + \frac{e^2}{8}\left(-4k^2C_1^{(k)} + D(C_1^{(k)}) + D^2(C_1^{(k)})\right)\right]k\sin kS \\ &- \frac{e}{2}\left[2kC_1^{(k)} + D(C_1^{(k)})\right](k-1)\sin(kS+M) \\ &+ \frac{e^2}{8}\Big[4k^2C_1^{(k)} - 5kC_1^{(k)} + (4k-3)D(C_1^{(k)}) \\ &+ D^2(C_1^{(k)})\Big](k-2)\sin(kS+2M)\Bigg\} - \frac{\mu a}{a_2^2}\left[\left(1-\frac{e^2}{2}\right)\sin S\right. \\ &\left.+ e\sin(S-l+\omega) - \frac{e^2}{8}\sin(S+2l-2\omega) + \frac{9e^2}{8}\sin(S-2l+2\omega)\right],\end{aligned} \tag{3.2.4}$$

where

$$C_1^{(k)} = \frac{2(2k-1)!!}{(2k)!!}\alpha^k F\left(\frac{1}{2},\ k+\frac{1}{2},\ k+1;\ \alpha^2\right); \tag{3.2.5}$$

$$D^{(s)}(C_1^{(k)}) = \frac{d^s(C^{(k)})}{d(\ln\alpha)^s},\qquad s = 1, 2, \tag{3.2.6}$$

$$S = l_2 - l_1,\quad l_1 = M_1 + \omega_1, \tag{3.2.7}$$

$F(1/2,\ k+1/2,\ k+1;\ \alpha^2)$ is a hypergeometric series (see Duboshin [1]), a_2, l_2 are respectively the major semiaxis (radius of the circular orbit in the bounded circular problem) and the average longitude of the disturbing planet. Using Newcomb's recurrence relations (see Duboshin [1]) it is possible to express the derivatives $D^{(s)}(C_1^{(k)})$ also by means of hypergeometric series:

$$D^{(s+1)}(C_1^{(k)}) = kD^{(s)}(C_1^{(k)}) + D^{(s)}(C_1^{(k,1)}), \tag{3.2.8}$$

where

$$C_1^{(k,1)} \equiv \frac{2(2k+1)!!}{(2k+2)!!}\,\alpha^{k+2}F\Big(\frac{3}{2},\,k+\frac{3}{2},\,k+2;\,\alpha^2\Big). \tag{3.2.9}$$

From (3.2.8) it is easy to derive formulae for the derivatives

$$\begin{aligned} D^{(1)}\big(C_1^{(k)}\big) &= kC_1^{(k)} + C_1^{(k,1)},\\ D^{(2)}\big(C_1^{(k)}\big) &= k^2C_1^{(k)} + (k+3)C_1^{(k,1)} + C_1^{(k,2)}, \end{aligned} \tag{3.2.10}$$

$$C_1^{(k,2)} \equiv \frac{6(2k+3)!!}{(2k+4)!!}\,\alpha^{k+4}F\Big(\frac{5}{2},\,k+\frac{5}{2},\,k+3;\,\alpha^2\Big). \tag{3.2.11}$$

Substituting the above relations into the main inequality (3.2.3), and performing the necessary computations we obtain the inequality

$$\big|K_0 + eK_1 + e^2K_2\big| > 0, \tag{3.2.12}$$

where

$$\begin{aligned} K_0 = &-\left[\frac{1\cdot 3}{2\cdot 4}\alpha^3 + \frac{1\cdot 3\cdot 3\cdot 5}{2\cdot 4\cdot 4\cdot 6}\alpha^5 + \cdots\right]\sin\left(M_1+\tilde{\omega}\right)\\ &-\sum_{k=2}^{\infty} kC_1^{(k)}\sin\left(kM_1+k\tilde{\omega}\right), \quad \tilde{\omega} = \omega_1 - l_2, \end{aligned} \tag{3.2.13}$$

$$\begin{aligned} K_1 = &\ \alpha\sin\left(2M_1+\tilde{\omega}\right)\\ &+\frac{1}{2}\sum_{k=-\infty}^{\infty}\big(3kC_1^{(k)} + C_1^{(k,1)}\big)(k-1)\sin\left[(k-1)M_1+k\tilde{\omega}\right], \end{aligned} \tag{3.2.14}$$

$$\begin{aligned} K_2 = &-\frac{\alpha}{2}\sin\left(M_1+\tilde{\omega}\right) + \frac{\alpha}{8}\sin\left(M_1-\tilde{\omega}\right) + \frac{9\alpha}{8}\sin\left(3M_1+\tilde{\omega}\right)\\ &-\frac{1}{8}\sum_{k=-\infty}^{\infty}\Big\{\Big[(-3k^2+1)C_1^{(k)} + (k+4)C_1^{(k,1)} + C_1^{(k,2)}\Big]k\sin\left(kM_1+k\tilde{\omega}\right)\\ &+\Big[(9k^2-8k)C_1^{(k)} + 5kC_1^{(k,1)} + C_1^{(k,2)}\Big](k-2)\sin\left[(k-2)M_1+k\tilde{\omega}\right]\Big\}. \end{aligned} \tag{3.2.15}$$

Inequality (3.2.12) should be satisfied for all M_1, ω_1, $l_2 = n_2t + l_{2,0}$, provided that the major axis a_1 and eccentricity e_1 of the orbit of the disturbed planet P_1 belong to the domain G_2 of analyticity of the disturbed function R_1. Such a domain may be the direct Cartesian product of the two segments: $A = [\beta a_2,\,\delta a_2]$, $0 < \beta < \delta < 1$, and $E = [0,\,\varepsilon]$, $0 \le e \le \varepsilon \le 1$. This is clearly a ring with radii βa_2 and δa_2, with nearly all orbits located within it, and the motion of the disturbed planet is such that within a sufficiently long time interval (of order $1/\mu$) the functions M_1 and $\tilde{\omega}(t)$ do not become zero at the same time. If the functions $M_1(t)$ and $\tilde{\omega}$ simultaneously become zero, then inequality (3.2.12) is not true whatever the domain G_2 is.

Now we pass on to the mathematical justification of the first method of Delaunay–Hill as applied to the plane bounded circular three-body problem. Let the average motions (frequencies) of the disturbed and the disturbing bodies satisfy the condition of exact resonance

$$k_1^* n_1^{(0)} - k_2^* n_2^{(0)} = 0, \tag{3.2.16}$$

where $k^* = (k_1^*, k_2^*)$ is a concrete integer vector with norm $\|k^*\| \neq 0$. It is obvious that in this case there exists an infinite number of collinear vectors sk^* satisfying the resonance correlation (3.2.16), in particular, with $s = \pm 1, \pm 2, \ldots$. Now introducing the Delaunay anomaly by the formula

$$D = k_1^* M + k_2^* \tilde{\omega}, \tag{3.2.17}$$

it is easy to verify that in undisturbed motion the Delaunay anomaly is constant because

$$\begin{aligned} D^{(0)} &= k_1^* M_1^{(0)} + k_2^* \tilde{\omega}^{(0)} \\ &= k_1^* \left(n_1^{(0)} t + M_{10}^{(0)}\right) + k_2^* \left(\omega_1^{(0)} - n_2^{(0)} t\right) \\ &= k_1^* M_{10}^{(0)} + k_2^* \omega_1^{(0)}. \end{aligned}$$

Substitution of the Delaunay anomaly into the expansion of the disturbing function R_1 constructed by Le Verrier [1] generates the following Fourier series:

$$R_1 = \mu \sum_{k_1=-\infty}^{\infty} \sum_{k_2=-\infty}^{\infty} C_{k_1,k_2}(a,e) \exp\left\{ i \left(\frac{k_1 k_2^* - k_2 k_1^*}{k_2^*} M_1 + \frac{k_2}{k_2^*} D \right) \right\}. \tag{3.2.18}$$

So, using relation (3.2.17), we have excluded one of the fast variables $\tilde{\omega}$ from the expansion of the disturbing functions. Further, in the process of averaging for continuous disturbances the fast variable (the average anomaly M) is replaced by the expression $M_1 = n_1 t + M_{10}$ and integrated with respect to t, and all other phase variables (a, e, D) in the two-body problem are considered constant.

In (3.2.18) the average values of terms are non-zero only for those with $k_1 k_2^* - k_2 k_1^* = 0$. From this it follows that $k_1 = sk_1^*$, $k_2 = sk_2^*$, and so the result of the averaging of (3.2.18) generates the function

$$\overline{R}_1(a, e, D) = \mu \sum_{s=-\infty}^{\infty} C_{sk_1^*, sk_2^*}(a, e) e^{isD}. \tag{3.2.19}$$

The right-hand members of the differential equations of the plane bounded circular three-body problem (they are particular cases of equations (3.2.3) if it is assumed that $i \equiv \Omega \equiv 0$) are analytical functions of phase variables in the whole Keplerian phase space with the exception of those points that correspond to the collision of an infinitely small mass with one of the attracting bodies (in this case one of the mutual distances approaches zero, whereas the attracting force and Newtonian potential become arbitrary large). In the bounded circular three-body problem connected domains of three-dimensional configuration space (and consequently, six-dimensional Keplerian space as well, because the change from rectangular Cartesian coordinates to Keplerian elements is performed by means of

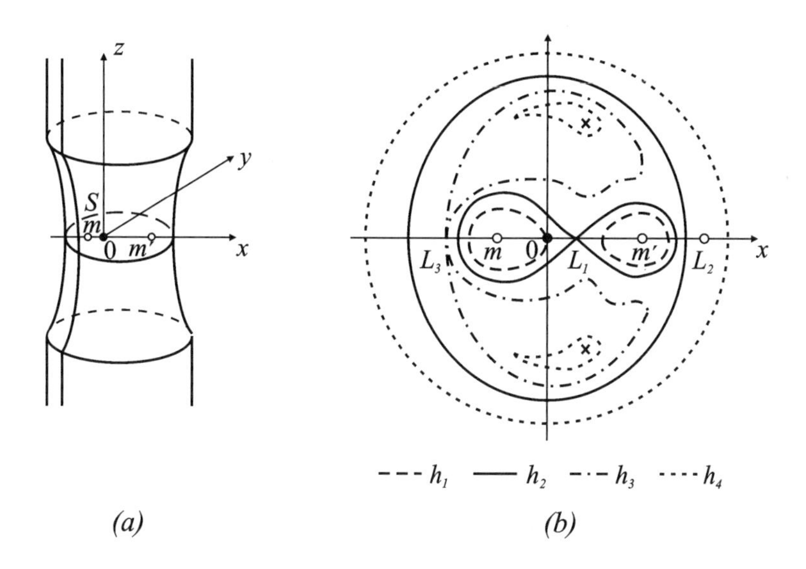

FIG. 3.1.1. Quasiquadratic Hill surface (a); projection of the Hill surface onto the main coordinate plane, $h_1, ..., h_4$ are constants of the Jacobi integral, corresponding to the boundary Hill curves (b).

a nondegenerate transformation) exist in which the motion of the three bodies is performed eternally, without collisions. For instance, such a domain (Fig. 3.2.1) is a part of three-dimensional space, located outside the quasicylindrical Hill surface S (see Duboshin [1]) serving as the geometric image of the Jacobi integral — the only known first integral of the differential equations of the bounded circular three-body problem. This surface S corresponds to the external closed contour of the double connected curve h_1 (in Fig. 3.1, h_1, h_2, h_3, h_4 denote different numerical values of the Jacobi constant h, and the corresponding Hill boundary curves are constructed); for more details refer to Grebenikov and Ryabov [1], and Duboshin [1]. This part of Euclidean space can be taken as the domain G_{m+n} appearing in the theorems from Sections 2.3, 2.6, 2.9, 2.10.

Indeed, all conditions of the next theorem are satisfied in this domain. This is typical in the justification of the averaging method over some of fast variables.

Theorem (see Grebenikov and Ryabov [1]) *Let:*

(1) *the partial derivatives of the disturbing function R_1 with respect to all variables (in total there are six phase variables $a_1, \dots, M_1$) be analytic and 2π-periodic with respect to M_1, $l_2 = n_2 t + l_{20}$ in the domain G_{4+2};*
(2) *the average motions n_1, n_2 be analytic in G_{4+2};*
(3) *the norms of all partial derivatives of the disturbing function $\|\partial R_1/\partial a_1\|$, $\|\partial R_1/\partial l_1\|$, ..., $\|\partial R_1/\partial M_1\|$, and also n_1, n_2 be bounded in G_{4+2};*

(4) *the following condition be satisfied:*

$$C_1^{-1}\mu < \left| k_1 \frac{dn_1}{dt} + k_2 \frac{dn_2}{dt} \right| < C_1\mu, \qquad (3.2.20)$$

where $1 \leq \|k\| = |k_1| + |k_2| \leq N$.

Then there exists $\mu_0 \in [0,\, 1]$, *such that for all* $0 \leq \mu \leq \mu_0$, $0 \leq t \leq \mu^{-1}$ *the following estimation is true*

$$|a_1(t) - \bar{a}_1(t)| + |e_1(t) - \bar{e}_1(t)| + \left|i_1(t) - \bar{i}_1(t)\right| < C_2\sqrt{\mu}\ln^2\mu, \qquad (3.2.21)$$

where $\bar{a}_1(t)$, $\bar{e}_1(t)$, $\bar{i}_1(t)$ *are projections of the solution of the averaged system of differential equations of the bounded circular three-body problem.*

The detailed proof for the case of $(m+n)$-dimensional multifrequency system can be found in Grebenikov and Ryabov [1]; therefore here we will comment upon the conditions of the theorem for the bounded circular three-body problem. Conditions (1) – (3) are evidently satisfied outside the quasicylindrical Hill surface because this domain does not contain bodies with nonzero masses. Consequently, in this part of configuration space the Newtonian potential of the problem (the function $m_0/r_0 + \mu/\Delta_{12}$) is a harmonic function, and the disturbing function R_1 is an analytic function of six variables a_1, e_1, i_1, ω_1, M_1, $\tilde{l}_2$, hence it follows that its partial derivatives $\partial R_1/\partial M_1, \ldots, \partial R_1/\partial\tilde{l}_2$ are also analytical functions of their own arguments. The following quantity can be taken as a constant C bounding the norms $\|\partial R_1/\partial M_1\|, \ldots, \|\partial R_1/\partial\tilde{l}_2\|$, $|n_1| + |n_2|$ (see Grebenikov and Ryabov [1]):

$$C = \frac{3\sqrt{3}}{\inf\limits_{s\in S} \rho[S,\mu]}$$

where the denominator denotes the exact lower edge of the distance between any point of the surface $s \in S$ and the attracting body with mass μ.

Then, the motion of an infinitely small mass continues perpetually outside the surface S if it was there at the initial time, and consequently, the condition $(a_1(t), \ldots, M_1(t)) \in G_{4+2}$ with $t \in (-\infty, \infty)$, necessary for the proof of the theorem, is also satisfied. It remains to check condition (4) of the theorem.

First of all note that in the bounded three-body problem $n_2 = \text{const}$, so $dn_2/dt = 0$. In view of this, the condition (4) can be written as

$$C_1^{-1}\mu < \left| k_1 \frac{dn_1}{dt} \right| < C_1\mu, \quad 1 \leq k_1 \leq N. \qquad (3.2.22)$$

Since $n_1 = a_1^{-3/2}$, then $dn_1/dt = -\dfrac{3}{2}a_1^{-5/2}da_1/dt$, so inequality (3.2.22) with consideration for the first equation of system (3.2.3) will take the form

$$C_1^{-1}\mu < \frac{3}{2}\,a_1^{-2}\left| k_1 \frac{\partial R_1}{\partial M_1} \right| < C_1\mu. \qquad (3.2.23)$$

But for the domain G_{4+2} the inequality $a_1 > a_2$ is obvious, i.e. the major semiaxis of the orbit of an infinitely small mass is always larger than the major semiaxis of the orbit of a disturbing body with mass μ; therefore condition (3.2.23) is always satisfied provided N is a finite number.

So we have shown that the Jacobi integral of the bounded circular three-body problem (not only in the plane version, but in the space version as well) allows us to divide the three-dimensional Euclidean configuration space into connected three-dimensional domains, including those for which all the conditions of the above theorem are satisfied. Consequently for such domains the solutions of equations averaged as per Delaunay–Hill with $\mu \to 0$ are arbitrarily close in norm to the solutions of the exact equations.

For this problem the following result is also correct. For some commensurabilities of average motions within an infinite time interval, and consequently even more so in the interval $0 \le t \le \mu^{-1}$, the average motion (frequency $n_1(\tilde{a}_1)$) differs from its initial values by a value of order μ:

$$n_1(\tilde{a}_1) = n_1\big(a_1^{(0)}\big) + \mu\beta(\bar{a}_1(t)), \tag{3.2.24}$$

where β is a function bounded for all t. Condition (3.2.24) means that in the plane bounded circular three-body problem the norm $|\partial n_1/\partial \tilde{a}_1|$ may be a value of order μ.

We calculate these resonances. The plane bounded circular problem averaged as per Delaunay–Hill has first integral (see Grebenikov and Ryabov [1])

$$\bar{a}_1(t) = a_{10}\frac{\left(k_2^* - k_1^*\sqrt{1-e_{10}^2}\right)^2}{\left(k_2^* - k_1^*\sqrt{1-\bar{e}_1^2(t)}\right)^2}, \tag{3.2.25}$$

where k_1^*, k_2^* are positive integers satisfying the resonance condition $k_1^* n_1(a_{10}) - k_2^* n_2 = 0$ (a_{10}, e_{10} are the initial values of the major semiaxis and the eccentricity, corresponding to the resonance). In addition, if the motion of an infinitely small mass is stable in the sense of Lagrange–Hill (see Grebenikov and Ryabov [1]), i.e. the conditions $h < h_{L_2}$, $a_{10} < R_{L_2}$ are satisfied (h is a constant of the Jacobi integral for the given orbit, h_{L_2} is a constant of the Jacobi integral for the libration point L_2, and R_{L_2} is the barycentric distance of the point L_2), then within an infinite interval the function $\tilde{a}_1(t)$ is bounded, and

$$\sup_{t\in[0,\infty)} \bar{a}_1(t) = a_{10}\frac{\left(k_2^* - k_1^*\sqrt{1-e_{10}^2}\right)^2}{\left(k_2^* - k_1^*\right)^2},$$

$$\inf_{t\in[0,\infty)} \bar{a}_1(t) = a_{10}\frac{\left(k_2^* - k_1^*\sqrt{1-e_{10}^2}\right)^2}{k_2^{*2}}.$$

From this it follows that the maximum amplitude of the change of the major axis with $t \in [0, \infty)$ equals

$$\sup_{t\in[0,\infty)} \Delta\bar{a}(t) = \frac{a_0}{2}\left[\frac{\left(k_2^* - k_1^*\sqrt{1-e_0^2}\right)^2}{(k_2^* - k_1^*)^2} - \frac{\left(k_2^* k_1^*\sqrt{1-e_0^2}\right)^2}{{k_2^*}^2}\right].$$

From the above formula it follows that there exists a resonance for which we always have:

$$\sup_{t\in[0,\infty)} \Delta\bar{a}(t) = O(\mu).$$

This is true if $k_1^*/k_2^* \sim \mu$. For such k_1^* and k_2^*, and all $t \geq 0$, $\tilde{a}(t) = a_0 + O(\mu)$, and consequently

$$n_1(\tilde{a}(t)) = n_1(a_0) + O(\mu). \tag{3.2.26}$$

For these resonances the system of differential equations of the plane bounded circular three-body problem does not contain fast variables (of type y in the standard notation of multifrequency systems), i.e. it is a standard system in Bogolyubov's sense. And this in its turn means that for the justification of the averaging principle in the case in question it is possible to use numerous well-known results (see e.g. Bogoliubov and Mitropolsky [1], Bogoliubov, *et al.* [1], Grebenikov and Ryabov [1], Martynyuk [1], Volosov and Morgunov [1], etc.).

3.3 Construction of Explicit Solutions of Averaged Differential Equations of the Bounded Three-Body Problem in the Case of Resonance

In Section 3.1 we gave the complete system of first integrals (3.1.23) of averaged differential equations of the fourth-order plane bounded circular three-body problem. Relations (3.1.23) express the implicit dependence of the sought variables $\bar{a}(t)$, $\bar{p}(t)$ (or $\bar{e}(t)$), $\bar{\omega}(t)$ and $D(t)$ on time, and for the construction of disturbed asymptotic theory it is necessary to have the variables $\bar{a}(t)$, $\bar{p}(t)$, $\bar{\omega}(t)$, $D(t)$ in the form of explicit functions of t, i.e. it is necessary to know $\bar{a} = \bar{a}(t)$, $\bar{p} = \bar{p}(t)$, $\bar{\omega} = \bar{\omega}(t)$, $D = D(t)$.

To solve this problem we will first of all denote the "focal parameter" $\bar{p}$ by the major semiaxis $\bar{a}$, using the first equality of system (3.1.23):

$$\bar{p} = \frac{{k_2^*}^2}{{k_1^*}^2}\left(\bar{a} + C_1^2 - 2C_1\sqrt{\bar{a}}\right). \tag{3.3.1}$$

Substituting this expression into the second relation of system (3.1.23), we obtain

$$F(\bar{a}, D) = C_2\bar{a}, \tag{3.3.2}$$

where $F(\bar{a}, D)$ is an even function with respect to D (a trigonometric series of multiple argument D, with its coefficients being power series of $\sqrt{\bar{a}}$). This can be written in the form

$$F(\bar{a}, D) = \sum_{s=0}^{\infty} F_s(\bar{a}) \cos sD, \tag{3.3.3}$$

where the coefficients $F_s(\bar{a})$ are analytic functions with respect to $\sqrt{\bar{a}}$ and are expressed via so-called Laplace coefficients (see Duboshin [1]). In this case equality (3.3.2) may be given the form

$$\sum_{s=0}^{\infty} F_s(\bar{a}) \cos sD = C_2 \bar{a}. \tag{3.3.4}$$

We describe two methods of solving problem (3.3.4) with respect to $\bar{a}$.

Method I Write $\bar{a}$ in the form of a trigonometric series

$$\bar{a}(D) = \sum_{s=0}^{\infty} a_s \cos sD \tag{3.3.5}$$

with unknown coefficients a_s. Substituting (3.3.5) into (3.3.4) and using binomial expansions for the functions $\bar{a}^{r/2}$ (r is an odd number), it is possible to deduce an infinite system of nonlinear algebraic equations for the unknown coefficients $a_0, a_1, \ldots$. To solve this system in an analytic form seems practically impossible, so the most effective method is an iteration one. Due to the awkwardness of the system we do not give it here, and the interested reader is referred to the literature (see Karaganchu [1], Prikhod'ko [1]).

Method II Write $\bar{a}$ as a power series of the small parameter μ (here the disturbing mass of Jupiter is usually considered as such):

$$\bar{a}(D) = \sum_{s=0}^{\infty} \mu^s b_s(D). \tag{3.3.6}$$

The coefficient b_0 is calculated from

$$m_0 + \frac{2k_2^* n_2^{(0)}}{k_1^*} \sqrt{m_0} b_0^{3/2} - 2C_2 b_0 = 0; \tag{3.3.7}$$

therefore it does not depend on D. The other coefficients $(b_1, b_2, \ldots)$ are calculated successively in the form of trigonometric series of multiples of D. After that the series (3.3.6) can be reconstructed in the form (3.3.5). This allows us to conclude that the form (3.3.5) is preferable.

Supposing that the coefficients of the series (3.3.5) have already been found, we state the algorithm for calculation of other sought functions in an explicit form. If we can write the focal parameter $\bar{p}$ as the trigonometric series

$$\bar{p}(D) = \sum_{s=0}^{\infty} p_s \cos sD \tag{3.3.8}$$

then the coefficients p_s are determined from the system

$$
\begin{aligned}
k_1^{*2} p_0 &= k_2^{*2} a_0 + k_2^{*2} C_1 - 2K_2^{*2} a_0^{1/2} + \dots, \\
k_1^{*2} p_1 &= k_2^{*2} a_1 - k_2^{*2} C_1 a_0^{-1/2} + \dots, \\
k_1^{*2} p_2 &= k_2^{*2} a_2 - k_2^{*2} C_1 a_0^{-1/2} + \cdots, \\
&\dots\dots\dots\dots\dots\dots
\end{aligned}
\tag{3.3.9}
$$

Equations (3.3.9) are obtained from the first integral (3.3.1) if $\bar{a}$ is replaced by series (3.3.5) and identical equalities are written for the coefficients at equal trigonometric functions. System (3.3.9) has actually been solved with respect to the unknown $p_0, p_1, \dots$ and so their calculation is trivial. Using the correlation $\bar{p} = \bar{a}(1 - \bar{e}^2)$, it is possible to express the "eccentricity" $\bar{e}$ as a trigonometric series of multiples of D:

$$
\bar{e}(D) = \sum_{s=0}^{\infty} e_s \cos sD. \tag{3.3.10}
$$

The coefficients e_s are calculated from the system of algebraic equations

$$
\begin{aligned}
&a_0^{1/2} e_0 + \frac{a_0^{-1/2} a_1}{4} e_1 + \frac{a_0^{-1/2} a_2}{4} e_2 + \dots = (a_0 - p_0)^{1/2}, \\
&\frac{a_0^{-1/2} a_1}{4} e_0 + a_0^{1/2} e_1 + \frac{a_0^{-1/2} a_1}{4} e_2 + \dots = \frac{1}{2}(a_0 - p_0)^{-1/2}(a_1 - p_1), \\
&\frac{a_0^{-1/2} a_2}{4} e_0 + \frac{a_0^{-1/2} a_1}{4} e_1 + a_0^{1/2} e_2 + \dots = \frac{1}{2}(a_0 - p_0)^{-1/2}(a_2 - p_2), \\
&\dots\dots\dots\dots\dots\dots\dots\dots\dots\dots
\end{aligned}
\tag{3.3.11}
$$

The coefficients n_s of the trigonometric series

$$
\bar{n}(D) = \sum_{s=0}^{\infty} n_s \cos sD \tag{3.3.12}
$$

for the average motion (frequency) $\bar{n}$ of the disturbed body are determined in a similar way. Formulae are easily written for them, if the known relation $\bar{n} = m_0 \bar{a}^{-3/2}$ is used:

$$
\begin{aligned}
n_0 &= \frac{m_0}{a_0^{3/2}} \left[1 + \frac{15}{16a_0^2} \left(a_1^2 + a_2^2 + \cdots \right) + \cdots \right], \\
n_1 &= \frac{m_0}{a_0^{3/2}} \left[-\frac{3a_1}{2a_0} + \frac{15a_1 a_2}{8a_0^2} + \cdots \right], \\
n_2 &= \frac{m_0}{a_0^{3/2}} \left[-\frac{3a_2}{2a_0} + \frac{15a_1^2}{16a_0^2} + \dots \right], \\
&\dots\dots\dots\dots\dots\dots\dots\dots
\end{aligned}
\tag{3.3.13}
$$

Thus, construction of the explicit analytic dependences $\bar{a}$ and $\bar{p}$ (and consequently $\bar{e}$, $\bar{n}$ as well) on the Delaunay anomaly is the result of the inversion of the first two integrals of system (3.1.23). Now in the third integral of system (3.1.23), we pass from the integration variable t to the new variable D, using the fourth integral:

$$\bar{\omega} - \bar{\omega}_0 = -2 \int_{D_0}^{D} \frac{\sqrt{\bar{a}^3 \bar{p}}\, \dfrac{\partial \bar{\bar{R}}}{\partial \bar{p}}\, dD}{k_1^* m_0 - k_2^* n_2^{(0)} \sqrt{\bar{a}^3 m_0} - 2k_1^* \bar{a}^2 \dfrac{\partial \bar{\bar{R}}}{\partial \bar{a}} - 2k_2^* \sqrt{\bar{a}^3 \bar{p}}\, \dfrac{\partial \bar{\bar{R}}}{\partial \bar{p}}}. \tag{3.3.14}$$

The integrand can be written as the ratio of two trigonometric cosine series:

$$\bar{\omega} - \bar{\omega}_0 = \int_{D_0}^{D} \frac{\sum\limits_{s=0}^{\infty} c_s \cos sD}{\sum\limits_{s=0}^{\infty} d_s \cos sD}\, dD. \tag{3.3.15}$$

Performing the operation of division of a series by a series and integrating the result, we obtain

$$\bar{\omega} - \bar{\omega}_0 = A_0^{(0)} + A_0 D + A_1 \sin D + A_2 \sin 2D + \ldots, \tag{3.3.16}$$

where the coefficients $A_0, A_1, A_2, \ldots$ are determined from the system of equations

$$\begin{aligned} d_0 A_0 + \frac{d_1}{2} A_1 + \frac{d_2}{2} A_2 + \ldots &= c_0, \\ d_1 A_0 + \left(d_0 + \frac{d_2}{2}\right) A_1 + \frac{d_1}{4} A_2 + \ldots &= c_1, \\ d_2 A_0 + \frac{d_1}{2} A_1 + \frac{d_0}{2} A_2 + \ldots &= c_2, \\ \ldots\ldots\ldots\ldots\ldots\ldots\ldots\ldots\ldots\ldots\ldots \end{aligned} \tag{3.3.17}$$

When the coefficients A_s are determined, the absolute term is determined from the initial conditions

$$A_0^{(0)} = -A_0 D_0 - A_1 \sin D_0 - A_2 \sin 2D_0 - \cdots \tag{3.3.18}$$

Finally, the integrand in the fourth equation of (3.1.23) can be written as the trigonometric series

$$t - t_0 = \int_{D_0}^{D} \sum_{s=0}^{\infty} B_s \cos sD\, dD. \tag{3.3.19}$$

If equations (3.3.17) for the coefficients A_s are already available, they can also be used for the calculation of the coefficients B_s. For this it is sufficient to replace the quantities $c_0, c_1, \ldots$ in (3.3.17) by the coefficients of the expansion $\bar{a}^{3/2}$ into a trigonometric series of multiple argument D, and the quantities $A_0, A_1, \ldots$ by $B_0, B_1, \ldots$ respectively.

Integrating (3.3.19), we obtain

$$t - t_0 = B_0^{(0)} + B_0 D + B_1 \sin D + B_2 \sin 2D + \cdots \qquad (3.3.20)$$

Relation (3.3.20) can be called a generalized Kepler equation for an averaged version (by Delaunay and Hill) of the plane bounded circular three-body problem, because, if one substitutes into it the small parameter $\mu = 0$ (the mass of the disturbing body, Jupiter) and changes D by the average anomaly $\overline{M}$, it is finally reduced to the Kepler equation. The average anomaly $\overline{M}$ of an averaged system is determined from the equality

$$\overline{M} = \frac{D}{k_1^*} - \frac{k_2^*}{k_1^*}(\bar{\omega} - l_2) . \qquad (3.3.21)$$

Using formulae (3.3.16), (3.3.20) and the linear dependence of the average longitude of Jupiter l_2 on time t, we finally obtain the explicit dependence of the average anomaly $\overline{M}$ on the argument D:

$$\overline{M} = M_0^{(0)} + M_0 D + M_1 \sin D + M_2 \sin 2D + \cdots \qquad (3.3.22)$$

Thus, the complete system of first integrals for (3.1.23) allows us to construct in an explicit and analytic form the exact solution of the averaged equations of the bounded circular three-body problem, when all three bodies (the central body with mass m_0, the disturbing body with mass μ, and the disturbed body with zero mass) always move in one and the same fixed plane (the so-called plane version of the bounded three-body problem).

The analytic representations (3.3.5), (3.3.8), (3.3.10) show that the phase variables $\bar{a}(D)$, $\bar{p}(D)$, $\bar{e}(D)$ determining the form and dimensions of the averaged (staging) orbit are changed in an oscillatory, but not a secular way, because there are no terms proportional to D in them. This circumstance is essential for asymptotic perturbation theory in celestial mechanics, and it allows us to draw the conclusion that asymptotic methods based on the application of averaging schemes for obtaining the first approximation are more effective in comparison with other methods of perturbation theory. Indeed, if one constructs an asymptotic theory of the motion of celestial bodies on the basis of the two-body problem as a first approximation, then in the next approximation the eccentricity $e(t)$ will contain a secular term proportional to t, which seems illogical.

In the general case $A_0 \neq 0$, therefore the phase variable $\bar{\omega}$ that determines the orientation of the staging orbit in coordinate space, as (3.3.16) shows, changes in a secular way, and an oscillation is superposed on this secular shift of the major axis of the trajectory. This oscillation is periodic with respect to D.

From this it follows that the staging orbit (first approximation) has a "peripleg-matic" nature and everywhere densely fills (for nearly all initial conditions in the sense of Lebesgue) a ring of radii $\inf \bar{a}(t)$ and $\sup \bar{a}(t)$ determined in Section 3.2.

In conclusion note that the explicit dependences of the phase coordinates $\bar{a}$, $\bar{p}$, $\bar{\omega}$, $\overline{M}$ on the Delaunay anomaly D can also be used for the study of the dynamics of concrete asteroids with their orbits having a small inclination, if in the first approximation it is considered that $i = i_0$ and $\Omega = \Omega_0$. Then for the phase variables $\bar{\pi}$ and $\bar{\varepsilon}$ that may replace the coordinates $\bar{\omega}$ and $\overline{M}$ we have

$$\bar{\pi} = \bar{\omega} + \Omega_0, \qquad \bar{\varepsilon} = \bar{\omega} + \overline{M} - \int_{t_0}^{t} \bar{n}\, dt + \Omega_0.$$

In these formulae, $\bar{\omega}, \overline{M}$ and $\bar{n}$ should be replaced by the corresponding series of the multiple argument D.

3.4 Quasi-Periodic Solutions of Resonance Hamiltonian Systems

Let the motion of a dynamical model be described by the canonical (Hamiltonian) system of differential equations

$$\frac{dp}{dt} = -\frac{\partial H}{\partial q}, \quad \frac{dq}{dt} = \frac{\partial H}{\partial p}, \tag{3.4.1}$$

where the Hamiltonian H has the form

$$H(p, q, \mu) = H_0(p) + \mu H_1(p.q), \tag{3.4.2}$$

$p = (p_1, \dots, p_n)$, $q = (q_1, \dots, q_n)$ are n-dimensional vectors, μ is a small positive parameter, and the function $H_1(p, q)$ is 2π-periodic with respect to $q_1, \dots, q_n$. Suppose that the Hamiltonian H is analytic in a complex $2n$-dimensional domain

$$G_{2n} = \{\operatorname{Re} p \in P_n, \quad \|\operatorname{Im} p\| < \rho, \quad \|\operatorname{Im} q\| < \sigma\}. \tag{3.4.3}$$

In addition, let the functions H_0 and H_1 be such that they take real values at real values of the arguments.

In mechanics (see Poincaré [1], and Ter Chaar [1]) the impulse vector p is often interpreted as a vector of positional variables, and the vector of generalized Lagrange coordinates q as a vector of angular velocities. Many problems of non-linear mechanics (including celestial mechanics) are related to so-called degenerate case, when the undisturbed part of the Hamiltonian — the function H_0 — only depends on the part n_0 $(n_0 \leq n)$ of positional variables $p_1, p_2, \dots, p_{n_0}$. Such degeneration complicates the tools of asymptotic theory, because some angle variables are slow (like the impulses p). Indeed, dividing the vectors p and q into components

$$p_0 = (p_1, \dots, p_{n_0}), \qquad q_0 = (q_1, \dots, q_{n_0}),$$
$$p^* = (p_{n_0+1}, \dots, p_n), \quad q^* = (q_{n_0+1}, \dots, q_n),$$

we write the Hamiltonian H as

$$H(p, q) = H_{00}(p_0) + \mu H_1(p, q). \tag{3.4.4}$$

Then, obviously,

$$\det\left(\frac{\partial^2 H_{00}}{\partial p^2}\right) \equiv 0, \tag{3.4.5}$$

which corresponds to the case of so-called proper degeneration (see Arnol'd [1], and Grebenikov and Ryabov [1]). Suppose that in G_{2n}

$$\det\left(\frac{\partial^2 H_{00}}{\partial p_0^2}\right) \neq 0, \quad \det\left(\frac{\partial^2 H_1}{\partial p^2}\right) \neq 0.$$

Taking the above equation into account, (3.4.1) can be written as

$$\begin{aligned} \frac{dp_0}{dt} &= -\mu\frac{\partial H_1}{\partial q_0}, & \frac{dq_0}{dt} &= \frac{\partial H_{00}}{\partial p_0} + \mu\frac{\partial H_1}{\partial p_0}, \\ \frac{dp^*}{dt} &= -\frac{\partial H_1}{\partial q^*}, & \frac{dq^*}{dt} &= \mu\frac{\partial H_1}{\partial p^*}. \end{aligned} \tag{3.4.6}$$

From (3.4.6) it is clear that the vector q^* is a vector of slow angle variables, because $\|\dot{q}^*\| \sim \mu$. Let the disturbing part $H_1(p,q)$ of the Hamiltonian $H(p,q,\mu)$ be expanded into the cosine Fourier series

$$H_1(p,q) = \sum_{\|k\|>0} h_k(p)\cos\left[(k_0,q_0) + (k^*,q^*) + l_k\right], \tag{3.4.7}$$

where $k = (k_0,k^*)$, $k_0 = (k_1,\dots,k_{n_0})$, $k^* = (k_{n_0+1},\dots,k_n)$, $\|k\| = |k_1| + \cdots + |k_n|$, $k_s = 0,1,2,\dots$, and l_k is a constant.

The problem we have to solve is finding quasi-periodic and stationary solutions of systems (3.4.1) or (3.4.6) in the case when there is a resonance among the basic frequencies $\partial H_{00}/\partial p_1$, $\partial H_{00}/\partial p_2$, $\dots$, $\partial H_{00}/\partial p_{n_0}$ of an undisturbed system (with $\mu = 0$). Note that if $n_0 = n$, all angle variables q are fast phase variables, and then the matter concerns the resonance of all undisturbed frequencies (in the usual notation $\omega_1 = \partial H_{00}/\partial p_1$, $\omega_2 = \partial H_{00}/\partial p_2$, $\dots$, $\omega_n = \partial H_{00}/\partial p_n$).

For this purpose we will apply the well-known method of canonical substitutions of variables (see Arnol'd [1], and Grebenikov and Ryabov [1]), and at each stage of the transformations we will consider the presence of resonances found by means of the averaging operator for continuous disturbances, or, which is the same in this case, by means of Delaunay–Hill averaging.

First we will construct a canonical substitution of variables such that the disturbing part H_1 of the Hamiltonian in the new variables has the special form

$$H_1(p,q) = \overline{H}_1(p,q^*) + \widetilde{H}_1(p,q), \tag{3.4.8}$$

where $\overline{H}_1(p,q^*)$ is an integral average value of the function H_1 over the rapid angle variables q_0 within a parallelepiped of periods $2\pi \times \cdots \times 2\pi$, and $\widetilde{H}_1(p,q)$ is the periodic part H_1 depending on the angle variables q:

$$\overline{H}_1(p,q^*) = \frac{1}{(2\pi)^{n_0}} \int_0^{2\pi} \cdots \int_0^{2\pi} H_1(p,q)\,dq_0. \tag{3.4.9}$$

Formula (3.4.9) gives the averaged value of H_1 taking no account of frequency commensurability; therefore we should take this fact into consideration in the analytic structure. From $\widetilde{H}_1$ extract the corresponding partial sum $[\widetilde{H}_1]_N$ such that

$$\widetilde{H}_1 = [\widetilde{H}_1]_N + R_N H_1,$$

where $R_N H_1$ is the remainder term of the trigonometric series, and N is a positive integer bounding the norm $\|k\|$. We obtain

$$H_1(p,q) = \overline{H}_1(p,q^*) + [\widetilde{H}_1(p,q)]_N + R_N H_1, \tag{3.4.10}$$

where

$$\overline{H}_1(p,q^*) = \sum_{1\le\|k\|\le N} h_k(p)\cos\left[(k^*,q^*) + l_k\right], \tag{3.4.11}$$

$$[\widetilde{H}_1(p,q)]_n = \sum_{1\le\|k\|\le N} h_k(p)\cos\left[(k_0,q_0) + (k^*,q^*) + l_k\right], \tag{3.4.12}$$

$$R_N H_1 = \sum_{\|k\|>N} h_k(p)\cos\left[(K_0,q_0) + (k^*,q^*) + l_k\right]. \tag{3.4.13}$$

In accordance with the general idea of the method of successive canonical transformation, during the construction of quasi-periodic solutions it is necessary to introduce a canonical substitution of variables $(p,q)\to(P,Q)$ such that in the new variables the disturbing part of the Hamiltonian — the function H_1 — should take the form

$$H_1 = \overline{H}_1 + R_N H_1, \tag{3.4.14}$$

i.e. in the new variables the periodic part $[\widetilde{H}_1]_N$ should become zero. This can be achieved through a canonical substitution determined, e.g., by the formulae

$$p = P + \frac{\partial S}{\partial q}, \qquad Q = q + \frac{\partial S}{\partial P}, \tag{3.4.15}$$

where the generating function $S(p,q)$ will be chosen in a special way. According to canonical transformations theory, equation (3.4.1) can be replaced by the new canonical system

$$\frac{dP}{dt} = -\frac{\partial K}{\partial Q}, \qquad \frac{dQ}{dt} = \frac{\partial K}{\partial P} \tag{3.4.16}$$

with the Hamiltonian

$$K \equiv H = H_0 + \mu\overline{H}_1 + \mu[\widetilde{H}_1]_N + R_N H_1, \tag{3.4.17}$$

where the substitution of variables $(p,q)\to(P,Q)$ is performed. The generating function $S(P,q)$ can be chosen by an infinite number of methods, e.g. in such a way that

$$\left(\frac{\partial H_0^{(1)}}{\partial P}, \frac{\partial S}{\partial q}\right) + \mu[\widetilde{H}_1(P,q)]_N = 0, \tag{3.4.18}$$

$$H_0^{(1)} = H_{00}(P_0) + \mu\overline{H}_1(P,q^*). \tag{3.4.19}$$

One of the solutions of equation (3.4.18) has the form

$$S(P,q) = \mu \sum_{1 \le \|k\| \le N} S_k(P) \sin\left[(k_0, q_0) + (k^*, q^*) + l_k\right], \tag{3.4.20}$$

$$S_k(P) = -\frac{h_k(P)}{(k,\, \partial H_0^{(1)}/\partial P)}. \tag{3.4.21}$$

For a system with resonances, the denominators in (3.4.21) for some k can be very small or even zero, and we get stuck on the problem of "small denominators". Therefore in the case of sharp commensurability the substitution of variables (3.4.15) with the generating function (3.4.20), (3.4.21) does not bring desirable results, i.e. without additional rather strict conditions it is impossible to transform the disturbing part of the Hamiltonian into the form (3.4.8) – (3.4.13) and destroy the periodic part $\widetilde{H}_1$ of the Hamiltonian.

So the widespread method of the construction of successive canonical substitutions with the use of averaging over fast angle variables as a preliminary transformation can be found unfit in the case of resonance. This difficulty could be overcome if, for example, one follows the following algorithm.

Let the system in question be such that with some fixed vector $\overline{k}$ (where $\overline{k} = (\overline{k}_0, \overline{k}^*)$, $\overline{k}_0 = (\overline{k}_1, \ldots, \overline{k}_{n_0})$, $\overline{k}^* = (\overline{k}_{n_0+1}, \ldots, \overline{k}_n)$) with its norm not exceeding N, there is a resonance between the derivatives of the components of fast angle variables q_0 which are linear time functions in the undisturbed motion. Introduce a new angle variable — the so-called critical argument (known in celestial mechanics as the Delaunay anomaly) — by the formula

$$d = \left(\overline{k}_0, q_0\right). \tag{3.4.22}$$

It can be shown that with this substitution of variables when instead of one fast angle variable (e.g. component q_0) a slow variable d is introduced, the canonical nature of system (3.4.1) remains, but there is a redistribution between the number of fast and slow angle variables.

For the sake of simplicity assume that $\overline{k}_1 \neq 0$. Then from (3.4.22) it is possible to express the fast component q_1 in the form

$$q_1 = \frac{d}{\overline{k}_1} - \frac{1}{\overline{k}_1}\left(\overline{k}_2 q_2 + \cdots + \overline{k}_{n_0} q_{n_0}\right). \tag{3.4.23}$$

Excluding the component q_1 from the Hamiltonian H_1, we obtain

$$\begin{aligned} H_1(p,q) &= H_1^*(p, d, q_2, \ldots, q_n) \\ &= \overline{H}_1^*(p, d, q^*) + [\widetilde{H}_1^*(p, d, q_2, \ldots, q_n)]_N + R_N H_1^*. \end{aligned} \tag{3.4.24}$$

Here the "secular" part $\overline{H}_1^*$ is calculated as an integral average over the variables $q_2, q_3, \ldots, q_{n_0}$, i.e.

$$\overline{H}_1^*(p, d, q^*) = \frac{1}{(2\pi)^{n_0 - 1}} \int_0^{2\pi} \cdots \int_0^{2\pi} H_1^*(p, d, q_2, \ldots, q_n)\, dq_2 \cdots dq_{n_0}. \tag{3.4.25}$$

Now we will show that the introduction of the critical argument d does not upset the canonical character of the system (3.4.1) or (3.4.6). For this purpose, separate the equations for q_1 and for the corresponding impulse p_1, i.e. rewrite system (3.4.6) in the form

$$\begin{aligned} \frac{dp_1}{dt} &= -\frac{\partial H}{\partial q_1}, & \frac{dq_1}{dt} &= \frac{\partial H}{\partial p_1}, \\ \frac{dp_0''}{dt} &= -\frac{\partial H}{\partial q_0''}, & \frac{dq_0''}{dt} &= \frac{\partial H}{\partial p_0''}, \\ \frac{dp^*}{dt} &= -\frac{\partial H}{\partial q^*}, & \frac{dq^*}{dt} &= \frac{\partial H}{\partial p^*}, \end{aligned} \tag{3.4.26}$$

where $p_0'' = (p_2, \dots, p_{n_0})$, $q_0'' = (q_2, \dots, q_{n_0})$.

Perform the substitution of variables $(p \equiv (p_1, p_0'', p^*),\ q \equiv (q_1, q_0'', q^*)) \to (P \equiv (P_1, P_0'', P^*),\ Q \equiv (Q_1, Q_0'', Q^*))$, using the generating function

$$V_2(P, q) = \sum_{s=1}^{n} P_s f_s, \tag{3.4.27}$$

where

$$f_1 = d = \left(\overline{k}_0, q_0\right), \quad f_m = q_m, \quad m = 2, \dots, n. \tag{3.4.28}$$

This generating function determines the canonical transformation if

$$\det\left(\frac{\partial^2 V_2}{\partial P_s \partial q_s}\right) = \det\left(\frac{\partial f_s}{\partial q_s}\right) \neq 0. \tag{3.4.29}$$

Calculations show that

$$\left\|\frac{\partial f_s}{\partial q_s}\right\| = \left|\overline{k}_a\right| \neq 0. \tag{3.4.30}$$

Then according to the theorem on canonical transformations by Poincaré [1] the explicit form of a canonical transformation is determined by the relations

$$\begin{aligned} Q_s &= \frac{\partial V_2}{\partial P_s} = f_s(q_1, \dots, q_n), \\ p_s &= \sum_{j=1}^{n} P_j \frac{\partial f_j}{\partial q_s}, \quad s = 1, \dots, n. \end{aligned} \tag{3.4.31}$$

From equalities (3.4.31) it is possible to express new canonical variables (P, Q) through the old (p, q):

$$\begin{aligned} &Q_1 = d, \quad Q_m = q_m, \quad m = 2, \dots, n, \\ &P_1 = \frac{p_1}{\overline{k}_1}, \quad P_j = p_j - \frac{\overline{k}_j}{\overline{k}_1} p_1, \quad j = 2, \dots, n_0, \\ &P_s = p_s, \quad s = n_0 + 1, \dots, n. \end{aligned} \tag{3.4.32}$$

So the system of canonical equations (3.4.26) is transformed by the substitution of variables (3.4.22) into the new Hamiltonian system

$$\frac{dP_s}{dt} = -\frac{\partial R}{\partial Q_s}, \qquad \frac{dQ_s}{dt} = \frac{\partial R}{\partial P_s}, \tag{3.4.33}$$

where $R(P, Q) = H(p(P), q(Q))$.

Considering the above transformation fulfilled, retaining the old symbols for the new canonical variables, we write the new system in the expanded form:

$$\begin{aligned} \frac{dp_1}{dt} &= -\frac{\partial H^*}{\partial d}, & \frac{dd}{dt} &= \frac{\partial H^*}{\partial p_1}, & \frac{dp_0''}{dt} &= -\frac{\partial H^*}{\partial q_0''}, \\ \frac{dq_0''}{dt} &= \frac{\partial H^*}{\partial p_0''}, & \frac{dp^*}{dt} &= -\frac{\partial H^*}{\partial q^*}, & \frac{dq^*}{dt} &= \frac{\partial H^*}{\partial p^*}. \end{aligned} \tag{3.4.34}$$

Now we will show that the introduction of a critical argument allows the removal of some small denominators. The explicit expression for the periodic part $[\widetilde{H}_1^*(p, d, q_2, \ldots, q_n)]_N$ has the form

$$[\widetilde{H}_1^*(p, d, q_2, \ldots, q_n)]_N = \sum_{1 \le \|k\| \le N}'' h_k(p) \cos\,[k_1'' d + (k_0'', q_0'') + (k^*, q^*) + l_k]. \tag{3.4.35}$$

Primes on the sum mean, as before, that the vector k passes "non-resonance" values only. Introduce new canonical variables $\widetilde{P} = (P_1, P_0'', P^*)$, $\widetilde{Q} = (D, Q_0'', Q^*)$ by the formulae

$$\begin{aligned} p_1 &= P_1 + \frac{\partial S}{\partial d}, & D &= d + \frac{\partial S}{\partial P_1}, \\ p_0'' &= P_0'' + \frac{\partial S}{\partial q_0''}, & Q_0'' &= q_0'' + \frac{\partial S}{\partial P_0''}, \\ p^* &= P^* + \frac{\partial S}{\partial q^*}, & Q^* &= q^* + \frac{\partial S}{\partial P^*}, \end{aligned} \tag{3.4.36}$$

where $S(\widetilde{P}, \widetilde{q})$ is the generating function of this transformation, $\widetilde{q} = (d, q_0'', q^*)$. This is determined from the partial differential equation

$$\left(\frac{\partial H_0^{*(1)}}{\partial \widetilde{P}}, \frac{\partial S}{\partial \widetilde{q}}\right) + \mu[\widetilde{H}_1^*(\widetilde{P}, d, q_2, \ldots, q_n)]_N = 0 \tag{3.4.37}$$

with the notation

$$H_0^{*(1)} = H_0(\widetilde{P}) + \widetilde{H}_1^*(\widetilde{P}, d, q^*). \tag{3.4.38}$$

One of the solutions of equation (3.4.37) has the form

$$S(\widetilde{P}, \widetilde{q}) = \mu \sum_{1 \le \|k\| \le N}'' S_k(\widetilde{P}) \sin\,[k_1'' d + (k_0'', q_0'') + (k^*, q^*) + l_k], \tag{3.4.39}$$

$$S_k(\widetilde{P}) = -\frac{h_k(\widetilde{P})}{(k,\ \partial H_0^{*(1)}/\partial \widetilde{P})}. \tag{3.4.40}$$

Since in the latter equation the vector k takes nonresonance values only, the denominator $\left(k,\, \partial H_0^{*(1)}/\partial\widetilde{P}\right)$, generally speaking, is not small. Now average out the Hamiltonian system (3.4.34) taking account of frequency commensurability:

$$\begin{aligned} &\frac{d\bar{p}_1}{dt} = -\mu\frac{\partial\overline{H}_1^*}{\partial\bar{d}}, && \frac{d\bar{d}}{dt} = \frac{\partial H_{00}^*}{\partial\bar{p}_1} + \mu\frac{\partial\overline{H}_1^*}{\partial\bar{p}_1}, && \frac{d\bar{p}_0''}{dt} = 0, \\ &\frac{d\bar{q}_0''}{dt} = \frac{\partial H_{00}^*}{\partial\bar{p}_0''} + \mu\frac{\partial\overline{H}_1^*}{\partial\bar{p}_0''}, && \frac{d\bar{p}^*}{dt} = -\mu\frac{\partial\overline{H}_1^*}{\partial\bar{q}^*}, && \frac{d\bar{q}^*}{dt} = \mu\frac{\partial\overline{H}_1^*}{\partial\bar{p}^*}. \end{aligned} \tag{3.4.41}$$

As will be shown below, the construction of quasi-periodic solutions is constructively possible when the disturbing part of the Hamiltonian function, averaged as per Delaunay–Hill, either does not depend on the slow angle variables q^*, or can be represented in the form

$$\overline{H}_1^*\left(\widetilde{p}, d, q^*\right) = H_{01}\left(\widetilde{p}\right) + H_{01}^{(\mathrm{res})}\left(\widetilde{p}, d\right) + H_{11}\left(\widetilde{p}, d, q^*\right), \tag{3.4.42}$$

where $H_{01} \sim 1$, $H_{01}^{(\mathrm{res})} \sim 1$, $H_{11} \sim \mu$ and the resonance part of equations (3.4.41) (the first couple for $\bar{p}_1$ and $\bar{d}$, split out from the other equations after truncation of H_{11}) has the stationary solution

$$\bar{p}_1 = p_{10} = \mathrm{const}, \quad \bar{d} = d_{00} = \mathrm{const} \tag{3.4.43}$$

for arbitrary values of the impulses $\bar{p}_0''$ and $\bar{p}^*$.

Let H_{11} be independent of q^*. Then $\partial\overline{H}_1^*/\partial\bar{q}^* \equiv 0$, therefore

$$\bar{p}_0'' = p_{00}'' = \mathrm{const}, \quad \bar{p}^* = p_{00}^* = \mathrm{const}.$$

In this case the first two equations really split out from all the rest, and for the existence of (3.4.43) it is necessary that

$$\begin{aligned} &\frac{\partial\overline{H}_{01}^*}{\partial\bar{d}} = 0, \qquad \frac{\partial H_{00}^*}{\partial\bar{p}_1} + \mu\frac{\partial\overline{H}_{01}^*}{\partial\bar{p}_1} = 0, \\ &\overline{H}_{01}^* = H_{01}\left(\widetilde{p}\right) + H_{01}^{(\mathrm{res})}\left(\widetilde{p}, d\right). \end{aligned} \tag{3.4.44}$$

After that the integration of the remaining equations (3.4.41) amounts to simple quadratures. Finally we obtain the following stationary (equilibrium) solution of system (3.4.41):

$$\begin{aligned} &\bar{p}_1 = p_{10} = \mathrm{const}, && \bar{d} = d_{00} = \mathrm{const}, \\ &\bar{p}_0'' = p_{00}'' = \mathrm{const}, && \bar{p}^* = p_{00}^* = \mathrm{const}, \quad \bar{q}_0'' = q_{00}'' + \omega t, \\ &\omega = \left(\omega_2, \ldots, \omega_{n_0}\right), && \omega = \left.\frac{\partial}{\partial p_0''}\left(H_{00}^* + \mu\overline{H}_{01}^*\right)\right|_{\widetilde{p}=\widetilde{p}_{00}}, \\ &\bar{q}^* = q_{00}^* + \omega^* t, && \omega^* = \left(\omega_{n_0+1}^*, \ldots, \omega_n^*\right), \\ &\omega^* = \left.\mu\frac{\partial\overline{H}_{01}^*}{\partial\bar{p}^*}\right|_{\widetilde{p}=\widetilde{p}_{00}}, && \widetilde{p} = \left(p_1, p_0'', p^*\right). \end{aligned} \tag{3.4.45}$$

The resonance solution (3.4.45) can be given a geometric interpretation similar to that described in the study of Van der Pol equations. Indeed, in two-dimensional phase space (p, q) we introduce n-dimensional tori (or more exactly, toroidal surfaces) $C_1 \times \cdots \times C_n$, considering the quantities $|p_{10}|$, $\|p''_{00}\| = |p''_{1,00}| + \cdots + |p''_{n_0,00}|$, $\|p^*_{00}\| = |p^*_{n_0+1,00}| + \cdots + |p^*_{n,00}|$ as the radii of circumferences $C_1, \dots, C_n$. On these tori we introduce n "geographic" coordinates (longitude and latitude): $\bar{d}$, $\bar{q}''_0$, $\bar{q}^*$. As one of the angular coordinates on the torus is constant ($d = \text{const}$), the motion of an imaginary point is performed either along a generalized parallel on the torus, or along a generalized meridian. The other angle variables ($\bar{q}''_0$ and $\bar{q}^*$) are linear functions of time ($\bar{q}''_0$ are fast functions, $\bar{q}^*$ slow functions).

If the frequencies ω and ω^* are rationally commensurable, then on the $(n-1)$-dimensional manifold (generalized parallel and meridian) we will obtain a periodic solution. If there is no rational commensurability, then the trajectory of a point is a quasi-periodic function of time.

The following question emerges: do quasi-periodic solutions of the initial Hamiltonian system exist such that they are close to the stationary solution of type (3.4.45) of the averaged Hamiltonian system? There is no rigorous mathematical proof of this statement, but the existence of approximate quasi-periodic solutions follows from the construction method itself.

The construction scheme for these solutions is as follows. If resonances occur at any step of the transformation, then the averaged system should be constructed by means of the averaging operator for continuous disturbances (i.e. by introducing the Delaunay anomaly), then at this step a canonical substitution of variables is constructed in order to destroy some of the terms of the disturbing function $H_1(p, q)$. If at the given iteration step there are no frequency resonances, then the averaged Hamiltonian system can be obtained by means of the operator of averaging over all fast angle variables. At such a step, instead of the Hamiltonian system (3.4.41) we obtain the system

$$\begin{aligned} \frac{d\bar{p}''_0}{dt} &= 0, & \frac{d\bar{q}''_0}{dt} &= \frac{\partial H^*_{00}}{\partial \bar{p}''_0} + \mu \frac{\partial \overline{H}^*_1}{\partial \bar{p}''_0}, \\ \frac{d\bar{p}^*}{dt} &= -\mu \frac{\partial \overline{H}^*_1}{\partial \bar{q}^*}, & \frac{d\bar{q}^*}{dt} &= \mu \frac{\partial \overline{H}^*_1}{\partial \bar{p}^*}, \end{aligned} \tag{3.4.46}$$

i.e. the resonance part of the system is lacking, and the vectors do not have the dimensions $(n_0 - 1)$, but n_0.

So at each step a simplified, averaged system of the first approximation (in the accepted terms) is constructed which is studied with the purpose of integration and finding an equilibrium (stationary) solution. Under certain conditions (occurring at each iteration step) it is possible to theoretically construct a chain

$$(\widetilde{p}, \widetilde{q}) \leftrightarrows (\widetilde{P}^{(0)}, \widetilde{Q}^{(0)}) \leftrightarrows (\widetilde{P}^{(1)}, \widetilde{Q}^{(1)}) \leftrightarrows \cdots \leftrightarrows (\widetilde{P}^{(\infty)}, \widetilde{Q}^{(\infty)}) \tag{3.4.47}$$

of analytic and one-to-one canonical substitutions of variables, allowing us to obtain closer and closer approximations to the sought quasi-periodic solutions. For this scheme to be mathematically justified, it is necessary to find the conditions under which the infinite procedures converge in one or another sense. This question requires special study.

Now we pass on to a more detailed statement of the described algorithm. Let there be a rigorous commensurability between the basic frequencies for the fast angle variables q_0, i.e. the following inequality is satisfied (see Arnold [1]):

$$|(k_0, \omega)| < K(\omega)||k_0||^{-(n_0+1)}. \tag{3.4.48}$$

Introduce a critical argument d by formula (3.4.22), assuming that k_0 is chosen from the condition

$$|(\overline{k}_0, \omega)| = \inf_{1 \le ||k_0|| \le N} |(k_0, \omega)|. \tag{3.4.49}$$

In the disturbing part H_1 of the Hamiltonian H we separate its secular part $\overline{H}_1$, partial sum $[\widetilde{H}_1]_N$ and the remainder term $R_N H_1$ by the formulae

$$\overline{H}_1(p, d, q^*) = \sum_{1 \le ||k|| \le N}{}' h_k(p) \cos\left[\overline{k}_1 d + (k^*, q^*) + l_k\right], \tag{3.4.50}$$

$$[\widetilde{H}_1]_N = \sum_{1 \le ||k|| \le N}{}'' h_k(p) \cos\left[k_1'' d + (k_0'', q_0'') + (k^*, q^*) + l_k\right], \tag{3.4.51}$$

$$R_N H_1 = \sum_{||k|| > N} h_k(p) \cos\left[k_1'' d + (k_0'', q_0'') + (k^*, q^*) + l_k\right]. \tag{3.4.52}$$

Introduce the new canonical variables $\widetilde{P} = (P_1, P_0'', P^*)$, $\widetilde{Q} = (D, Q_0'', Q^*)$ by the formulae

$$\begin{aligned} p_1 &= P_1 + \mu \sum_{||k|| \ge 1}{}'' k S_k(\widetilde{P}) \cos\left[k_1'' d + (k_0'', q_0'') + (k^*, q^*) + l_k\right], \\ D &= d + \mu \sum_{||k|| \ge 1}{}'' S_k'(\widetilde{P}) \sin\left[k_1'' d + (k_0'', q_0'') + (k^*, q^*) + l_k\right]; \end{aligned} \tag{3.4.53}$$

$$\begin{aligned} p &= P + \mu \sum_{||k|| \ge 1}{}'' k S_k(\widetilde{P}) \cos\left[k_1'' d + (k_0'', q_0'') + (k^*, q^*) + l_k\right], \\ Q &= q + \mu \sum_{||k|| \ge 1}{}'' \widetilde{S}_k'(\widetilde{P}) \sin\left[k_1'' d + (k_0'', q_0'') + (k^*, q^*) + l_k\right]; \end{aligned} \tag{3.4.54}$$

$$\begin{aligned} S_k' &= \frac{\partial S_k}{\partial P_1}, & \overline{S}_k' &= \frac{\partial S_k}{\partial P}, & p &= (p_0'', p^*), \\ q &= (q_0'', q^*), & P &= (P_0'', P^*), & Q &= (Q_0'', Q^*). \end{aligned} \tag{3.4.55}$$

Since the generating function $S(P, q)$ is determined by (3.4.39), it does not contain, generally speaking, small denominators, therefore relations (3.4.53) – (3.4.54) do not have peculiar properties. From equalities (3.4.53), (3.4.54) one can obtain explicit formulae connecting the old $(\widetilde{p}, d, q)$ and the new $(\widetilde{P}, D, Q)$ variables.

Concrete dependences can be constructed by the method of step-by-step approximations. These relations have the form

$$\begin{aligned}
p_1 &= P_1 + \mu \sum_{\|k\|\geq 0}'' f_{p_1 k}(\widetilde{P}) \cos\left[k_1'' D + (k_0'', Q_0'') + (k^*, Q^*) + l_k\right], \\
d &= D + \mu \sum_{\|k\|\geq 1}'' f_{dk}(\widetilde{P}) \sin\left[k_1'' D + (k_0'', Q_0'') + (k^*, Q^*) + l_k\right], \\
p &= P + \mu \sum_{\|k\|\geq 1}'' f_{pk}(\widetilde{P}) \cos\left[k_1'' D + (k_0'', Q_0'') + (k^*, Q^*) + l_k\right], \\
q &= Q + \mu \sum_{\|k\|\geq 1}'' f_{qk}(\widetilde{P}) \sin\left[k_1'' D + (k_0'', Q_0'') + (k^*, Q^*) + l_k\right], \\
P_1 &= p_1 + \mu \sum_{\|k\|\geq 0}'' \varphi_{p_1 k}(\widetilde{p}) \cos\left[k_1'' d + (k_0'', q_0'') + (k^*, q^*) + l_k'\right], \\
D &= d + \mu \sum_{\|k\|\geq 1}'' \varphi_{dk}(\widetilde{p}) \sin\left[k_1'' d + (k_0'', q_0'') + (k^*, q^*) + l_k'\right], \\
P &= p + \mu \sum_{\|k\|\geq 0}'' \varphi_{pk}(\widetilde{p}) \cos\left[k_1'' d + (k_0'', q_0'') + (k^*, q^*) + l_k'\right], \\
Q &= q + \mu \sum_{\|k\|\geq 1}'' \varphi_{qk}(\widetilde{p}) \sin\left[k_1'' d + (k_0'', q_0'') + (k^*, q^*) + l_k'\right],
\end{aligned} \tag{3.4.56}$$

where $f_{p_1 k}, \ldots, \varphi_{qk}$ are some analytic functions of their own variables.

The equations for the new variables P_1, D, P, Q will again be canonical:

$$\frac{dP_1}{dt} = -\frac{\partial H'}{\partial D}, \quad \frac{dD}{dt} = \frac{\partial H'}{\partial P_1}, \quad \frac{dP}{dt} = -\frac{\partial H'}{\partial Q}, \quad \frac{dQ}{dt} = \frac{\partial H'}{\partial P}, \tag{3.4.57}$$

where $P = (P_0'', P^*)$, $Q = (Q_0'', Q^*)$, and $H' = H'(P_1, D, P, Q)$ is a new Hamiltonian obtained from the old Hamiltonian $H(p_1, d, p, q)$ by a substitution of variables (p_1, d, p, q), using relations (3.4.56). The new Hamiltonian has the form

$$H'(P_1, D, P, Q) = H_{00}'(P_1) + \mu \overline{H}_1(P_1, D, P, Q^*) + \mu \widetilde{H}_1(P_1, D, P, Q), \tag{3.4.58}$$

where $\mu\widetilde{H}_1$ is the set of terms not accounted for at the previous step of the transformation, plus the remainder term $\mu R_N H_1$. From the condition of the construction of a generating function $S(P_1, P, q)$ and the choice of N it follows that $\mu\widetilde{H}_1 \sim \mu^2$ in the domain of substitution of the new variables P_1, D, P, Q, provided that the old variables p_1, d, p, q change in the initial domain G_{2n}.

Then, using the same method, it is possible to transform the canonical equations (3.4.57), first having written in explicit form the dependence of the new Hamiltonian H' on the new canonical variables P_1, D, P, Q. Finding resonance terms of the Hamiltonian at each step and including them into its secular part (through the introduction of a critical argument and averaging for continuous disturbances), we again construct a canonical substitution of variables that allows us to decrease the norm of the remaining part of the Hamiltonian (the construction of higher approximations is described in Arnol'd [1], and Grebenikov and Ryabov [1]). During this construction, frequencies of quasi-periodic solutions (including the small frequencies ω^* proportional to μ) are also defined more exactly at each step.

If $\widetilde{p}^{(s)}$, $d^{(s)}$, $\widetilde{q}^{(s)}$ denote an s-th approximation of the initial canonical variables $\widetilde{p}$, d, $\widetilde{q}$, they are expressed through the canonical variables of the previous $(s-1)$-th approximation $\widetilde{P}^{(s-1)}$, $D^{(s-1)}$, $\widetilde{Q}^{(s-1)}$ by means of the equalities

$$\begin{aligned}
\widetilde{p}^{(s)}(t,\mu) &= \widetilde{P}_{00}^{(s-1)} + \mu \sum_{\|k\|\geq 0} f_{pk}^{(s)}(\widetilde{P}_{00}^{(s-1)}t) \\
&\quad \times \cos\,[k_1'' D_{00}^{(s-1)} + (k_0'',\, \omega_j^{(s)} t + Q''^{(s-1)}_{00}) + (k^*,\, \omega^{*(s)}_i t + Q^{*(s-1)}_{00} t) + l_k], \\
d^{(s)}(t,\mu) &= D_{00}^{(s-1)} + \mu \sum_{\|k\|\geq 0} f_{dk}^{(s)}(\widetilde{P}_{00}^{(s-1)}) \\
&\quad \times \sin\,[k_1'' D_{00}^{(s-1)} + (k_0'',\, \omega_j^{(s)} t + Q''^{(s-1)}_{00}) + (k^*,\, \omega^{*(s)}_i t + Q^{*(s-1)}_{00}) + l_k], \\
\widetilde{q}^{(s)} &= \widetilde{Q}_{00}^{(s-1)} + \mu \sum_{\|k\|\geq 0} f_{qk}^{(s)}(\widetilde{P}_{00}^{(s-1)}) \\
&\quad \times \sin\,[k_1'' D_{00}^{(s-1)} + (k_0'',\, \omega_j^{(s)} t + Q''^{(s-1)}_{00}) + (k^*,\, \omega^{*(s)}_i t + Q^{*(s-1)}_{00}) + l_k],
\end{aligned} \tag{3.4.59}$$

where $f_{pk}^{(s)}$, $f_{dk}^{(s)}$, $f_{qk}^{(s)}$ are analytic functions of their own variables, and $\widetilde{P}_{00}^{(s-1)}$, $D_{00}^{(s-1)}$, $\widetilde{Q}_{00}^{(s-1)}$ (the initial values of the variables $\widetilde{P}^{(s-1)}$, $D^{(s-1)}$, $\widetilde{Q}^{(s-1)}$) are connected with the initial conditions $\bar{p}_{00}$, d_{00}, $\bar{q}_{00}$ by the relations described in Arnol'd [1], and Zhuravlev [1].

Formulae (3.4.59) show that the approximate solution of the initial Hamiltonian system $\widetilde{p}^{(s)}(t,\mu)$, $d^{(s)}(t,\mu)$, $\widetilde{q}^{(s)}(t,\mu)$ is usually quasi-periodic with respect to time t. It can move to a periodic solution only for "exotic" initial conditions, when the exact commensurability of fast and slow frequencies of the s-th approximation $\omega_j^{(s)}$, $\omega^{*(s)}_i$ is possible.

3.5 Motion of a Geostationary Satellite

One of the most interesting resonance problems of celestial mechanics is the problem of the motion of an artificial celestial body with its period of rotation around the Earth's center approximately equal to 24 hours. Such a body moves in the equatorial plane of the Earth with an angular velocity approximately equal to the velocity of the Earth's rotation around the polar axis. Due to this, the satellite "hangs" above the same point of the Equator; therefore such satellites are extremely important for global TV broadcasts. These objects move approximately along a circular orbit with its center at the Earth's center, at a height of about 36 000 km above the Earth. Naturally, such a description is quite approximate, and for the exact prediction of the motion of a geostationary satellite it is necessary to study a very intricate nonlinear system of differential equations, which is a mathematical model of its dynamics.

From the point of view of dynamics a geostationary Earth satellite is a very compound object for the following two reasons: first, it is a typical resonance object, as the angular velocities of rotation of the Earth around its axis and rotation of the satellite around the Earth's center are approximately equal, and consequently, there is a resonance of lowest order (in the usual notation $\omega_1 : \omega_2 = 1 : 1$); and second, disturbances in its motion due to the Earth's polar compression on the one hand and by the influence of the Moon and the Sun on the other are values of the same order of infinitesimals and therefore should be determined together, and not separately, in order not to lose accuracy.

These two factors were accounted for by S.G. Zhuravlev in the construction of an asymptotic theory of the motion of a daily satellite with the use of the averaging method (see Zhuravlev [1, 2]). Below we will adhere in the main to the results of his research.

The differential equations of the motion of such a body, written with the use of a special system of phase coordinates (see Zhuravlev [1]), have the form

$$
\begin{aligned}
\frac{d\tilde{n}}{d\tau} &= -3\mu\tilde{n}^{4/3}\frac{\partial U}{\partial \lambda},\\
\frac{dh}{d\tau} &= \frac{\mu\tilde{n}^{1/3}}{\sqrt{1-q^2-k^2}}\left[-\sqrt{h^2-l^2}\,\frac{\partial U}{\partial l} - h\left(k\frac{\partial U}{\partial q} - q\frac{\partial U}{\partial k} + \frac{\partial U}{\partial \lambda}\right)\right],\\
\frac{dl}{d\tau} &= \frac{\mu\tilde{n}^{1/3}}{\sqrt{1-q^2-k^2}}\left[\sqrt{h^2-l^2}\,\frac{\partial U}{\partial k} - l\left(k\frac{\partial U}{\partial q} - q\frac{\partial U}{\partial k} + \frac{\partial U}{\partial \lambda}\right)\right],\\
\frac{dq}{d\tau} &= \mu\tilde{n}^{1/3}\sqrt{1-q^2-k^2}\left[\frac{\partial U}{\partial k} - \frac{q}{1+\sqrt{1-q^2-k^2}}\frac{\partial U}{\partial l}\right.\\
&\quad \left. + \frac{k}{1-q^2-k^2}\left(h\frac{\partial U}{\partial h} + l\frac{\partial U}{\partial l}\right)\right],\\
\frac{dk}{d\tau} &= \mu\tilde{n}^{1/3}\sqrt{1-q^2-k^2}\left[\frac{\partial U}{\partial q} - \frac{k}{1+\sqrt{1-q^2-k^2}}\frac{\partial U}{\partial \lambda}\right.\\
&\quad \left. - \frac{q}{1-q^2-k^2}\left(h\frac{\partial U}{\partial h} + l\frac{\partial U}{\partial l}\right)\right],\\
\frac{d\lambda}{d\tau} &= \tilde{n} + 3\mu\tilde{n}^{4/3}\frac{\partial U}{\partial \tilde{n}} + \mu\tilde{n}^{1/3}\sqrt{1-q^2-k^2}\\
&\quad \times\left[\frac{1}{1+\sqrt{1-q^2-k^2}}\left(q\frac{\partial U}{\partial k} + k\frac{\partial U}{\partial q}\right)\right.\\
&\quad \left. + \frac{1}{1-q^2-k^2}\left(h\frac{\partial U}{\partial h} + l\frac{\partial U}{\partial l}\right)\right],
\end{aligned}
\tag{3.5.1}
$$

where τ, $\widetilde{n}$, U, μ are respectively the dimensionless time, the average motion (average angular velocity of the satellite), the disturbing function and a small parameter, calculated by the formulae

$$\tau = tp_0^{-3/2}\sqrt{\mu_E}, \qquad \widetilde{n} = \left(1-e^2\right)^{3/2} p_0^{-3/2},$$
$$U = \frac{p_0}{\mu\mu_E}V, \qquad \mu = -3c_{20}\left(\frac{r_0}{p_0}\right)^2, \tag{3.5.2}$$

where t is time, p_0 the focal parameter of the orbit at an initial point of time, μ_E the gravitational potential of the Earth, V the disturbing function of the problem, r_0 the largest equatorial radius of the Earth, c_{20} the coefficient characterizing the polar compression of the Earth (see Samoilenko and Ronto [1]), and e the orbital eccentricity.

If in the disturbing function U we take account of the influence of the polar and equatorial compression of the Earth and the attraction of the Moon and the Sun, in the variables $\widetilde{n}$, h, l, q, k, λ it will have the form (see Zhuravlev [1])

$$\begin{aligned} U &= R^3\widetilde{n}^2\left[\frac{1-6k}{6} + \frac{q^2+k^2}{2} + h\cos 2u\right] \\ &+ \frac{\bar{m}_L R_L^3 \widetilde{n}_L^2 (1-q^2-k^2)}{R^2\widetilde{n}^{4/3}\left(1-e_L^2\right)^3}\left[\sum_{\nu=0}^{2}(\bar{a}_\nu^{(L)}\cos\nu u + \bar{b}_\nu^{(L)}\sin\nu u)\right. \\ &+ \left.\frac{R_L\widetilde{n}_L^{2/3}(1-q^2-k^2)}{R\widetilde{n}^{4/3}\left(1-e_L^2\right)}\sum_{\nu=1}^{3}(\overline{A}_\nu\cos\nu u + \overline{B}_\nu\sin\nu u) + \frac{3R_L^2\widetilde{n}_L^{4/3}(1-q^2-k^2)}{4R^2\widetilde{n}^{4/3}\left(1-e_L^2\right)^2}\right] \\ &+ \frac{\bar{m}_s R_s^3\widetilde{n}_s^2(1-q^2-k^2)}{R^2\widetilde{n}^{4/3}(1-e_s^2)^3}\sum_{\nu=0}^{2}(\bar{a}_\nu^{(s)}\cos\nu u + \bar{b}_\nu^{(s)}\sin\nu u) \\ &+ \frac{\mu b\widetilde{n}^2R^3}{(1-q^2-k^2)^3}\left\{\sqrt{h^2-l^2}\sin 2\bar{\omega}_3\tau - l\cos 2\bar{\omega}_3\tau\right. \\ &- \left.\frac{1-2h}{2}\left[\frac{\sqrt{h^2-l^2}}{h}\sin 2(\bar{\omega}_3\tau - u) - \frac{l}{h}\cos 2(\bar{\omega}_3\tau - u)\right]\right\}, \end{aligned} \tag{3.5.3}$$

where $\bar{\omega}_3$ is a dimensionless angular velocity of the Earth's rotation, calculated by the formula $\bar{\omega}_3 = \omega_3\sqrt{p_0^3/\mu_E}$, and ω_3 is the angular velocity of the Earth's rotation. In (3.5.3) the following symbols are also used:

$$\bar{m}_j = \left(\frac{p_{0j}}{r_0}\right)^3 m_j, \qquad \mu m_j = \frac{\mu_j}{\mu_E}\left(\frac{r_0}{p_{0j}}\right)^3,$$

$$b = 6\left(\frac{r_0}{p_0}\right)^2\frac{\sqrt{c_{22}^2+d_{22}^2}}{\mu^2}, \qquad R_j = 1 + q_j\cos u_j + k_j\sin u_j,$$

$$q_j = e_j\cos\omega_j, \quad k_j = e_j\sin\omega_j, \quad u_j = v_j + \omega_j, \quad j = L, S,$$

$$R = 1 + \left(q\sqrt{\frac{h+l}{2h}} + k\sqrt{\frac{h-l}{2h}}\right)\cos u + \left(q\sqrt{\frac{h-l}{2h}} - k\sqrt{\frac{h+l}{2h}}\right)\sin u, \qquad u = v + \omega,$$

where v is the true anomaly of the satellite, v_L, v_S the true anomalies of the Moon and the Sun, p_{0L}, p_{0S} the focal parameters of the orbits of the Moon and the Sun at the initial time, μ_L, μ_S the gravitational potential of the Moon and the Sun, and c_{22}, d_{22} the coefficients characterizing the equatorial compression of the Earth (numerical values of c_{20}, c_{22}, d_{22} can be found e.g. in Duboshin [1]).

The connection between the phase variables $\widetilde{n}$, h, q, k, l, λ and the classical Keplerian elements a, e, i, ω, Ω, M is expressed by well-known formulae, see Duboshin [1]; and in Zhuravlev [1] one can find analytic expressions for the coefficients $\bar{a}_\nu^{(j)}$, $\bar{b}_\nu^{(j)}$, $\overline{A}_\nu$, $\overline{B}_\nu$ of series (3.5.3).

So the equations of the motion of a geostationary satellite (3.5.1) are a multi-frequency system with five slow variables $(\widetilde{n}, h, q, k, l)$ determined by its first five equations (their right-hand members are proportional to the small parameter μ) and one fast variable λ determined by the last, the sixth equation. Therefore one might think that the system (3.5.1) is not in resonance, but actually this is not so. The point is that due to the presence of the equatorial compression (terms with $\bar{\omega}_3$) it is not autonomous, i.e. its right-hand members explicitly depend on t. In an extension to seventh order (if another additional fast variable is introduced, e.g. by the formula $y = \bar{\omega}_3 \tau$) the new seventh-order system will be autonomous, but with two fast variables (λ and y).

For the construction of an approximate solution of equations (3.5.1) we will use the averaging method in accordance with which we will search for a substitution of variables

$$\begin{aligned} &P = \overline{P} + \mu p_1 + \mu^2 p_2 + \cdots, \qquad P = (\widetilde{n}, h, l, q, k), \\ &\lambda = \overline{\lambda} + \mu\lambda_1 + \mu^2\lambda^2 + \cdots, \end{aligned} \tag{3.5.4}$$

that would transform equations (3.5.1) into the equations

$$\begin{aligned} \frac{d\overline{P}}{d\tau} &= \mu A_{1p}(\overline{P}) + \mu^2 A_{2p}(\overline{P}) + \cdots, \\ \frac{d\overline{\lambda}}{d\tau} &= \widetilde{n} + \mu B_{1\lambda}(\overline{P}) + \mu^2 B_{2\lambda}(\overline{P}) + \cdots, \end{aligned} \tag{3.5.5}$$

where A_{1p}, $A_{2p}, \ldots,$ $B_{1\lambda}$, $B_{2\lambda}, \ldots$ are still unknown functions of the averaged slow variables $\overline{P}$.

In the accepted terminology, equations (3.5.5) are averaged equations of any approximation, and the algorithm for finding the unknown functions $p_1, p_2, \ldots,$ $\lambda_1, \lambda_2, \ldots,$ $A_{1p}, A_{2p}, \ldots,$ $B_{1\lambda}, B_{2\lambda}, \ldots$ has been given more than once in the previous section, so we will implement this algorithm in this specific case.

To obtain explicit expressions for the functions A_{1p}, $A_{2p}, \ldots$ it is necessary to have an averaged value of the disturbing problem, taking into consideration that the ratio of the period of the orbital motion of the satellite and the period of rotation of the Earth around the axis is 1:1. If as an averaging operator we use the operator for continuous disturbances (a particular case of this for problems of celestial mechanics is the Delaunay–Hill scheme), then the averaged disturbing functions with consideration for the terms up to μ^2 inclusive will have the form

$$\begin{aligned}\overline{U} &= \frac{\widetilde{n}^2}{6}\left[1 - 6h + \frac{3}{2}(q^2 + k^2)\right] + \sum_j \bar{m}_j \frac{\widetilde{n}_j^2}{\widetilde{n}^{4/3}} \bar{a}_{00}^{(j)} + \sum_j A^{(j)} \\ &\quad + B^{(L)} + \frac{\mu b \widetilde{n}^2}{2}(1 - 2h)\left(\frac{l}{h}\cos 2D + \frac{\sqrt{h^2 + l^2}}{h}\sin 2D\right).\end{aligned} \tag{3.5.6}$$

Here

$$\begin{aligned}A^{(j)} &= \frac{\bar{m}_j \widetilde{n}_j^2}{\widetilde{n}^{4/3}}\left\{\frac{3}{2}[e_j^2 + (q^2 + k^2)]\bar{a}_{00}^{(j)} + 5\left[\frac{q^2 - k^2}{2}\frac{\sqrt{h^2 - l^2}}{h} - \frac{kql}{h}\right]\bar{b}_{20}^{(j)}\right. \\ &\quad \left. + \frac{5}{2}\left[\frac{(q^2 - k^2)l}{h} + \frac{2qk\sqrt{h^2 - l^2}}{h}\right]\bar{a}_{20}^{(j)}\right\}, \\ B^{(L)} &= -\frac{5\bar{m}_L \widetilde{n}_L^{8/3} e_L}{2\widetilde{n}^2}\left[\left(q\sqrt{\frac{h + l}{2h}} + k\sqrt{\frac{h - l}{2h}}\right)(\bar{\alpha}_{11}\cos\omega_L + \bar{\beta}_{11}\sin\omega_L)\right. \\ &\quad \left. + \left(q\sqrt{\frac{h - l}{2h}} - k\sqrt{\frac{h + l}{2h}}\right)(\bar{\gamma}_{11}\cos\omega_L + \bar{\delta}_{11}\sin\omega_L)\right] + \frac{3\bar{m}_L \widetilde{n}_L^{10/3}}{4\widetilde{n}^{8/3}}, \\ &\quad j = L, S, \\ D &= \lambda - \bar{\omega}_3 \tau,\end{aligned}$$

where ω_L is the angular distance of the pericenter of the Moon's orbit, and D is the Delaunay anomaly.

Substituting the partial derivatives of the averaged disturbing function $\bar{U}$ with respect to the corresponding arguments, by means of the method described, e.g., in Sections 2.3 and 2.6, it is possible to find all values included in the structure of the averaged equations of the second approximation (accurate to μ^2 inclusive):

$$\begin{aligned}A_{1\widetilde{n}} &= 0, \qquad A_{1h} = -\sum_j \frac{\bar{m}_j \widetilde{n}_j^2 \sqrt{h^2 - l^2}}{\widetilde{n}} \frac{\partial a_{00}^{(j)}}{\partial l}, \\ A_{1l} &= \sum_j \frac{\bar{m}_j \widetilde{n}_j^2 \sqrt{h^2 - l^2}}{\widetilde{n}} \frac{\partial \bar{a}_{00}^{(j)}}{\partial h} - \sqrt{h^2 - l^2}\,\widetilde{n}^{7/3}, \\ A_{1q} &= \frac{k\widetilde{n}^{7/3}}{2} + \sum_j \frac{\bar{m}_j \widetilde{n}_j^2}{\widetilde{n}} \frac{\partial \overline{A}^{(j)}}{\partial k} - \frac{\bar{m}_L \widetilde{n}_L^2}{\widetilde{n}} \frac{\partial \overline{B}^{(L)}}{\partial k}, \\ A_{1k} &= -\frac{q\widetilde{n}^{7/3}}{2} - \sum_j \frac{\bar{m}_j \widetilde{n}_j^2}{\widetilde{n}} \frac{\partial \overline{A}^{(j)}}{\partial q} + \frac{\bar{m}_L \widetilde{n}_L^2}{\widetilde{n}} \frac{\partial \overline{B}^{(L)}}{\partial q}, \\ B_{1\lambda} &= \widetilde{n}^{7/3} - \sum_j 4\bar{m}_j \frac{\widetilde{n}_j^2}{\widetilde{n}} \bar{a}_{00}^{(j)}, \\ A_{2\widetilde{n}}^{(L)} &= 0, \qquad A_{2\widetilde{n}}^{(S)} = 0,\end{aligned}$$

$$A_{2\widetilde{n}}^{(E)} = -3b\widetilde{n}^{10/3}\left(\frac{l}{h}\sin 2D - \frac{\sqrt{h^2-l^2}}{h}\cos 2D\right),$$

$$A_{2l}^{(E)} = -\frac{b\widetilde{n}^{7/3}}{2}\left(\frac{l^2}{h^2}\sin 2D - \frac{l\sqrt{h^2-l^2}}{h^2}\cos 2D\right),$$

$$A_{2l}^{(L)} = \frac{\bar{m}_L\widetilde{n}_L^2\sqrt{h^2-l^2}}{\mu\widetilde{n}}\left[\frac{\partial\overline{A}^{(L)}}{\partial h} + \frac{\partial\overline{B}^{(L)}}{\partial h} - \frac{q^2+k^2}{2}\frac{\partial\bar{a}_{00}^{(L)}}{\partial h}\right],$$

$$A_{2l}^{(S)} = \frac{\bar{m}_S\widetilde{n}_S^2\sqrt{h^2-l^2}}{\mu\widetilde{n}}\left[\frac{\partial\overline{A}^{(S)}}{h} - \frac{q^2+k^2}{2}\frac{\partial\bar{a}_{00}^{(S)}}{\partial h}\right],$$

$$A_{2h}^{(E)} = -\frac{b\widetilde{n}^{7/3}}{2}\left(\frac{l}{h}\sin 2D - \frac{\sqrt{h^2-l^2}}{h}\cos 2D\right),$$

$$A_{2h}^{(L)} = -\frac{\bar{m}_L\widetilde{n}_L^2\sqrt{h^2-l^2}}{\mu\widetilde{n}}\left[\frac{\partial\overline{A}^{(L)}}{\partial l} + \frac{\partial\overline{B}^{(L)}}{\partial l} + \frac{q^2+k^2}{2}\frac{\partial\bar{a}_{00}^{(L)}}{\partial l}\right],$$

$$A_{2h}^{(S)} = -\frac{\bar{m}_S\widetilde{n}_S^2\sqrt{h^2-l^2}}{\mu\widetilde{n}}\left[\frac{\partial\overline{A}^{(S)}}{\partial l} + \frac{q^2+k^2}{2}\frac{\partial\bar{a}_{00}^{(S)}}{\partial l}\right],$$

$$A_{2q}^{(E)} = -\frac{qb\widetilde{n}^{7/3}}{2}\left(\frac{l}{h}\sin 2D - \frac{\sqrt{h^2-l^2}}{h}\cos 2D\right) - \frac{kh\widetilde{n}^{7/3}}{\mu} - \frac{k(q^2+k^2)\widetilde{n}^{7/3}}{4\mu},$$

$$A_{2q}^{(L)} = \frac{\bar{m}_L\widetilde{n}_L^2}{\widetilde{n}}\left[-\frac{q^2+k^2}{2\mu}\left(\frac{\partial\overline{A}^{(L)}}{\partial k} - \frac{\partial\overline{B}^{(L)}}{\partial k}\right) + \frac{k}{\mu}\left(h\frac{\partial\bar{a}_{00}^{(L)}}{\partial h} + l\frac{\partial\bar{a}_{00}^{(L)}}{\partial l}\right)\right],$$

$$A_{2q}^{(S)} = \frac{\bar{m}_S\widetilde{n}_S^2}{\widetilde{n}}\left[-\frac{q^2+k^2}{2\mu}\frac{\partial\overline{A}^{(S)}}{\partial k} + \frac{k}{\mu}\left(h\frac{\partial\bar{a}^{(S)}}{\partial h} + l\frac{\partial\bar{a}_{00}^{(S)}}{\partial l}\right)\right],$$

$$A_{2k}^{(E)} = -\frac{kb\widetilde{n}^{7/3}}{2}\left(\frac{l}{h}\sin 2D - \frac{\sqrt{h^2-l^2}}{h}\cos 2D\right) + \frac{qh\widetilde{n}^{7/3}}{\mu} + \frac{(q^2+k^2)q\widetilde{n}^{7/3}}{4\mu},$$

$$A_{2k}^{(L)} = \frac{\bar{m}_L\widetilde{n}_L^2}{\widetilde{n}}\left[\frac{q^2+k^2}{2\mu}\left(\frac{\partial\overline{A}^{(L)}}{\partial q} - \frac{\partial\overline{B}^{(L)}}{\partial q}\right) - \frac{q}{\mu}\left(h\frac{\partial\bar{a}_{00}^{(L)}}{\partial h} + l\frac{\partial\bar{a}_{00}^{(L)}}{\partial l}\right)\right],$$

$$A_{2k}^{(S)} = \frac{\bar{m}_S\widetilde{n}_S^2}{\widetilde{n}}\left[\frac{q^2+k^2}{\mu}\frac{\partial\overline{A}^{(S)}}{\partial q} - \frac{q}{\mu}\left(h\frac{\partial\bar{a}_{00}^{(S)}}{\partial h} + l\frac{\partial\bar{a}_{00}^{(S)}}{\partial l}\right)\right],$$

$$B_{2\lambda}^{(E)} = -3b\widetilde{n}^{7/3}\left(\frac{l}{h}\cos 2D + \frac{\sqrt{h^2-l^2}}{h}\sin 2D\right) + \frac{\widetilde{n}^{7/3}}{2\mu}[3(q^2+k^2) - 13h + qk],$$

$$B_{2\lambda}^{(L)} = \frac{\bar{m}_L \tilde{n}_L}{\tilde{n}} \left[\frac{1}{\mu}(4\overline{A}^{(L)} + 6\overline{B}^{(L)}) + \frac{q}{2\mu}\left(\frac{\partial \overline{A}^{(L)}}{\partial k} + \frac{\partial \overline{B}^{(L)}}{\partial k} \right) \right.$$
$$+ \frac{k}{2\mu}\left(\frac{\partial \overline{A}^{(L)}}{\partial q} + \frac{\partial \overline{B}^{(L)}}{\partial q} \right) + \frac{h}{2\mu}\left(\frac{\partial \bar{a}_{00}^{(L)}}{\partial h} + \frac{\partial \overline{A}^{(L)}}{\partial h} + \frac{\partial \overline{B}^{(L)}}{\partial h} \right)$$
$$\left. + \frac{l}{2\mu}\left(\frac{\partial \bar{a}_{00}^{(L)}}{\partial l} + \frac{\partial \overline{A}^{(L)}}{\partial l} + \frac{\partial \overline{B}^{(L)}}{\partial l} \right) \right],$$

$$B_{2\lambda}^{(S)} = \frac{\bar{m}_S \tilde{n}_S^2}{\tilde{n}} \left[\frac{4\overline{A}^{(S)}}{\mu} + \frac{q}{2\mu}\frac{\partial \overline{A}^{(S)}}{\partial k} + \frac{k}{2\mu}\frac{\partial \overline{A}^{(S)}}{\partial q} \right.$$
$$\left. + \frac{h}{2\mu}\left(\frac{\partial \bar{a}_{00}^{(S)}}{\partial h} + \frac{\partial \overline{A}^{(S)}}{\partial h} \right) + \frac{l}{2\mu}\left(\frac{\partial \bar{a}_{00}^{(S)}}{\partial l} + \frac{\partial \overline{A}^{(S)}}{\partial l} \right) \right],$$

$$\overline{A}^{(j)} = \frac{\tilde{n}^{4/3}}{\bar{m}_j \tilde{n}_j^2} A^{(j)}, \quad j = L, S, \quad \overline{B}^{(L)} = \frac{\tilde{n}^{4/3}}{\bar{m}_L \tilde{n}_L^2} B^{(L)},$$

$$A_{2p}^{(E)} + A_{2p}^{(L)} + A_{2p}^{(S)} = A_{2p}, \quad p = h, l, q, k, \tilde{n},$$

$$B_{2\lambda}^{(E)} + B_{2\lambda}^{(L)} + B_{2\lambda}^{(S)} = B_{2\lambda}.$$

If in equations (3.5.5) the terms of order μ^2 are neglected, then the averaged equations of the first approximation can be integrated in the same way as the similar equations of the first approximation for a multifrequency rotary system of general form (see Sections 2.3, 2.6, 2.8). In this case we will have

$$\begin{aligned} \bar{\tilde{n}} &= \tilde{n}_0 + \mu\delta_1\bar{\tilde{n}}, & \bar{l} &= l_0 + \mu\delta_1\bar{l}, \\ \bar{q} &= q_0 + \mu\delta_1\bar{q}, & \bar{k} &= k_0 + \mu\delta_1\bar{k}, \\ \bar{h} &= h_0 + \mu\delta_1\bar{h}, & \overline{\lambda} &= \lambda_0 + \tilde{n}_0\tau + \mu\delta_1\overline{\lambda}; \end{aligned} \tag{3.5.7}$$

the functions $\delta_1\bar{\tilde{n}}, \ldots, \delta_1\overline{\lambda}$ are expressed by the formulae

$$\delta_1\bar{\tilde{n}} = 0,$$

$$\delta_1\bar{l} = \left[-\sqrt{h_0^2 - l_0^2}\,\tilde{n}_0^{4/3} + \frac{3\sqrt{2}}{8}\sum_j m_j\left(\frac{\tilde{n}_j}{\tilde{n}_0}\right)^2 \sin 2i_j \right.$$
$$\left. \times \left(\sqrt{h_0 - l_0}\sin\Omega_j + \sqrt{h_0 + l_0}\cos\Omega_j \right) \right] \tilde{n}_0\tau,$$

$$\delta_1\bar{q} = \left\{ \frac{k_0\tilde{n}_0^{4/3}}{2} + \sum_j \bar{m}_j\left(\frac{\tilde{n}_j}{\tilde{n}_0}\right)^2 \left[3k_0\bar{a}_{00}^{(j)} - 5\bar{a}_{20}^{(j)}\left(\frac{k_0 l_0}{h_0} - \frac{q_0\sqrt{h_0^2 - l_0^2}}{h_0} \right) \right.\right.$$
$$\left. - 5\bar{b}_{20}^{(j)}\left(\frac{k_0\sqrt{h_0^2 - l_0^2}}{h_0} + \frac{q_0 l_0}{h_0} \right) \right] - \frac{5\bar{m}_L e_L}{2}\left(\frac{\tilde{n}_L}{\tilde{n}_0}\right)^{8/3} \left[\sqrt{\frac{h_0 - l_0}{2h_0}} \right.$$
$$\left.\left. \times (\bar{\alpha}_{11}\cos\omega_L + \bar{\beta}_{11}\sin\omega_L) - \sqrt{\frac{h_0 + l_0}{2h_0}}(\bar{\gamma}_{11}\cos\omega_L + \bar{\delta}_{11}\sin\omega_L) \right] \right\} \tilde{n}_0\tau,$$

$$\delta_1\bar{h} = -\frac{3\sqrt{2}}{8}\widetilde{n}_0\tau\sum_j \bar{m}_j\left(\frac{\widetilde{n}_j}{\widetilde{n}_0}\right)^2 \sin 2i_j\left(\sqrt{h_0-l_0}\sin\Omega_j - \sqrt{h_0+l_0}\cos\Omega_j\right),$$

$$\delta_1\bar{k} = \left\{-\frac{q_0\widetilde{n}_0^{4/3}}{2} + \sum_j \bar{m}_j\left(\frac{\widetilde{n}_j}{\widetilde{n}_0}\right)^2\left[3q_0\bar{a}_{00}^{(j)}\right.\right.$$

$$\left. + 5\bar{a}_{20}^{(j)}\left(\frac{q_0 l_0}{h_0} + \frac{k_0\sqrt{h_0^2-l_0^2}}{h_0}\right) + 5\bar{b}_{20}^{(j)}\left(\frac{q_0\sqrt{h_0^2-l_0^2}}{h_0} - \frac{k_0 l_0}{h_0}\right)\right]$$

$$+ \frac{5\bar{m}_L e_L}{2}\left(\frac{\widetilde{n}_L}{\widetilde{n}_0}\right)^{8/3}\left[\sqrt{\frac{h_0+l_0}{2h_0}}(\bar{\alpha}_{11}\cos\omega_L + \bar{\beta}_{11}\sin\omega_L)\right.$$

$$\left.\left. + \sqrt{\frac{h_0-l_0}{2h_0}}(\bar{\gamma}_{11}\cos\omega_L + \bar{\delta}_{11}\sin\omega_L)\right]\right\}\widetilde{n}_0\tau,$$

$$\delta_1\bar{\lambda} = \left[\widetilde{n}_0^{4/3} - 4\sum_j \bar{m}_j\left(\frac{\widetilde{n}_j}{\widetilde{n}_0}\right)^2 \bar{a}_{00}^{(j)}\right]\widetilde{n}_0\tau, \qquad j = L, S.$$

In the expressions for the coefficients $\bar{a}_{00}^{(j)}$, $\bar{a}_{20}^{(j)}$, $\bar{b}_{20}^{(j)}$, $\bar{\alpha}_{11}$, $\bar{\beta}_{11}$, $\bar{\gamma}_{11}$, $\bar{\delta}_{11}$ it is necessary to replace the osculating elements by their initial values. It is clear that in the first approximation the polar compression of the Earth, as well as the attraction of the Moon and the Sun, do not generate secular disturbances in the element $\widetilde{n}$.

This statement is an analogue of the classical Laplace theorem on the absence of first-order secular disturbances in the major semiaxes of the orbits of the three-body problem (see Duboshin [1]). Using the formulae given in Zhuravlev [1]

$$\delta i = \frac{\delta h}{\sqrt{h(1-h)}}, \qquad \delta e = \frac{1}{e}(q\delta k + k\delta q),$$

$$\delta\pi = \frac{1}{e}(\delta q\cos\pi - \delta k\sin\pi),$$

we can calculate the first-order disturbances in the inclination $\delta_1 i$, eccentricity $\delta_1 e$ and the longitude of the pericenter $\delta_1\pi$. From these relations it follows that the polar compression of the Earth does not cause secular disturbances in the eccentricity and the inclination, and the Moon and the Sun generate them in all elements except $\widetilde{n}$. The so-called parallactic terms in the Moon's potential (see Hill [1]) generate secular disturbances in the elements q and k, and consequently in the elements e and π of the geostationary satellite's orbit. S.G. Zhuravlev also constructed an asymptotic theory of the second approximation, i.e. integrating equations (3.5.5) with consideration for the terms proportional to μ^2 and found (see Zhuravlev [1]) second-order disturbances $\delta_2\bar{\widetilde{n}}$, $\delta_2\bar{l}$, $\delta_2\bar{q}$, $\delta_2\bar{k}$, $\delta_2\bar{h}$, $\delta_2\bar{\lambda}$, i.e. analytic expressions

$$\begin{aligned} \bar{\widetilde{n}} &= \widetilde{n}_0 + \mu\delta_1\bar{\widetilde{n}} + \mu^2\delta_2\bar{\widetilde{n}}, & \bar{h} &= h_0 + \mu\delta_1\bar{h} + \mu^2\delta_2\bar{h}, \\ \bar{l} &= l_0 + \mu\delta_1\bar{l} + \mu^2\delta_2\bar{l}, & \bar{k} &= k_0 + \mu\delta_1\bar{k} + \mu^2\delta_2\bar{k}, \\ \bar{q} &= q_0 + \mu\delta_1\bar{q} + \mu^2\delta_2\bar{q}, & \bar{\lambda} &= \lambda_0 + \widetilde{n}_0\tau + \mu\delta_1\bar{\lambda} + \mu^2\delta_2\bar{\lambda}. \end{aligned} \tag{3.5.8}$$

Here we cannot write the formulae for $\delta_2\bar{\bar{n}}, \ldots, \delta_2\overline{\lambda}$, but their analysis shows that equatorial compression of the Earth in the second approximation already generates secular disturbances in the element $\bar{\bar{n}}$, and this is an essential difference of the problem of the motion of a geostationary satellite from the Newtonian three-body problem for which the Poisson theorem is true: second-order secular disturbances in major semiaxes of orbits are zero.

S.G. Zhuravlev has also performed a comparison of the asymptotic theory of the motion of a geostationary satellite, constructed by himself, with the results related to the concrete satellite Early Bird and published by Arnaud [1].

According to Arnaud [1], the drift of that satellite along the longitude was $\dot{\lambda} = 0.02$ degree/day. Calculation by asymptotic formulae from the papers of Zhuravlev [1, 2] gave 0.0206 deg/day, that due to the polar compression of the Earth $\dot{\lambda}_{C_{20}} = 0.0268$ deg/day, that due to the attraction of the Moon $\dot{\lambda}_L = -0.0045$ deg/day, that due to the attraction of the Sun $\dot{\lambda}_S = -0.0022$ deg/day and that due to the equatorial compression of the Earth $\dot{\lambda}_{C_{22}} = 0.0005$ deg/day. The value of the small parameter of the problem is of order 10^{-4}.

3.6 Averaging Method in the Theory of Partial Differential Equations

The averaging method in conjunction with the method of separation of variables (Fourier method) allows us in many cases to study oscillatory processes in systems with distributed parameters. These questions were investigated in numerous works, including the monographs by Mitropolsky and Moiseenkov [1]. The essence of this approach is that one partial differential equation is reduced by the Fourier method to an infinite-dimensional system of ordinary differential equations, to which one or another averaging method can be applied.

Consider the quasi-linear equation of hyperbolic type with constant coefficients

$$A\frac{\partial^2 u}{\partial t^2} + 2B\frac{\partial^2 u}{\partial t \partial x} + C\frac{\partial^2 u}{\partial x^2} + D\frac{\partial u}{\partial t} + E\frac{\partial u}{\partial x} + Gu = \mu F\left(t, x, u, \frac{\partial u}{\partial t}, \frac{\partial u}{\partial x}\right), \quad (3.6.1)$$

where μ is a small parameter, F is a function nonlinear with respect to its arguments, and the constants A, B, C satisfy the condition of hyperbolicity

$$B^2 - AC > 0. \quad (3.6.2)$$

By means of known substitutions of variables (see Vladimirov [1]) equation (3.6.1) can be transformed to a simpler form (for the new variables the old symbols u, t, x, F are kept):

$$\frac{\partial^2 u}{\partial t^2} - a^2\frac{\partial^2 u}{\partial x^2} = \lambda u + \mu F\left(t, x, u, \frac{\partial u}{\partial t}, \frac{\partial u}{\partial x}\right). \quad (3.6.3)$$

Let it be required to find a solution of this equation under the following initial and boundary conditions:

$$u(0,t) = u(l,t) = 0, \tag{3.6.4}$$

$$u|_{t=0} = f(x), \quad \left.\frac{\partial u}{\partial t}\right|_{t=0} = \Phi(x), \tag{3.6.5}$$

where $f(x)$, $\Phi(x)$ are functions of the degree of smoothness ensuring the existence of a solution of the problem (3.6.3) – (3.6.5).

First consider the undisturbed equation obtained from (3.6.3) at $\mu = 0$:

$$\frac{\partial^2 u}{\partial t^2} - a^2\frac{\partial^2 u}{\partial x^2} = \lambda u, \tag{3.6.6}$$

with the previous initial and boundary conditions. Supposing that $(n\pi a l^{-1})^2 - \lambda > 0$ for $n = 1, 2, \dots,$ with the help of the Fourier method we find the solution of equation (3.6.6.) in the form of the series

$$u(t,x) = \sum_{n=1}^{\infty}(A_n\cos\omega_n t + B_n\sin\omega_n t)\,\sin\frac{n\pi x}{l}, \tag{3.6.7}$$

where $\omega_n = \sqrt{(n\pi a l^{-1})^2 - \lambda}$ are frequencies of normal oscillations; the constants A_n and B_n are determined from the initial conditions.

Proceeding from the analytical structure of solution (3.6.7) and supposing that with sufficiently small μ the forms of normal oscillations in the presence of disturbances are determined with sufficient accuracy by the same functions $\sin(n\pi x/l)$ $(n = 1, 2, \dots)$, we will now seek the solution of the initial equation (3.6.3) in the form of the series

$$u(t,x,\mu) = \sum_{n=1}^{\infty} z_n(t,\mu)\sin\frac{n\pi x}{l}, \tag{3.6.8}$$

where $z_n(t,\mu)$ are unknown functions. The analytic form (3.6.8) of the sought solution is typical for the well-known Boubnov–Galyorkin method. Substituting (3.6.8) into (3.6.3) and (3.6.5) and integrating with respect to x from 0 to 1, we obtain for the calculation of $z_n(t,\mu)$ an infinite countable system of ordinary differential equations

$$\frac{d^2 z_n}{dt^2} + \omega_n^2 z_n = \mu F_n(t, z_1, z_2, \dots, \dot z_1, \dot z_2, \dots), \quad n = 1, 2, \dots, \tag{3.6.9}$$

with initial conditions

$$z_n|_{t=0} = f_n, \quad \left.\frac{dz_n}{dt}\right|_{t=0} = \Phi_n, \tag{3.6.10}$$

where f_n, Φ_n are the Fourier coefficients of the functions $f(x)$, $\Phi(x)$ respectively.

Now rearrange the system of equations (3.6.9) into standard form. Instead of z_n and $\dot{z}_n$ introduce new variables x_n and x_{-n}, using the equalities

$$z_n = x_n e^{i\omega_n t} + x_{-n} e^{-i\omega_n t}, \tag{3.6.11}$$

$$\frac{dz_n}{dt} = i\omega_n x_n e^{i\omega_n t} - i\omega_n x_{-n} e^{-i\omega_n t}. \tag{3.6.12}$$

From these formulae it follows that x_n and x_{-n} are slowly changing complex conjugate variables. Upon the necessary transformation one can obtain the system

$$\begin{aligned} &\frac{dx_n}{dt} e^{i\omega_n t} + \frac{dx_{-n}}{dt} e^{-i\omega_n t} = 0, \\ &i\omega_n \frac{dx_n}{dt} e^{i\omega_n t} - i\omega_n \frac{dx_{-n}}{dt} e^{-i\omega_n t} = \mu F_n. \end{aligned} \tag{3.6.13}$$

It can be proved that $F_{-n} = F_n$. Therefore, solving (3.6.13) with respect to $\dot{x}_n$ and $\dot{x}_{-n}$, we obtain an infinite system of ordinary differential equations in standard form, equivalent to system (3.6.9):

$$\frac{dx_n}{dt} = \mu X_n(t, x_1, \dots ; x_{-1}, \dots), \qquad n = \pm 1, \pm 2, \dots , \tag{3.6.14}$$

where

$$\begin{aligned} &X_n(t, x_1, \dots ; x_{-1}, \dots) \\ &\quad = \frac{e^{i\omega_n t}}{2i\omega_n} F_n(t,\ x_1 e^{i\omega_1 t} + x_{-1} e^{-i\omega_1 t}, \dots ;\ i\omega_1 x_1 e^{i\omega_1 t} - i\omega_1 x_{-1} e^{-i\omega_1 t}, \dots), \\ &\quad n = \pm 1, \pm 2, \dots \end{aligned}$$

Then using expressions (3.6.11), (3.6.12) and initial conditions (3.6.10), we obtain the initial conditions for (3.6.14):

$$x_n|_{t=0} = \frac{\Phi_n + i\omega_n f_n}{2i\omega_n}, \qquad n = \pm 1, \pm 2, \dots , \tag{3.6.15}$$

where it is assumed that $f_{-n} = f_n$ and $\Phi_{-n} = \Phi_n$.

We apply the method of averaging with respect to t to system (3.6.14) by means of operator M_t, and obtain the following averaged system of the first approximation:

$$\frac{d\bar{x}_n}{dt} = \mu M_t \left[X_n \left(t, \bar{x}_1, \dots ; \bar{x}_{-1}, \dots \right) \right], \quad n = \pm 1, \pm 2, \dots \tag{3.6.16}$$

with the initial conditions

$$\bar{x}_n|_{t=0} = x_n|_{t=0} \, . \tag{3.6.17}$$

Thus the initial problem with distributed parameters finally turns to the autonomous infinite-dimensional system (3.6.16) with initial conditions (3.6.17). For the justification of this algorithm it is necessary to have theorems on ε-promixity of infinite-dimensional vectors $x = (x_1, \dots ; x_{-1}, \dots)$ and $\bar{x} = (\bar{x}_1, \dots ; \bar{x}_{-1}, \dots)$. It should be noted that at the moment there is no complete mathematical justification as there is for the finite-dimensional case (see Grebenikov and Tunusbaev [1]). Nevertheless for the approximate calculation of oscillation amplitudes this method is effective enough. Let us give two examples.

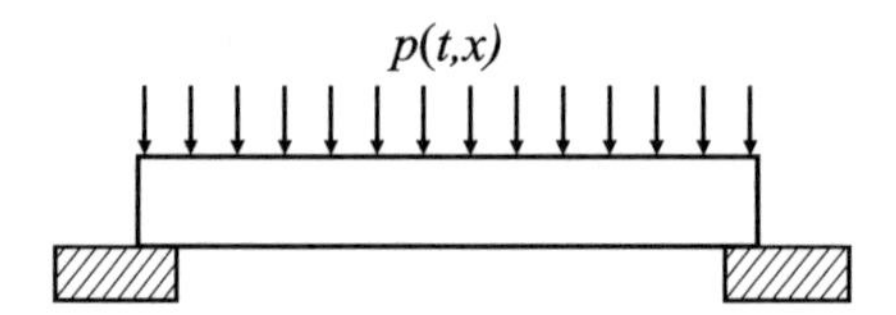

FIG. 3.6.1. Horizontal beam under a shear load $p(t, x)$.

Example 1 We study the transverse oscillation of a beam of length l fixed at the ends and under the shear load $p(t, x)$ (Fig. 3.6.1).

The equation of transverse oscillation of the beam is a particular case of the Kirchhoff equations, and in this case it has the form

$$\rho \frac{\partial^2 u}{\partial t^2} + EJ \frac{\partial^4 u}{\partial x^4} + ku = p(t, x) + \mu \varphi \left(u, \frac{\partial u}{\partial t} \right) + \mu N(u) \frac{\partial^2 u}{\partial x^2}, \tag{3.6.18}$$

where ρ is the linear density of the beam (assumed thin enough), J the moment of inertia of the cross-section, E the modulus of elasticity of the material, φ the reaction of the base of the beam, N the longitudinal force occurring in the beam as a result of the neutral line elongation and equal to

$$N(u) = \frac{ES}{2l} \int_0^l \left(\frac{\partial u}{\partial x} \right)^2 dx, \tag{3.6.19}$$

S is the area of the beam cross-section, k = const, and μ is a small positive parameter.

In equation (3.6.18) the sought function is $u(t, x)$ — the cross-motion of the beam at point x at the moment of time t.

From (3.6.18) and (3.6.19) it is clear that the oscillatory process is nonlinear, and its study in the general case is impossible. Therefore assume that the disturbing load has the infinitesimal order μ, i.e.

$$p(t, x) = \mu q(t, x). \tag{3.6.20}$$

If we now apply to equation (3.6.18) the method of Boubnov–Galerkin and seek a solution in the form of series (3.6.8), then we obtain the following countable system:

$$\begin{gathered} \frac{d^2 z_n}{dt^2} + \omega_n^2 z_n = \mu Z_n(t, z_1, z_2, \ldots; \dot{z}_1, \dot{z}_2, \ldots), \qquad n = 1, 2 \\ \rho Z_n = q_n(t) + \varphi_n(z_1, \ldots; \dot{z}_1, \ldots) - ES \frac{(n\pi)^2}{l^2} z_n \sum_{s=1}^{\infty} \frac{(s\pi)^2 z_s^2}{4l^2}, \end{gathered} \tag{3.6.21}$$

where $q_n(t)$, $\varphi_n(z_1, \ldots; \dot{z}_1, \ldots)$ are coefficients in the Fourier series for functions $q(t, x)$ and $\varphi(u, \dot{u})$ with respect to the fundamental functions $\sin(n\pi x/l)$, $n = 1, 2, \ldots$

Using the above method it is possible to rearrange equations (3.6.21) into the standard form

$$\begin{aligned} \frac{dx_n}{dt} &= \frac{\mu e^{-i\omega_n t}}{2i\omega_n} X_n\left(t,\, x_1 e^{\omega_1 t} + x_{-1} e^{-i\omega_1 t},\, \ldots\right), \\ \frac{dx_{-n}}{dt} &= \frac{\mu e^{i\omega_n t}}{2i\omega_n} X_{-n}\left(t,\, x_1 e^{i\omega_1 t} + x_{-1} e^{-i\omega_1 t},\, \ldots\right). \end{aligned} \tag{3.6.22}$$

Assume that the function $\widetilde{\varphi} = q + \varphi$ has cubic nonlinearity:

$$\widetilde{\varphi}(t, x, u, \dot{u}) = q(t, x) + (\alpha + \beta u + \gamma u^2)\dot{u}$$

with specified constant parameters α, β, γ. Such a concrete definition is necessary for the completion of the analytic computation. Then the Fourier coefficient φ_n can be represented by the formula

$$\varphi_n = J'_n + J''_n + J'''_n,$$

where

$$J'_n = \frac{2\alpha}{l} \int_0^l \sum_{k=1}^{\infty} \dot{z}_1 \sin\frac{k\pi x}{l} \sin\frac{n\pi x}{l}\, dx,$$

$$\begin{aligned} J''_n &= \frac{2\beta}{l} \int_0^l \sum_{m,s=1}^{\infty} z_m \dot{z}_s \sin\frac{m\pi x}{l} \sin\frac{s\pi x}{l} \sin\frac{n\pi x}{l}\, dx \\ &= \frac{\beta}{2l} \int_0^l \Bigg[\sum_{k=0}^{\infty} a_k \left(\sin\frac{(k-n)\pi x}{l} + \sin\frac{(k+n)\pi x}{l} \right) \\ &\quad - \sum_{j=2}^{\infty} b_j \left(\sin\frac{(j-n)\pi x}{l} + \sin\frac{(j+n)\pi x}{l} \right) \Bigg] dx, \end{aligned}$$

$$a_k = \sum_{s=1}^{\infty} z_{k+s} \dot{z}_s, \qquad b_j = \sum_{s=1}^{j-1} z_{j-s} \dot{z}_s,$$

$$\begin{aligned} J'''_n &= \frac{\gamma}{4l} \int_0^l \Bigg[\sum_{k_1=0}^{\infty} A_{k_1} \left(\cos\frac{(k_1-n)\pi x}{l} - \cos\frac{(k_1+n)\pi x}{l} \right) \\ &\quad + \sum_{k_2=1}^{\infty} B_{k_2} \left(\cos\frac{(k_2-n)\pi x}{l} - \cos\frac{(k_2+n)\pi x}{l} \right) \\ &\quad - \sum_{k_2=1}^{\infty} C_{k_3} \left(\cos\frac{(k_3-n)\pi x}{l} - \cos\frac{(k_3+n)\pi x}{l} \right) \\ &\quad - \sum_{k_4=1}^{\infty} D_{k_4} \left(\cos\frac{(k_4-n)\pi x}{l} - \cos\frac{(k_4+n)\pi x}{l} \right) \Bigg] dx, \end{aligned}$$

$$A_{k_1} = \sum_{k,s=1}^{\infty} z_{k_1+k+s} z_k \dot{z}_s, \qquad B_{k_2} = \sum_{k=1}^{\infty} \sum_{s=1}^{k_2-1} z_{k_2+k-s} z_k \dot{z}_s,$$

$$C_{k_3} = \sum_{s=1}^{\infty} \sum_{k=1}^{k_3+s-1} z_{k_3-k+s} z_k \dot{z}_s, \qquad D_{k_4} = \sum_{s=1}^{k_4-1} \sum_{k=1}^{k_4-s-1} z_{k_4-k-s} z_k \dot{z}_s.$$

Consider in more detail the expressions for J_n', J_n'', J_n'''. Due to the orthogonality of trigonometric functions, we find that $J_n' = 0$ at $k \neq n$ and $J_n' = \alpha z_n$ at $k = n$. Let $k \pm n$ and $j \pm n$ be odd numbers, i.e. suppose that

(a) if n is odd, and k and j are even, then

$$J_n'' = \frac{4\beta}{l} \sum_{k=1}^{\infty} \frac{k(a_{2k} - b_{2k})}{4k^2 - (2n-1)^2};$$

(b) if n is odd, k and j are even, then

$$J_n'' = \frac{2\beta}{l} \left[\sum_{k=2}^{\infty} \frac{(2k-1)(a_{2k-1} - b_{2k-1})}{(2k-2n-1)(2k+2n-1)} + \frac{a_1}{1-4n^2} \right].$$

With even values of $k \pm n$ and $j \pm n$ we have $J_n'' = 0$. If $n > 2$, then $J_n''' = \gamma(A_n + B_n - C_n - D_n)/4$. For $n = 1, 2$, $J_n''' = \gamma(A_n + B_n - C_n)/4$.

If we use the expression for $\widetilde{\varphi}$, system (3.6.21) may be given the form

$$\frac{d^2 z_n}{dt^2} + \omega_n^2 z_n = \frac{\mu}{\rho} \left[q_n(t) + \alpha \dot{z}_n + J_n'' + J_n''' - ES \frac{n^2 \pi^4}{4l^4} z_n \sum_{k=1}^{\infty} (k z_k)^2 \right], \tag{3.6.23}$$

and equations (3.6.22) in this case will have the form

$$\begin{aligned} \frac{dx_n}{dt} = \frac{\mu}{\rho} \Bigg[& \frac{q_n(t) e^{-i\omega_n t}}{2i\omega_n} + \frac{\alpha}{2} \left(x_n - x_{-n} e^{-2i\omega_n t} \right) + \frac{e^{-i\omega_n t}}{2i\omega_n} (J_n'' + J_n''') \\ & - \frac{ES(n\pi)^2}{8il^2\omega_n} \sum_{k=1}^{\infty} \left(\frac{k\pi}{ld} \right)^2 (x_k^2 x_n e^{2i\omega_n t} + 2x_k x_{-k} x_n \\ & + x_{-k}^2 x_n e^{-2i\omega_n t} + x_k^2 x_{-n} e^{2i(\omega_k - \omega_n)t} \\ & + 2x_k x_{-k} x_{-n} e^{-2i\omega_n t} + x_{-k}^2 x_{-n} e^{-2i(\omega_k + \omega_n)t}) \Bigg] \end{aligned} \tag{3.6.24}$$

$$\begin{aligned} \frac{dx_{-n}}{dt} = -\frac{\mu}{\rho} \Bigg[& \frac{q_n(t) e^{i\omega_n t}}{2i\omega_n} + \frac{\alpha}{2} \left(x_n e^{2i\omega_n t} - x_{-n} \right) + \frac{e^{i\omega_n t}}{2i\omega_n} (J_n'' + J_n''') \\ & - \frac{ES(n\pi)^2}{8il^2\omega_n} \sum_{k=1}^{\infty} \left(\frac{k\pi}{l} \right)^2 (2x_k x_{-k} x_{-n} + x_k^2 x_n e^{2i(\omega_k + \omega_n)t} \\ & + 2x_k x_{-k} x_n e^{2i\omega_n t} + x_{-k}^2 x_n e^{2i(\omega_n - \omega_k)t} \\ & + x_k^2 x_{-n} e^{2i\omega_k t} + x_k^2 x_{-n} e^{-2i\omega_n t}) \Bigg]. \end{aligned} \tag{3.6.25}$$

Now suppose that the linear disturbance μX_n is a periodic function t, and consider two cases.

1. Let the Fourier coefficients $q_n(t)$ of the external load have the form $q_n(t) = q_n^{(0)} \sin \nu t$, where ν does not depend on n. Then, averaging equations (3.6.24) and (3.6.25) over t, we obtain

$$\begin{aligned}\frac{d\bar{x}_n}{dt} = \frac{\mu \bar{x}_n}{\rho}\Bigg[\frac{\alpha}{2} - \frac{ES(n\pi)^2}{8il^2\omega_n}\bar{x}_n\bar{x}_{-n} - \frac{\gamma}{8}\bar{x}_n\bar{x}_{-n} \\ - \frac{ES(n\pi)^2}{4il^2\omega_n}\sum_{k=1}^{\infty}\left(\frac{k\pi}{l}\right)^2 \bar{x}_k\bar{x}_{-k}\Bigg] - \mu a_n,\end{aligned} \tag{3.6.26}$$

$$\begin{aligned}\frac{d\bar{x}_n}{dt} = -\frac{\mu \bar{x}_{-n}}{\rho}\Bigg[-\frac{\alpha}{2} - \frac{ES(n\pi)^2}{8il^2\omega_n}\bar{x}_n\bar{x}_{-n} + \frac{\gamma}{8}\bar{x}_n\bar{x}_{-n} \\ - \frac{ES(n\pi)^2}{4il^2\omega_n}\sum_{k=1}^{\infty}\left(\frac{k\pi}{l}\right)^2 \bar{x}_k\bar{x}_{-k}\Bigg] - \mu a_{-n},\end{aligned} \tag{3.6.27}$$

where $a_n = 0$ at $\omega_n \neq 0$ and $a_n = q_n/(4\rho\omega_n)$, if $\omega_n = \nu$ with the value of n equal to n_0. The condition $\omega_{n_0} = \nu$ shows the presence of resonance between the frequency ω_{n_0} of the beam's natural oscillations and the frequency ν of the applied load.

In the case of zero resonance, equations (3.6.26) and (3.6.27) can be integrated. For that, multiply (3.6.26) by $\bar{x}_{-n}$, (3.6.27) by $\bar{x}_n$, and sum the products:

$$\bar{x}_{-n}\frac{d\bar{x}_n}{dt} + \bar{x}_n\frac{d\bar{x}_{-n}}{dt} = \frac{\mu\alpha\bar{x}_n\bar{x}_{-n}}{\rho} - \frac{\mu\gamma\left(\bar{x}_n\bar{x}_{-n}\right)^2}{4\rho}.$$

This is the well-known Bernoulli equation (see Smirnov [1]) for the function $\bar{x}_n\bar{x}_{-n}$. Solving it with the specified initial conditions for $\bar{x}_n(0)\bar{x}_{-n}(0)$, we obtain

$$\bar{x}_n(t)\bar{x}_{-n}(t) = 4\alpha\bar{x}_n(0)\bar{x}_{-n}(0)e^{\mu\alpha t/\rho}\left[4\alpha - \gamma\bar{x}_n(0)\bar{x}_{-n}(0)\left(e^{\mu\alpha t/\rho} - 1\right)\right]^{-1}.$$

Upon substitution of this expression into the right-hand member of equations (3.6.26) and (3.6.27) and upon integration we find

$$\bar{x}_n(t,\mu) = \bar{x}_n(0)e^{\lambda_n(t,\mu)}, \tag{3.6.28}$$

$$\bar{x}_{-n}(t,\mu) = \bar{x}_{-n}(0)e^{\lambda_{-n}(t,\mu)}. \tag{3.6.29}$$

In these formulae

$$\begin{aligned}\lambda_n(t,\mu) = \frac{\mu\alpha t}{2\rho} - \frac{1}{2}\left[1 + \frac{ES(n\pi)^2}{i\gamma l^2\omega_n}\right] \\ \times \ln\left[1 + \gamma\bar{x}_n(0)\bar{x}_{-n}(0)\frac{1}{4\alpha}(e^{\mu\alpha t/\rho} - 1)\right] \\ - \frac{ES(n\pi)^2}{i\gamma l^2\omega_n}\sum_{s=1}^{\infty}\left(\frac{s\pi}{l}\right)^2 \ln\left[1 + \gamma\bar{x}_s(0)\bar{x}_{-s}(0)\frac{1}{4\alpha}(e^{\mu\alpha t/\rho} - 1)\right],\end{aligned} \tag{3.6.30}$$

$$\lambda_{-n}(t,\mu) = \frac{\mu\alpha t}{2\rho} + \frac{1}{2}\left[-1 + \frac{ES(n\pi)^2}{i\gamma l^2\omega_n}\right]$$
$$\times \ln\left[1 + \gamma\bar{x}_n(0)\bar{x}_{-n}(0)\frac{1}{4\alpha}(e^{\mu\alpha t/\rho} - 1)\right] \tag{3.6.31}$$
$$+ \frac{ES(n\pi)^2}{i\gamma l^2\omega_n}\sum_{s=1}^{\infty}\left(\frac{s\pi}{l}\right)^2 \ln\left[1 + \gamma\bar{x}_s(0)\bar{x}_{-s}(0)\frac{1}{4\alpha}(e^{\mu\alpha t/\rho} - 1)\right],$$
$$n = 1, 2, \ldots$$

If $\gamma = 0$, i.e. the function $\varphi(u, \dot{u})$ contains quadratic nonlinearity, and the mentioned resonance is zero, the averaged system of the first approximation will take the form

$$\frac{d\bar{x}_n}{dt} = \frac{\mu\bar{x}_n}{\rho}\left[\frac{\alpha}{2} - \frac{ES(n\pi)^2}{8il^2\omega_n}\bar{x}_n\bar{x}_{-n} - \frac{ES(n\pi)^2}{4il^2\omega_n}\sum_{s=1}^{\infty}\left(\frac{s\pi}{l}\right)^2\bar{x}_s\bar{x}_{-s}\right],$$
$$\frac{d\bar{x}_{-n}}{dt} = \frac{\mu\bar{x}_{-n}}{\rho}\left[\frac{\alpha}{2} + \frac{ES(n\pi)^2}{8il^2\omega_n}\bar{x}_n\bar{x}_{-n} + \frac{ES(n\pi)^2}{4il^2\omega_n}\sum_{s=1}^{\infty}\left(\frac{s\pi}{l}\right)^2\bar{x}_s\bar{x}_{-s}\right]. \tag{3.6.32}$$

In this case we will obviously obtain the linear equation

$$\bar{x}_{-n}\frac{d\bar{x}_n}{dt} + \bar{x}_n\frac{d\bar{x}_{-n}}{dt} = \frac{\mu\alpha}{\rho}\bar{x}_n\bar{x}_{-n},$$

which will give upon integration

$$\bar{x}_n(t,\mu)\bar{x}_{-n}(t,\mu) = \bar{x}_n(0)\bar{x}_{-n}(0)e^{\mu\alpha t/\rho},$$

and then

$$\bar{x}_n(t,\mu) = \bar{x}_n(0)e^{\lambda'_n(t,\mu)}, \tag{3.6.33}$$
$$\bar{x}_{-n}(t,\mu) = \bar{x}_{-n}(0)e^{\lambda'_{-n}(t,\mu)}, \tag{3.6.34}$$

$$\lambda'_n(t,\mu) = \frac{\mu\alpha t}{2\rho} - \frac{ES(n\pi)^2}{8\rho il^2\omega_n}\bar{x}_n(0)\bar{x}_{-n}(0)(e^{\mu\alpha t/\rho} - 1)$$
$$- \frac{ES(n\pi)^2}{4\rho il^2\omega_n}(e^{\mu\alpha t/\rho} - 1)\sum_{k=1}^{\infty}\left(\frac{k\pi}{l}\right)^2\bar{x}_k(0)\bar{x}_{-k}(0),$$
$$\lambda'_{-n}(t,\mu) = \frac{\mu\alpha t}{2\rho} + \frac{ES(n\pi)^2}{8\rho il^2\omega_n}\bar{x}_n(0)\bar{x}_{-n}(0)(E^{\mu\alpha t/\rho} - 1)$$
$$+ \frac{ES(n\pi)^2}{4\rho il^2\omega_n}(e^{\mu\alpha t/\rho} - 1)\sum_{k=1}^{\infty}\left(\frac{k\pi}{l}\right)^2\bar{x}_k(0)\bar{x}_{-k}(0).$$

In the case of resonance $\omega_{n_0} = \nu$ the system of equations (3.6.26), (3.6.27) is not analytically integrable, therefore it should be split into two subsystems: nonresonance — a countable subsystem with the variable index $n = 1, \ldots, n_0 - 1, n_0 + 1, \ldots$; and resonance — a subsystem consisting only of two equations with unknown functions $\bar{x}_{n_0}(t,\mu)$ and $\bar{x}_{-n_0}(t,\mu)$. The first subsystem is integrated in compact form by the above method (products $\bar{x}_n\bar{x}_{-n}$ with $n \neq n_0$ are found). As regards the resonance subsystem, it finally takes the form

$$\frac{d\bar{x}_{n_0}}{dt} = \frac{\mu\bar{x}_{n_0}}{\rho}\left[\frac{\alpha}{2} - \bar{x}_{n_0}\bar{x}_{-n_0}\left(\frac{\gamma}{8} + \frac{3ES(n_0\pi)^2}{8il^2\nu}\right) + A(t)\right] - \frac{\mu q_{n_0}^{(0)}\sin\nu t}{4\rho\nu},$$

$$\frac{d\bar{x}_{-n_0}}{dt} = \frac{\mu\bar{x}_{-n_0}}{\rho}\left[\frac{\alpha}{2} + \bar{x}_{n_0}\bar{x}_{-n_0}\left(\frac{\gamma}{8} + \frac{3ES(n_0\pi)^2}{8il^2\nu}\right) + A(t)\right] - \frac{\mu q_{n_0}^{(0)}\sin\nu t}{4\rho\nu},$$

where $A(t)$ is a time function calculated through integration of the nonresonance subsystem. The latter is nonlinear, so it seems reasonable to find its solution by numerical methods.

2. Let the load $q(t,x)$ be a quasi-periodic time function with the frequency spectrum $\nu_1, \nu_2, \dots, \nu_s$. Then its Fourier coefficients have the form

$$q_j(t) = q_j^{(0)} e^{i\nu_j t}. \tag{3.6.35}$$

In this case system (3.6.26), (3.6.27) also splits into two subsystems (resonance and nonresonance), both of which can contain an infinite number of equations. If the nonresonance subsystem can also be integrated in analytic form, then the resonance subsystem can usually be investigated by numerical methods. Certainly, in practice it is essential at what values of n_0 a resonance may occur. If n_0 is equal to a very large integer, then the standard system of ordinary differential equations for $\bar{x}_n(t,\mu)$ and $\bar{x}_{-n}(t,\mu)$ should be considered practically nonresonance, and the above methods are applicable to it.

Example 2 Consider the nonlinear equation

$$\frac{\partial^4 u}{\partial x^4} + a^2\frac{\partial^2 u}{\partial t^2} = \mu f(t,u), \tag{3.6.36}$$

occurring in problems of flexural oscillations of a beam located on an elastic base under the influence of external disturbing forces (see Mitropolsky and Moiseenkov [1]). Below we describe the method of rearrangement of this equation to a countable system of ordinary differential equations.

Suppose that $a \neq 0$, $\mu > 0$ and $f(t,u)$ is a function analytic in some domain, that does not contain in its expansion any absolute terms with coefficients depending on t. Under these conditions it is required to find a solution of $u(t,x)$, satisfying the boundary and initial conditions

$$u(t,0) = u(t,l) = 0, \quad \left.\frac{\partial^2 u}{\partial x^2}\right|_{x=0} = \left.\frac{\partial^2 u}{\partial x^2}\right|_{x=l} = 0, \tag{3.6.37}$$

$$u|_{t=0} = f_1(x), \quad \left.\frac{\partial u}{\partial t}\right|_{t=0} = F(x). \tag{3.6.38}$$

Let the functions $f_1(x)$ and $F(x)$ be decomposable into absolutely convergent series

$$f_1(x) = \sum_{k=1}^{\infty} \alpha_k \sin \frac{k\pi x}{l}, \qquad F(x) = \sum_{k=1}^{\infty} \beta_k \sin \frac{k\pi x}{l},$$

$$\alpha_k = \frac{2}{l} \int_0^l f_1(x) \sin \frac{k\pi x}{l}\, dx, \qquad \beta_k = \frac{2}{l} \int_0^l F(x) \sin \frac{k\pi x}{l}\, dx,$$

$$|\alpha_k| \leq M/k^2, \quad |\beta_k| \leq M/k^2, \quad M = \text{const.}$$

As before, we will seek the solution of the formulated problem in the form of the series

$$u(t,x) = \sum_{n=1}^{\infty} z_n(t) \sin \frac{n\pi x}{l}.$$

For the unknown functions $z_n(t,\mu)$ $(n = 1,2,\dots)$ we obtain the countable system of equations

$$\frac{d^2 z_n}{dt^2} + \omega_n^2 z_n = \mu f_n(t, z_1, z_2, \dots), \qquad n = 1,2,\dots \tag{3.6.39}$$

with the initial conditions

$$z_n(0) = \alpha_n, \qquad \left.\frac{dz_n}{dt}\right|_{t=0} = \beta_n, \tag{3.6.40}$$

$$\omega_n = \frac{n\pi}{l\sqrt{a}}, \qquad f_n = \frac{2}{a^2 l} \int_0^l f\left(t, \sum_{m=1}^{\infty} z_m(t) \sin \frac{m\pi x}{l}\right) \sin \frac{n\pi x}{l}\, dx.$$

Rearrangement of system (3.6.39) into the standard form in order to be able to use the averaging method does not meet with any difficulties.

In conclusion we will only note that in principle, the coefficients f_n may contain frequencies resonant with eigenfrequencies of the differential operator $d^2/dt^2 + \omega_n^2$, and this may substantially complicate the study of not only the standard system, but also the corresponding averaged system of the first approximation.

3.7 Energy Method of Construction of Amplitude–Phase Equations

For the construction of averaged systems of any approximation by means of the Krylov–Bogolyubov transform we often tried to obtain an explicit form of equations for the oscillation amplitudes and phases (so-called amplitude–phase equations). But the Krylov–Bogolyubov method is not the only possible one for solving this problem. Here we will describe another method of the construction of amplitude–phase equations of the first approximation, based on application of a so-called energy method (for the detailes see Godunov and Riabenky [1], and Mitropolsky and Moiseenkov [1]) well known in the equations of mathematical physics. It originates from the variational principle of virtual displacements, one of the classical principles of engineering mechanics. According to this principle, the virtual work

δW is equal to the scalar product of the vector F of forces influencing a mechanical system and the virtual displacement vector δx:

$$\delta W = (F, \delta x). \tag{3.7.1}$$

If the mathematical model is described by N degrees of freedom, the forces F are the generalized forces Q_{s0} from the second-order Lagrange equations and virtual displacements correspond to the variations of amplitude and phase of the first normal oscillation with a frequency $\omega_1(\tau)$, $\tau = \mu t$ is slow time. Then the virtual work of the generalized forces is equal to

$$\delta W = \sum_{s=1}^{N} Q_{s0}\delta x_s, \quad \delta x_s = \varphi_s^{(1)}(\tau)\cos\xi\delta a - \varphi_s^{(1)}(\tau)a\sin\xi\delta\psi. \tag{3.7.2}$$

The symbol δ, like in all variational principles, in the general case denotes not a differential, but a universal deviation; $\xi = p\theta/q + \psi$; p, q are coprime numbers; a is an amplitude; ψ is a phase; and θ is an angular argument with respect to which the functions of the Krylov–Bogolyubov transform are 2π-periodic. Let $\overline{\delta W}$ denote the value of the average virtual work within a complete oscillation cycle within the period 2π:

$$\overline{\delta W} = \frac{1}{2\pi}\int_0^{2\pi} \delta W\, d\xi. \tag{3.7.3}$$

Upon construction of a Fourier series for Q_{r0} $(r = 1, \dots, N)$ of arguments ξ and θ and further integration of the expression for δW over the complete phase ξ, we obtain

$$\overline{\delta W} = \frac{1}{4\pi^2}\sum_{s=-\infty}^{\infty} e^{isq\psi}\int_0^{2\pi}\int_0^{2\pi}\sum_{r=1}^{N} Q_{r0}\varphi_r^{(1)}e^{-irq\psi}\left[\cos\xi\delta a - a\sin\xi\delta\psi\right] d\theta\, d\xi. \tag{3.7.4}$$

The formulae show that the summation index s is connected with the indices n and m indices of the double Fourier series for Q_{r0} by the relations $n = -sp$, $m = sq + 1$. Expression (3.7.4) may be given the form

$$\overline{\delta W} = \sum_{s=-\infty}^{\infty} \overline{\delta W_s} = \sum_{s=-\infty}^{\infty}\left[\frac{\overline{\delta W_s}}{\delta a}\delta a + \frac{\overline{\delta W_s}}{\delta\psi}\delta\psi\right], \tag{3.7.5}$$

where

$$\frac{\overline{\delta W_s}}{\delta a} = \frac{1}{4\pi^2}e^{isq\psi}\int_0^{2\pi}\int_0^{2\pi}\sum_{r=1}^{N} Q_{r0}\varphi_r^{(1)}e^{-irq\psi}\cos\xi\, d\xi\, d\theta, \tag{3.7.6}$$

$$\frac{\overline{\delta W_s}}{\delta\psi} = -\frac{1}{4d\pi^2}e^{isq\psi}\int_0^{2\pi}\int_0^{2\pi}\sum_{r=0}^{N} Q_{r0}\varphi_r^{(1)}e^{-irq\psi}a\sin\xi\, d\xi\, d\theta. \tag{3.7.7}$$

If the amplitude–phase equations of the first approximation are written in the usual form

$$\begin{aligned}\frac{da}{dt} &= \mu A_1(\tau, a, \psi),\\ \frac{d\psi}{dt} &= \omega_1(\tau) - \frac{p\nu(\tau)}{q} + \mu B_1(\tau, a, \psi),\end{aligned} \tag{3.7.8}$$

then, according to the theorem given in Mitropolsky and Moiseenkov [1],

$$\begin{aligned}A_1(\tau, a, \psi) = &-\frac{a}{2m_1(\tau)\omega_1(\tau)}\frac{d[m_1(\tau)\omega_1(\tau)]}{d\tau}\\ &+\frac{2}{m_1(\tau)}\sum_{s=-\infty}^{\infty}\left[is(q\omega_1 - p\nu)\frac{\overline{\delta W_s}}{\delta a} + \frac{2\omega_1}{a}\frac{\overline{\delta W_s}}{\delta\psi}\right]\\ &\times\left[4\omega_1^2 - (q\omega_1 - p\nu)^2 s^2\right]^{-1},\\ B_1(\tau, a, \psi) = &\frac{2}{m_1(\tau)a}\sum_{s=-\infty}^{\infty}\left[(q\omega_1 - p\nu)\frac{is}{a}\frac{\overline{\delta W_s}}{\delta\psi} - 2\omega_1\frac{\overline{\delta W_s}}{\delta a}\right]\\ &\times\left[4\omega_1^2 - (q\omega_1 - p\nu)^2 s^2\right]^{-1},\\ m_1(\tau) = &\ 2T(\varphi^{(1)}(\tau)),\end{aligned} \tag{3.7.9}$$

where T is the kinetic energy of the undisturbed system.

$\overline{\delta W_s}$ is average virtual work that within an oscillation cycle would be performed by the disturbing force in a sinusoidal mode (the s-th term of its expansion into a Fourier series) over virtual displacements corresponding to variations of amplitude δa and phase $\delta\psi$ of the oscillation.

So the following rule can be formulated (see Mitropolsky and Moiseenkov [1]): for the construction of amplitude–phase equations of the first approximation it is necessary to find the value of the average virtual work that within an oscillation cycle would be performed by the disturbing force in a sinusoidal mode over virtual displacements corresponding to variations of amplitude and phase, expand the obtained expression into a Fourier series, and after that substitute the partial derivatives of the s-th term into (3.7.9).

This rule becomes much simpler if the oscillatory system is under the influence of potential disturbing forces only, because in this case

$$\delta W = -\delta V,$$

where V is the disturbed potential energy, i.e. the part of the potential energy that occurs due to the presence of a disturbance.

So the energy method allows us to derive approximate equations for the amplitudes and phases without preliminary construction of exact differential equations of the problem. Expressions for the work (or potential energy) and the kinetic energy are directly used for their construction. This circumstance makes it possible to

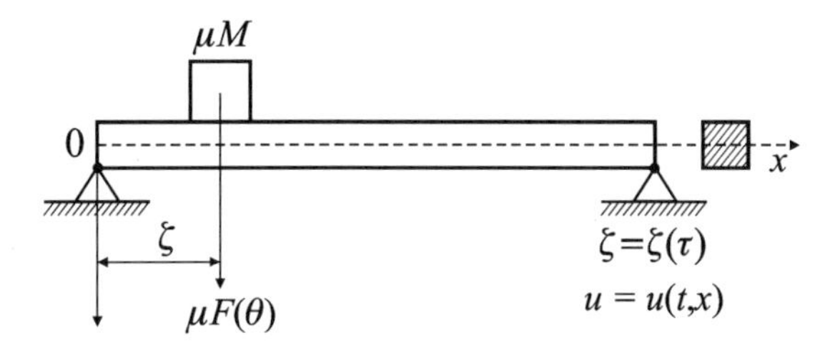

FIG. 3.7.1. Rod influenced by travelling load and pulsating force.

extend the energy method to the solving of problems of the theory of oscillations of systems with distributed parameters, described by partial differential equations.

As an example, consider the problem of transverse oscillations of a rod under the influence of a traveling load and a pulsating force (see Fig. 3.7.1).

Using the energy method, construct amplitude–phase equations for this problem. Let S be the area of the rod cross-section of dimensions small in comparison with its length l. Let us also assume that on the rod a mass μM is moving which is small compared with the mass M of the rod. In addition, let the rod be influenced by a vertical force $\mu F(\theta)$ periodic with respect to θ, with its point of application at any point of time coinciding with the center of mass μM. Let ρ, E be the density and Young's modulus of the material, and J be the moment of inertia of the rod's cross-section relative to the axis perpendicular to the flexure plane.

If the inertia of the cross-section rotation and transverse forces are neglected, then the expressions for the kinetic and potential energies of the rod–load system will have the form

$$
\begin{aligned}
T &= \frac{\rho S}{2}\int_0^l \left(\frac{\partial u}{\partial t}\right)^2 dx + \mu M \left(\frac{\partial u}{\partial t}\right)^2\bigg|_{x=\zeta} = T_0 + \mu T_1,\\
V &= \frac{EJ}{2}\int_0^l \left(\frac{\partial^2 u}{\partial x^2}\right)^2 dx - \mu M g \left(u|_{x=\zeta}\right) = V_0 + \mu V_1.
\end{aligned}
\tag{3.7.10}
$$

The equation for the natural oscillations of the rod (undisturbed motion) is determined by the undisturbed values of the kinetic and potential energies

$$
T_0 = \frac{\rho S}{2}\int_0^l \left(\frac{\partial u}{\partial t}\right)^2 dx, \qquad V_0 = \frac{EJ}{2}\int_0^l \left(\frac{\partial^2 u}{\partial x^2}\right)^2 dx
$$

and it can easily be deduced, e.g. from the variational principle of Hamilton–Ostrogradsky

$$
\delta \int (T_0 - V_0)\, dt = 0, \tag{3.7.11}
$$

wherefrom we obtain the equation

$$\frac{\partial^4 u}{\partial x^4} + \frac{\rho S}{EJ}\frac{\partial^2 u}{\partial t^2} = 0. \tag{3.7.12}$$

The boundary conditions for the simply supported or hinged ends of the rod are given by the equalities

$$u|_{x=0} = \left.\frac{\partial^2 u}{\partial x^2}\right|_{x=0} = u|_{x=l} = \left.\frac{\partial^2 u}{\partial x^2}\right|_{x=l} = 0. \tag{3.7.13}$$

We apply the Fourier method to equation (3.7.2). Then the equation for the eigenfunctions determining the forms of natural oscillations has the form

$$\frac{d^4 X}{dx^4} - k^4 X = 0, \qquad k^4 = \frac{\rho S \omega^2}{EJ}, \tag{3.7.14}$$

where ω is the frequency of the principal oscillation. Writing the general solution with the use of the formula

$$X(x) = C_1 \sin kx + C_2 \cos kx + C_3 \operatorname{sh} kx + C_4 \operatorname{ch} kx$$

and taking into consideration the boundary conditions written as

$$X|_{x=0} = \left.\frac{d^2 X}{dx^2}\right|_{x=0} = X|_{x=l} = \left.\frac{d^2 X}{dx^2}\right|_{x=l} = 0,$$

we obtain the basic frequency equation

$$\sin kl = 0, \tag{3.7.15}$$

with its solutions being $k_n = n\pi/l$ $(n = 1, 2, \ldots)$. Hence we determine the forms of normal (principal) oscillations of the undisturbed motion

$$X_n(x) \equiv \varphi_n(x) = \sin\frac{n\pi x}{l}$$

and their eigenfrequencies

$$\omega_n = \frac{n^2\pi^2}{l^2}\sqrt{\frac{EJ}{\rho S}}, \qquad n = 1, 2, \ldots,$$

required for the energy method application.

For definiteness we assume that $\mu F(\theta) = \mu F_0 \sin\theta$, and the instantaneous frequency of the disturbing force $d\theta/dt = \nu(\mu t)$ is positive. Suppose that in the time interval $0 \le t \le T$ the frequency of the vertical force $\nu(\mu t)$ takes a value equal to ω_1, i.e. the oscillatory system passes through principal resonance, because it is the most interesting case.

Taking into account the disturbing potential energy

$$\mu V_1 = -\mu M g \left(u|_{x=\zeta}\right),$$

(where $\zeta = v\tau$ is the current coordinate of the point of application of the disturbing force, v is the velocity of the motion of the load and pulsating force along the rod) and the disturbing kinetic energy

$$\mu T_1 = \mu M \left(\frac{\partial u}{\partial t} \right)^2 \bigg|_{x=\zeta}$$

the summed disturbing force can be presented as a sum of the pulsating force applied from the outside and the force of inertia, i.e.

$$\mu \Phi \left(\tau, \theta, \frac{\partial^2 u}{\partial t^2} \right) = \left[\mu F_0 \sin \theta + \mu M g u - \mu M \frac{\partial^2 u}{\partial t^2} \right]_{x=\zeta}. \tag{3.7.16}$$

From (3.7.16) it is clear that the beam flexure u should be taken as a generalized coordinate. Then the first approximation can be written as

$$u^{(1)} = a \sin \frac{\pi x}{l} \cos(\theta + \psi), \tag{3.7.17}$$

where the amplitude a and the phase ψ are determined from the equations of the first approximation (3.7.8) with functions (3.7.9) and with consideration for $\omega_1 \approx \nu(\tau)$.

Now, using the energy method, we deduce equations (3.7.8) and (3.7.9). First we find the expression for the disturbing force $\mu\Phi$ in a sinusoidal oscillation mode:

$$\begin{aligned} u^{(1)}(t, x) &= a \varphi^{(1)}(x) \cos(\theta + \psi), \\ \frac{\partial u^{(1)}}{\partial t} &= -a \omega_1 \varphi^{(1)}(x) \sin(\theta + \psi), \\ \frac{\partial^2 u^{(1)}}{\partial t^2} &= -a \omega_1^2 \varphi^{(1)}(x) \cos(\theta + \psi), \end{aligned}$$

where $\varphi^{(1)}(x) = \sin\left(\frac{\pi x}{l}\right)$. As a result we obtain

$$\mu \Phi_0^{(1)} = \mu \left[F_0 \sin \theta + M g + a \omega_1^2 M \sin \frac{\pi \zeta}{l} \cos(\theta + \psi) \right].$$

Then we write the expression for the virtual work corresponding to the variations of the amplitude and phase of the first normal oscillation:

$$\begin{aligned} \delta W = \mu \Phi_0^{(1)} \delta u^{(1)} = \mu & \left[F_0 \sin \theta + M g u + a \omega_1^2 M \sin \frac{\pi \zeta}{l} \cos(\theta + \psi) \right] \\ & \times \left[\sin \frac{\pi \zeta}{l} \cos(\theta + \psi) \delta a - a \sin \frac{\pi \zeta}{l} \sin(\theta + \psi) \delta \psi \right]. \end{aligned}$$

After that we find the average virtual work $\overline{\delta W}$ and its "partial derivatives"

$$\begin{aligned} \frac{\overline{\delta W_1}}{\delta a} &= \frac{iF_0}{4} e^{i\psi} \sin \frac{\pi\zeta}{l}, & \frac{\overline{\delta W_{-1}}}{\delta a} &= -\frac{iF_0}{4} e^{-i\psi} \sin \frac{\pi\zeta}{l}, \\ \frac{\overline{\delta W_1}}{\delta \psi} &= -\frac{aF_0}{4} e^{i\psi} \sin d\frac{\pi\zeta}{l}, & \frac{\overline{\delta W_{-1}}}{\delta \psi} &= -\frac{aF_0}{4} e^{-i\psi} \sin \frac{\pi\zeta}{l}, \\ \frac{\overline{\delta W_0}}{\delta a} &= \frac{a\omega_1^2 M}{2} \sin^2 \frac{\pi\zeta}{l}, & \frac{\overline{\delta W_0}}{\delta \psi} &= 0 \end{aligned} \tag{3.7.18}$$

and

$$m_1 = \rho S \int_0^l \sin^2 \frac{\pi x}{l}\, dx = \frac{\rho S l}{2}. \tag{3.7.19}$$

Substituting all these expressions into (3.7.9), we obtain amplitude–phase equations of the first approximation

$$\begin{aligned} \frac{da}{dt} &= -\frac{2\mu F_0}{\rho S l(\omega_1 + \nu)} \sin \frac{\pi\zeta}{l} \cos\psi, \\ \frac{d\psi}{dt} &= \omega_1 - \nu - \frac{\mu\omega_1 M}{\rho S l} \sin^2 \frac{\pi\zeta}{l} + \frac{2\mu F_0}{\rho S a l(\omega_1 + \nu)} \sin \frac{\pi\zeta}{l} \sin\psi, \end{aligned} \tag{3.7.20}$$

where $\zeta = \zeta(\tau)$, $\nu = \nu(\tau)$.

In conclusion we note that the amplitude–phase equations can also be given complex notation, using the substitution of variables $z = ae^{i\psi}$.

Then instead of system (3.7.20) obtain the equation

$$\frac{dz}{dt} = ip(\tau)z - \frac{2\mu F_0}{\rho S l(\omega_1 + \nu)} \sin \frac{\pi\zeta}{l}, \tag{3.7.21}$$

where

$$p(\tau) = \omega_1 - \nu - \frac{\mu\omega_1 M}{\rho S l} \sin^2 \frac{\pi\zeta}{l}. \tag{3.7.22}$$

3.8 Averaging Method and Maximum Principle in Boundary Value Problems

It seems logical that the asymptotic theory of differential equations, using the averaging method, has been widely adopted in optimal control theory, including its important part that is usually called the maximum principle (see Pontryagin, *et al.* [1]). Though there are numerous known results (see Avramchuk [1], Klick [1, 2], Akulenko and Chernousko [1], Chernousko [1], etc.) and a lot of publications on this subject, we will describe one of the first (chronologically) results obtained by Avramchuk and Klikh [1], and the mathematical issues of applicability of the averaging method in the maximum principle (see also Klikh [2]).

Consider the following time minimization problem. It is required to solve the problem of a material point's flight from a given initial point to a given end point within minimum time. The material point moves in the central field of forces under the influence of a small controlled traction.

The equations of the material point's motion in this model will have the form

$$\frac{d^2r}{dt^2} - r\left(\frac{d\theta}{dt}\right)^2 = -\frac{GM}{r^2} + w\cos\beta, \qquad \frac{d}{dt}\left(r^2\frac{d\theta}{dt}\right) = rw\sin\beta, \tag{3.8.1}$$

where r and θ are polar coordinates of the point, m is its mass, M the mass of the attracting center, G the gravity constant, and the acceleration constant in absolute value, caused by traction W is equal to

$$w = |W/m| \ll 1, \tag{3.8.2}$$

and β is the angle between the direction of traction and the radius-vector of the moving point.

The problem is to find an optimal control $\beta = \beta(t)$ of the traction vector, and of the corresponding path moving the phase point within minimum time from the initial position

$$r_0, \ \theta_0, \ \left.\frac{dr}{dt}\right|_{t=0}, \ \left.\frac{d\theta}{dt}\right|_{t=0} \tag{3.8.3}$$

to the given end position

$$r_k, \ \theta_k, \ \left.\frac{dr}{dt}\right|_{t=t_k}, \ \left.\frac{d\theta}{dt}\right|_{t=t_k}. \tag{3.8.4}$$

The authors proposed to seek for the optimal path in the form of an osculating ellipse (see Duboshin [1])

$$\rho = \frac{x_3}{1 + x_1\cos(\theta - x_2)}, \tag{3.8.5}$$

where $\rho = r/r_0$, $x_1 = a/h$ is the eccentricity of the orbit, $x_2 = \omega$ is the angular distance of the pericenter from the line of nodes, and $x_3 = 1/h$ is a focal parameter.

Passing to the new phase coordinates $x = (x_1, x_2, x_3)$ and to the new independent variable θ, we obtain

$$\begin{aligned}
\frac{dx_1}{d\theta} &= \mu x_3^2 \frac{2x_1 + 2\cos(\theta - x_2) - x_1\sin^2(\theta - x_2)}{[1 + x_1\cos(\theta - x_2)]^3}\sin\beta \\
&\quad + \mu x_3^2 \frac{\sin(\theta - x_2)}{[1 + x_1\cos(\theta - x_2)]^3}\cos\beta, \\
\frac{dx_2}{d\theta} &= \mu\frac{x_3^2}{x_1}\frac{[2 + x_1\cos(\theta - x_2)]\sin(\theta - x_2)}{[1 + x_1\cos(\theta - x_2)]^3}\sin\beta \\
&\quad - \mu\frac{x_3^2}{x_1}\frac{\cos(\theta - x_2)}{[1 + x_1\cos(\theta - x_2)]^3}\cos\beta, \\
\frac{dx_3}{d\theta} &= \mu\frac{2x_3^2}{[1 + x_1\cos(\theta - x_2)]^3}\sin\beta,
\end{aligned} \tag{3.8.6}$$

where $\mu = w$ is the small parameter of the problem. From the second equation of system (3.8.1) it follows that the control $\beta = \beta(\theta)$ should be chosen so as to minimize the integral

$$\int_{\theta_0}^{\theta_k} \frac{x_2^{3/2}}{[1 + x_1 \cos(\theta - x_2)]^2} \, d\theta. \tag{3.8.7}$$

The formulated problem will be solved with the use of the maximum principle.

First of all, construct the function

$$H = -f_0(x, \theta) + \mu \sum_{i=1}^{3} f_i(x, \theta, \beta) y_i + y_4, \tag{3.8.8}$$

where

$$f_0(x, \theta) = \frac{x_3^{3/2}}{[1 + x_1 \cos(\theta - x_2)]^2}, \tag{3.8.9}$$

f_i are the right-hand members of equations (3.8.6), and the auxiliary variables $y = (y_1, y_2, y_3)$ and y_4 are the solution of the conjugated system (for the detailes see Plotnikov [1], Pontryagin, *et al.* [1], ets).

The control $\beta = \beta(\theta)$ is now found from the condition of the existence of $\max H$, hence

$$\begin{aligned} \beta(\theta) = {} & \operatorname{arctg}\{x_1[2x_1 + 2\cos(\theta - x_2) - x_1 \sin^2(\theta - x_2)]y_1 \\ & + \sin(\theta - x_2)[2 + x_1 \cos(\theta - x_2)]y_2 + 2x_1 x_3 y_3\}/\{[1 \\ & + x_1 \cos(\theta - x_2)][x_1 \sin(\theta - x_2) y_1 - \cos(\theta - x_2) y_2]\}, \end{aligned} \tag{3.8.10}$$

Then, substituting $\beta(\theta)$ into (3.8.8), we obtain

$$H = - f_0(x, \theta) + \mu A(x, \theta) \sqrt{B(x, y, \theta)} + y_4, \tag{3.8.11}$$

$$A(x, \theta) = \frac{x_3^2}{x_1[1 + x_1 \cos(\theta - x_2)]^3}, \tag{3.8.12}$$

$$\begin{aligned} B(x, y, \theta) = {} & [x_1 \sin(\theta - x_2) y_1 - \cos(\theta - x_2) y_2]/[1 + x_1 \cos(\theta - x_2)]^2 \\ & + \{x_1[2x_1 + 2\cos(\theta - x_2) - x_1 \sin^2(\theta - x_2)]y_1 \\ & + \sin(\theta - x_2)[2 + x_1 \cos(\theta - x_2)]y_2 + 2x_1 x_3 y_3\}^2. \end{aligned} \tag{3.8.13}$$

As is known from Pontryagin, *et al.* [1], on the sought optimal path

$$- f_0(x(\theta), \theta) + \mu A(x(\theta), \theta) \sqrt{B(x(\theta), y(\theta), \theta)} + y_4(\theta) \equiv 0 \tag{3.8.14}$$

for any $\theta \in [\theta_0, \theta_k]$, and this relation is henceforth used as one of the system integrals. Also note that the function $A(x, \theta)$ essentially depends on x and θ, i.e. $\partial A / \partial x \not\equiv 0$, $\partial A / \partial \theta \not\equiv 0$.

Taking account of (3.8.14), the problem of finding an optimal trajectory turns to the following two-point boundary value problem (see Lans [1]): it is necessary to find the solution of the seventh-order system of differential equations

$$\frac{dx}{d\theta} = \mu^2 \frac{A^2(x,\theta)}{2[f_0(x,\theta) - y_4]} \frac{\partial B(x,y,\theta)}{\partial y},$$
$$\frac{dy}{d\theta} = \frac{\partial f_0(x,\theta)}{\partial x} - \frac{f_0(x,\theta) - y_4}{A(x,\theta)} \frac{\partial A(x,\theta)}{\partial x} - \mu^2 \frac{A^2(x,\theta)}{2[f_0(x,\theta) - y_4]} \frac{\partial B(x,y,\theta)}{\partial x},$$
$$\frac{dy_4}{d\theta} = \frac{\partial f_0(x,\theta)}{\partial \theta} - \frac{f_0(x,\theta) - y_4}{A(x,\theta)} \frac{\partial A(x,\theta)}{\partial \theta} - \mu^2 \frac{A^2(x,\theta)}{2[f_0(x,\theta) - y_4]} \frac{\partial B(x,y,\theta)}{\partial \theta} \tag{3.8.15}$$

with the boundary conditions

$$x(\theta_0) = x^0, \quad x(\theta_k) = x^k, \quad y_4(\theta_k) = 0. \tag{3.8.16}$$

The exact solution of the boundary value problem (3.8.15), (3.8.16) cannot be found in compact form, so for the construction of an approximate solution we will use the averaging method.

The generating system is presented by the equations

$$\frac{dx}{d\theta} = 0, \qquad x = \text{const},$$
$$\frac{dy}{d\theta} = \frac{\partial f_0(x,\theta)}{\partial x} - \frac{f_0(x,\theta) - y_4}{A(x,\theta)} \frac{\partial A(x,\theta)}{\partial x}, \tag{3.8.17}$$
$$\frac{dy_4}{d\theta} = \frac{\partial f_0(x,\theta)}{\partial \theta} - \frac{f_0(x,\theta) - y_4}{A(x,\theta)} \frac{\partial A(x,\theta)}{\partial \theta},$$

The last equation of (3.8.17) is easily integrable. Its first integral can be written in the form

$$\frac{f_0(x,\theta) - y_4}{A(x,\theta)} = \frac{f_0(x,\theta_0) - y_4(\theta_0)}{A(x,\theta_0)} = \text{const}. \tag{3.8.18}$$

If one uses equality (3.8.18), then the other equation of (3.8.17) can be easily integrated:

$$y_1 = \frac{x_3^{3/2} - (1 + x_1^2) y_4^0}{1 - x_1} \frac{\sin(\theta - x_2 - \theta_0 + x_2^0)}{[1 + x_1 \cos(\theta - x_2 - \theta_0 + x_2^0)]^3}$$
$$- \frac{2x_3^{3/2}(1 - x_1^2) - (1 + x_1)(1 + 4x_1^2)[x_3^{3/2} - (1 + x_1)^2 y_4^0]}{2(1 - x_1^2)^2}$$
$$\times \frac{\sin(\theta - x_2 - \theta_0 + x_2^0)}{[1 + x_1 \cos(\theta - x_2 - \theta_0 + x_2^0)]^2}$$

$$- \frac{2x_3^{3/2}(1-x_1^2)(1+2x_1^2) + (1+x_1)(1-16x_1^2)[x_3^{3/2} - (1+x_1)^2 y_4^0]}{2(1-x_1^2)^3}$$

$$\times \frac{\sin(\theta - x_2 - \theta_0 + x_2^0)}{1 + x_1 \cos(\theta - x_2 - \theta_0 + x_2^0)}$$

$$+ \frac{6x_3^{3/2} x_1^2 (1-x_1^2) + (1+x_1)(2 - 13x_1^2 - 4x_1^4)[x_3^{3/2} - (1+x_1)^2 y_4^0]}{x_1(1-x_1^2)^{7/2}}$$

$$\times \operatorname{arctg}\left(\frac{\sqrt{1-x_1^2}}{1+x_1} \operatorname{tg} \frac{\theta - x_2 - \theta_0 + x_2^0}{2}\right) + y_1^0,$$

$$y_2 = \frac{x_3^{3/2}}{\left[1 + x_1 \cos(\theta - x_2 - \theta_0 + x_2^0)\right]^2}$$

$$+ \frac{(1+x_1)\left[x_3^{3/2} - (1+x_1)^2\right]}{\left[1 + x_1 \cos(\theta - x_2 - \theta_0 + x_2^0)\right]^3} + y_2^0 + y_4^0,$$

$$y_3 = \frac{x_1\left[x_3^{3/2} - (1+x_1)^2 y_4^0\right]}{x_3(1-x_1)} \frac{\sin(\theta - x_2 - \theta_0 + x_2^0)}{\left[1 + x_1 \cos(\theta - x_2 - \theta_0 + x_2^0)\right]^2}$$

$$- \frac{3x_1\left\{x_3^{3/2}(1-x_1^2) - 2(1+x_1)\left[x_3^{3/2} - (1+x_1)^2 y_4^0\right]\right\}}{2x_3(1-x_1^2)^2}$$

$$\times \frac{\sin(\theta - x_2 - \theta_0 + x_2^0)}{1 + x_1 \cos(\theta - x_2 - \theta_0 + x_2^0)}$$

$$+ \frac{3x_3^{3/2}(1-x_1^2) - 2(2 + x_1^2)(1+x_1)\left[x_3^{3/2} - (1+x_1)^2 y_4^0\right]}{x_3(1-x_1)^{5/2}}$$

$$\times \operatorname{arctg}\left(\frac{\sqrt{1-x_1^2}}{1+x_1} \operatorname{tg} \frac{\theta - x_2 - \theta_0 + x_2^0}{2}\right) + y_3^0,$$

$$y_4 = - \frac{(1+x_1)\left[x_3^{3/2} - (1+x_1)^2 y_4^0\right]}{\left[1 + x_1 \cos(\theta - x_2 - \theta_0 + x_2^0)\right]^3} + \frac{x_3^{3/2}}{\left[1 + x_1 \cos(\theta - x_2 - \theta_0 + x_2^0)\right]^2}. \tag{3.8.19}$$

Here the notation $x_i^0 = x_i(\theta_0)$ is used. In formulae (3.8.19) the variable x_2 is separated so that the form of the solution should be such as required in accordance with the theorem of Volosov [1].

Then, the slow variables $x = (x_1, x_2, x_3)$ are determined from the system of three equations upon substitution of the fast variables y and y_4 in (3.8.15) by their values (3.8.19), and the averaging of the right-hand members over the angular argument θ. We then have

$$\begin{aligned}
\frac{d\bar{x}_1}{d\theta} &= -\mu^2 \frac{\bar{x}_3^4\big[(1-\bar{x}_1^2)\big(4-7\bar{x}_1^2-4\sqrt{1-\bar{x}_1^2}\big)y_1^0+6\bar{x}_1^3 y_3^0\big]}{2\bar{x}_1^2(1+\bar{x}_1)\,(1-\bar{x}_1^2)^{3/2}\big[\bar{x}_3^{3/2}-(1+\bar{x}_1)^2 y_4^0\big]}, \\
\frac{d\bar{x}_2}{d\theta} &= -\frac{\mu^2\bar{x}_3^4}{2\bar{x}_1^4(1+\bar{x}_1)\,[\bar{x}_3^{3/2}-(1+\bar{x}_1)^2 y_4^0]} \\
&\quad\times\Bigg[\frac{\bar{x}_3^{3/2}(2\bar{x}_1^4+11\bar{x}_1^2+4)}{(1-\bar{x}_1^2)^{5/2}}-\frac{4\bar{x}_3^{3/2}(2\bar{x}_1^2+1)}{(1-\bar{x}_1^2)^2} \\
&\quad+(y_2^0+y_4^0)\left(\frac{2\bar{x}_1^4+\bar{x}_1^2-4}{(1-\bar{x}_1^2)^{3/2}}+4\right) \\
&\quad-\frac{12\bar{x}_3^{3/2}\bar{x}_1^2}{(1-\bar{x}_1^2)^{5/2}}\ln\frac{2(1+\bar{x}_1)}{1+\sqrt{1-\bar{x}_1^2}}\Bigg]-\frac{\mu^2\bar{x}_3^4}{2\bar{x}_1^4}\Bigg[\frac{2(16\bar{x}_1^2-1)}{(1-\bar{x}_1^2)^3} \\
&\quad+\frac{42\bar{x}_1^6+40\bar{x}_1^5+269\bar{x}_1^4-96\bar{x}_1^3-316\bar{x}_1^2+32\bar{x}_1+16}{8(1-\bar{x}_1^2)^{9/2}} \\
&\quad+\frac{2(4\bar{x}_1^4+13\bar{x}_1^2-2)}{(1-\bar{x}_1^2)^{7/2}}\ln\frac{2(1+\bar{x}_1)}{1+\sqrt{1-\bar{x}_1^2}}\Bigg], \\
\frac{d\bar{x}_3}{d\theta} &= -\mu^2\frac{\bar{x}_3^5\,[3\bar{x}_1(1-\bar{x}_1^2)y_1^0-2\bar{x}_3(2+\bar{x}_1^2)y_3^0]}{(1+\bar{x}_1)\,(1-\bar{x}_1^2)^{5/2}\big[\bar{x}_3^{3/2}-(1+\bar{x}_1)^2 y_4^0\big]}.
\end{aligned} \tag{3.8.20}$$

So the initial problem turns to the integration (for the given initial conditions $x(\theta_0) = x^0$) of the three averaged equations (3.8.20). The initial values of the vector of fast variables $y^0 = (y_1^0, y_2^0, y_3^0)$ included in those equations as parameters should be chosen so that the boundary conditions $x(\theta_k) = x^k$ are satisfied, and the value of y_4^0 is considered as known, because it is determined from the equality

$$y_4^0 = f_0(x^k, \theta_0) - \frac{A(x^k, \theta_0)}{A(x^k, \theta_k)} f_0(x^k, \theta_k).$$

In the first approximation, the optimal control $\beta = \beta(\theta)$ determined by (3.8.10) where $x = (x_1(\theta), x_2(\theta), x_3(\theta))$ are replaced by the solution of the averaged system (3.8.20) at values of y_i^0 chosen from the boundary conditions, and the functions $y_1(\theta)$, $y_2(\theta)$, $y_3(\theta)$, $y_4(\theta)$ are determined by equalities (3.8.19) where x_1, x_2, x_3 in their turn are also replaced by the solution of equations (3.8.20), may not be close to the exact optimal control. However, as proved by Avramchuk and Klikh [1], in the interval $[\theta_0, \theta_k]$ of length $T \approx 1$ it is approximate in the sense of the minimized functional (3.8.7). These authors give a numerical example. Let $x_1^0 = 0.0005$, $x_2^0 = 0.0459$, $x_3^0 = 1.0005$, $\theta_0 = 0.2500$, $x_1^k = 0.0104$, $x_2^k = 1.0708 + 2k\pi$, $x_3^k = 1.1000$, $\theta_k = 0.5000 + 2k\pi$, $\mu = 0.001$.

Solving the differential equations for the optimal path (phase coordinates $x_1(\theta)$, $x_2(\theta)$, $x_3(\theta)$ and coordinate $\rho(\theta)$) on computer, we find the dependence of the above coordinates on θ (Fig. 3.8.1). The authors calculated the values of y_i^0 by

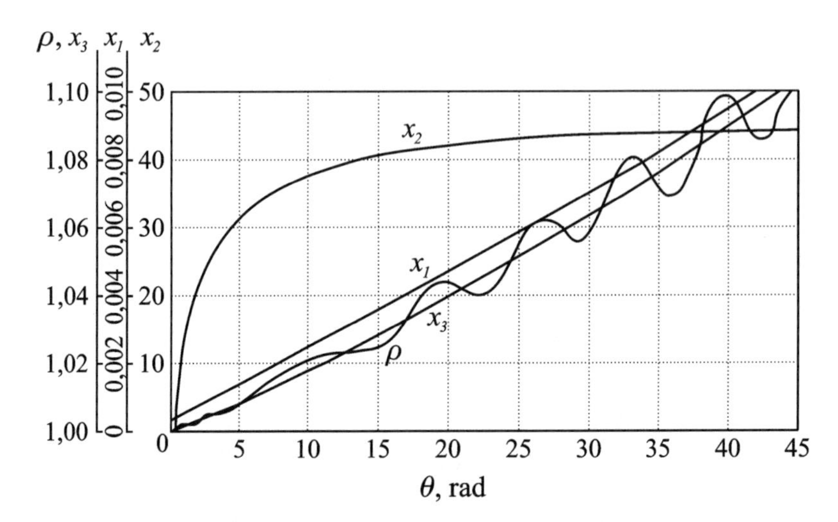

FIG. 3.8.1. Dependence of phase coordinates x_1, x_2, x_3, ρ of the optimal path on the polar angle θ.

means of successive refinement, minimizing the discrepancy at the end point (see Lans [1]), and they are as follows: $y_1^0 = 83.440$, $y_2^0 = 9.2355$, $y_3^0 = 457.9999$, $y_4^0 = 0.0082$. The optimal control $\beta = \beta(\theta)$ is given in Fig. 3.8.2. The authors have also calculated the minimum flight time by their approximate method and by the method of high accuracy numerical integration; the results were sufficiently close.

While using the averaging method for solving optimal control problems, a number of questions arise related to the mathematical justification of the averaging method. Among them is the question of ε-proximity of the exact and smoothed solutions of an optimal control problem within a sufficiently large time interval, the continuous dependence of a solution on the parameter, etc.

Here we will describe two such results. Let there be given a system of differential equations in standard form

$$\frac{dx}{dt} = \mu f(t, x, u) \tag{3.8.21}$$

and boundary conditions

$$x(t_0, \mu) = x_0, \quad h(t_1, x(t_1, \mu)) = 0, \tag{3.8.22}$$

where x, f are n-dimensional vectors, u is an r-dimensional control vector, μ is a small positive parameter, and t is the (real) time.

In addition, let the vector u be defined in a compact domain $U_r \subset R_r$ of Euclidean space, and the vector-function $f(t, x, u)$ in the $(n + r + 1)$-dimensional domain $G_{n+r+1} = \{x \in P_n,\ u \in U_r,\ t \in (t_0, \infty)\}$.

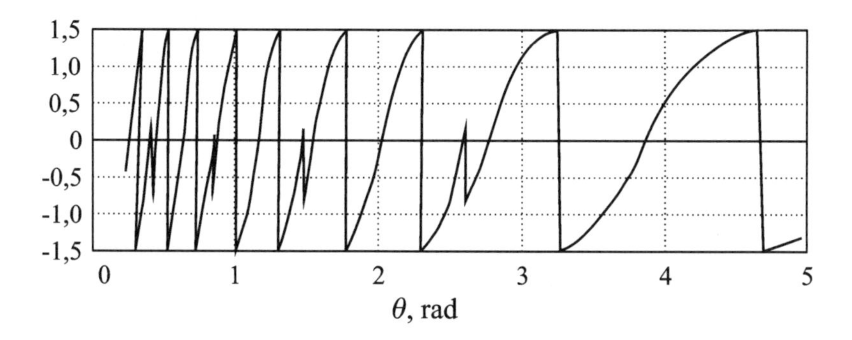

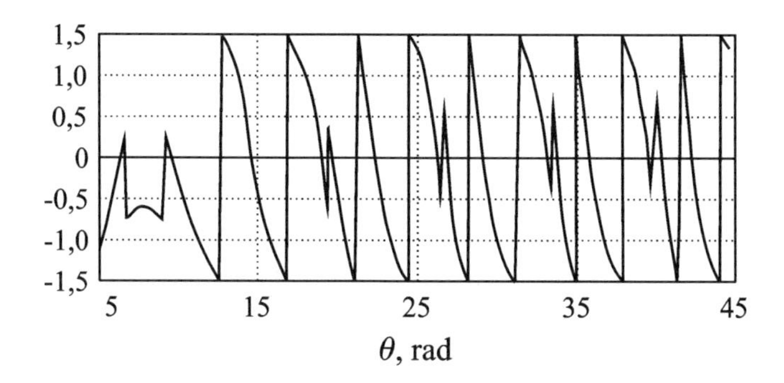

FIG. 3.8.2. Dependence of optimal control β on angle θ.

We will also suppose that f and h are defined and continuous with respect to their arguments, and f is continuously differentiable with respect to t and x in the corresponding domains.

Along with the problem (3.8.21), (3.8.22) consider the functional

$$J(x,u) = \mu \int_{t_0}^{t_1} f_0(x,t,u)\,dt. \tag{3.8.23}$$

It is required to find a control that would solve the boundary-value problem (3.8.21), (3.8.22) and minimize the functional (3.8.23). For the further computation assume that system (3.8.21) is controllable in P_n, the issue of optimal control has a solution in the interval $[t_0, t_1]$ for every $\mu \in [0, \mu^*]$, and the set of functions $V(t,x) = \{f_0(t,x,u), f(t,x,u)\}$, $u \in U_r$, is strictly convex for every $(n+1)$-dimensional vector (t,x). Under these conditions the control $u = u^*(t,x,\psi)$ solving the formulated problem is a single-valued function.

Substitute $u = u^*(t, x, \psi)$ into the right-hand members of system (3.8.21) and the system conjugated with the latter. If we denote

$$X_0(t, x, \psi) = f_0(t, x, u^*(t, x, \psi)),$$
$$X(t, x, \psi) = f(t, x, u^*(t, x, \psi)),$$
$$\Psi(t, x, \psi) = \frac{\partial f_0(t, x, u^*(t, x, \psi))}{\partial x} - \left(\frac{\partial f(t, x, u^*(t, x, \psi))}{\partial x}, \psi\right),$$

then we obtain a two-point boundary problem of the maximum principle:

$$\begin{aligned} \frac{dx}{dt} &= \mu X(t, x, \psi), \quad x(t_0, \mu) = x^0, \\ \frac{d\psi}{dt} &= \mu \Psi(t, x, \psi), \quad h(t_1, x(t_1, \mu)) = 0. \end{aligned} \tag{3.8.24}$$

We associate problem (3.8.24) with the two-point boundary problem with averaged equations

$$\begin{aligned} \frac{d\bar{x}}{dt} &= \mu \overline{X}(\bar{x}, \bar{\psi}), \quad \bar{x}(t_0, \mu) = x^0, \\ \frac{d\bar{\psi}}{dt} &= \mu \overline{\Psi}(\bar{x}, \bar{\psi}), \quad \bar{h}(t_1, \bar{x}(t_1, \mu)) = 0, \end{aligned} \tag{3.8.25}$$

using the notation

$$\overline{X}(\bar{x}, \bar{\psi}) = M_t[X(t, x, \psi)], \quad \overline{\Psi}(\bar{x}, \bar{\psi}) = M_t[\Psi(t, x, \psi)]. \tag{3.8.26}$$

Then for the two-point boundary problem of the maximum principle (3.8.4) the following theorem is true:

Theorem (see Klikh [1]) *Let:*

(1) *the vector-functions $X(t, x, \psi)$, $\Psi(t, x, \psi)$ be uniformly bounded and continuous with respect to x, ψ uniformly with respect to $t \in [t_0, \infty)$, $(x, \psi) \in P_{2n} \subset G_{2n}$;*

(2) *the function $h(t, x)$ be uniformly bounded and continuous with respect to $t \in [t_0, \infty)$, $x \in P_n \subset R_n$ with all $\mu \in [0, \mu^*]$;*

(3) *at each point $(x, \psi) \in P_{2n}$ there be limits*

$$\overline{X}(x, \psi) = \lim_{T \to \infty} \frac{1}{T} \int_{t_0}^{t_0+T} X(t, x, \psi)\, dt,$$

$$\overline{\Psi}(x, \psi) = \lim_{T \to \infty} \frac{1}{T} \int_{t_0}^{t_0+T} \Psi(t, x, \psi)\, dt;$$

(4) *for $\mu = 1$ the boundary problem (3.8.25) have the unique solution $(\bar{x}(t, \mu)$, $\bar{\psi}(t, \mu))$, determined for all $t \in [t_0, \infty)$ and lying in a bounded domain G_{2n}.*

Then for any $\varepsilon > 0$ *there exists* $\mu^0 \in [0, \mu^*]$, *such that with* $0 < \mu \leq \mu^0$ *the solution* $(x(t,\mu),\, \psi(t,\mu))$ *of the boundary problem (3.8.24) satisfying the boundary conditions*

$$x(t_0, \mu) = x^0, \quad h(t_0 + 1/\mu,\, x(t_0 + 1/\mu,\, \mu)) = 0,$$

and the solution $(\bar{x}(t,\mu),\, \bar{\psi}(t,\mu))$ *of the averaged boundary problem (3.8.25) in the same time interval* $[t_0, t_0 + \mu^{-1}]$ *are* ε*-close in their norm, i.e.*

$$\|x(t,\mu) - \bar{x}(t,\mu)\| < \varepsilon, \quad \|\psi(t,\mu) - \bar{\psi}(t,\mu)\| < \varepsilon, \quad t \in [t_0,\, t_0 + 1/\mu]. \tag{3.8.27}$$

Remark If furthermore the function $X_0(t, x, \psi)$ in the domain $G_{2n+1} = \{(x,\psi) \in P_{2n},\ t \in [t_0,\, t_0 + \mu^{-1}]\}$ satisfies with respect to $(x, \psi) \in P_{2n}$ the Lipschitz condition with the constant L, then we have the estimation for the functional

$$|J(x(t,\mu),\, u^*(t,x,\psi)) - J(\bar{x}(t,\mu),\, u^*(t,\bar{x},\bar{\psi}))| < 2L\varepsilon, \tag{3.8.28}$$

where $u^*(t,x,\psi)$, $u^*(t,\bar{x},\bar{\psi})$ are optimal controls of the corresponding boundary problems.

The second result relates to the continuous dependence of the solution of the two-point boundary problem of the maximum principle on the parameter that is in essence a generalization of the well-known results of Gikhman [1] and Krasnoselsky and Krein [1] for the case of optimal control.

This is as follows. Let there be a two-point boundary problem

$$\begin{aligned} \frac{dx}{d\tau} &= X(\tau, x, \psi, \mu), \quad x(0,\mu) = x^0, \\ \frac{d\psi}{d\tau} &= \Psi(\tau, x, \psi, \mu), \quad h(T, x(T,\mu), \mu) = 0, \end{aligned} \tag{3.8.29}$$

where

$$\begin{aligned} X(\tau, x, \psi, \mu) &= X(t_0 + \tau\mu^{-1}, x, \psi), \\ \Psi(\tau, x, \psi, \mu) &= \Psi(t_0 + \tau\mu^{-1}, x, \psi), \\ \tau = \mu(t - t_0), \quad h(\tau, x, \mu) &= h(t_0 + \tau\mu^{-1}, x). \end{aligned}$$

Under certain conditions imposed on the functions (3.8.29), Klikh [1] obtained the following result: for any $\varepsilon > 0$ there exists a sufficiently small neighborhood of the point μ_0 such that for all points of this neighborhood and for all solutions of $(x(\tau,\mu),\, \psi(\tau,\mu))$ of the boundary problem (3.8.29) defined for $0 \leq \tau \leq T$ and satisfying the boundary conditions

$$x(0,\mu) = x^0, \quad h(T, x(T,\mu), \mu) = 0,$$

the following ε-inequalities are true:

$$\|x(\tau,\mu) - x(\tau,\mu_0)\| < \varepsilon, \quad \|\psi(\tau,\mu) - \psi(\tau,\mu_0)\| < \varepsilon, \quad \tau \in [0,T].$$

In conclusion note that the real application of the averaging method in optimal control theory actually turns to the development of effective algorithms for computers, because in important applications of control theory it is hardly possible to manage without extensive computations.

3.9 Comments and References

This chapter described the solution of certain nonlinear problems of space dynamics and mechanics of continua, based on asymptotic approximations and the averaging principle. The contents of this chapter are to a certain extent an illustration of the mathematical results contained in Chapter 2. At the same time it should be noted that the field of application of the asymptotic theory of equations is significantly wider and varied. It is successfully used in high-energy physics (in the study of charged beam dynamics in accelerators), in aircraft navigation (in the study of automatic regimes of flight), in mathematical control theory (in solving boundary problems of differential equations), in mathematical biology, geology, and other areas of knowledge.

As the reader should have noticed, we tried to advance as far as possible in solving one or another problem with the help of mathematical analytic apparatus, though it is clear that on this path it is seldom possible to obtain a satisfactory solution of a real, but not simplified model problem. Only a sensible, well-considered synthesis of the modern mathematical arsenal (analytical, qualitative and numerical methods) and new computer technology give important practical recommendations and results.

It is also important to point out the following circumstance. In the previous chapters we only considered the Cauchy problem for ordinary differential equations (apart from problems of oscillations of a rod, which are typical mixed boundary problems). Actually the described mathematical theory in conjunction with mathematical methods of information processing can be successfully used (and has been used many times) in qualitative processing of discrete and continuous experimental information in problems of image processing, filtration problems, etc.

Section 3.3 Justification of the averaging method in the bounded three-body problem, using the averaging operator for continuous disturbances (the averaging schemes of Delaunay–Hill) by means of splitting three-dimensional configuration space into the domains of possibility and impossibility of motions, and detailed comments, can be found in Grebenikov and Ryabov [2].

Section 3.5 The averaging of Hamiltonians with consideration for frequency resonance not only in the first step of the iterations, but also in the second step, the third, etc., was first proposed in a paper by Grebenikov, *et al.* [1].

Section 3.7 The effectiveness of asymptotic theory based on averaging principles in quasi-linear partial differential equations was first proved in the book by Mitropolsky and Moseenkov [1, 2].

Section 3.8 The mathematical theory of the maximum principle widely used in applied problems of control theory is specified in the book by Pontryagin, *et al.* [1]).

4 Numerical–Analytic Methods

4.0 Introduction

This chapter is devoted to numerical–analytic methods of the construction of systems of ordinary differential equations in problems of analytical dynamics. Our first study in this field was performed in 1968 (see Ryabov [1]). Fundamental research in the application of numerical–analytic methods in problems of celestial mechanics was carried out in the 1970s (see Barton [1], Deprit [1], Hansen [1], Hearn [1], and Rom [1]). These methods continue to develop intensively in different directions (see Samoilenko and Ronto [1]) together with the development of computer networks and their mathematical software in the form of analytic programming packages.

In the monograph of Grebenikov and Ryabov [1] a number of examples are given of applications of these methods in problems of nonlinear mechanics, described by quasi-linear differential equations. This chapter is in a certain sense an addition to the material stated in Grebenikov and Ryabov [1], and its continuation. The essence of the methods described here is the fact that they are aimed at implementation of analytic algorithms arising from the iteration variant of the Poincaré–Lyapunov method of small parameters. We use iteration processes converging in the Cauchy sense (and not only asymptotically), which allows us to obtain the solution with a predefined accuracy.

We will also note that in this chapter we basically adhere to a simplified variant of numerical–analytic methods, when a result is presented not in a purely literal form, but in the form of a certain analytic structure (defined by one or another algorithm) with numerical coefficients. This allows us in a number of cases to reduce requirements to the capacity and operating speed of the computer, compared with the variant oriented to literal analytic transformations.

Sections 4.1 and 4.2 describe algorithms for the Lyapunov transform for linear systems of equations with periodic coefficients, and the construction of the Green and Lyapunov matrices for those systems; these play an important part in the analysis and construction of solutions to these systems. Some small examples are given to illustrate it.

In Section 4.3 an algorithm is proposed for direct construction of periodic solutions of linear heterogeneous systems with periodic coefficients.

In Section 4.4 an algorithm is considered for the construction of known periodic Hill's solutions in the problem of lunar motion, illustrated by a table for periodic solutions (in the form of Fourier polynomials) for several numerical values of the basic Hill's parameter.

In Section 4.5 there is a complete algorithm for Mathieu's construction in the form of Fourier polynomials, and a corresponding Quick-Basic program.

In Section 4.6 an algorithm for the construction of solutions of the plane bounded three-body problem is developed, an iteration process is proposed, corresponding to quadratic convergence. Of course, the direct programming of this algorithm and its numerical–analytic implementation are far from being simple, but the authors think that the difficulties are of a technical nature and are surmountable provided that a sufficiently high-powered and high-speed computational complex is available.

In Section 4.7 an algorithm is developed for implementation of the Krylov–Bogolyubov transform (see Bogolyubov [1] which plays an important part in the analysis of nonlinear multifrequency systems with fast and slow variables.

4.1 Construction of Lyapunov Transform for a Linear System with Periodic Coefficients

Consider the linear system

$$\frac{dx}{dt} = P(t)x, \tag{4.1.1}$$

where the matrix $P(t)$ is continuous or piecewise continuous, real and 2π-periodic with respect to t. As is known (see Erugin [1], and Lyapunov [1]), there exists a transform (called the Lyapunov transform)

$$x = L(t)y \tag{4.1.2}$$

with a nonsingular real periodic matrix $L(t)$, reducing (4.1.1) to the system

$$\frac{dy}{dt} = Wy \tag{4.1.3}$$

with constant real matrix W. The fundamental matrix $X(t)$ of system (4.1.1) will be expressed by the formula

$$X(t) = L(t)e^{Wt}. \tag{4.1.4}$$

Matrices $L(t)$ and W are called matrices of the Lyapunov transform and are connected by the equation

$$\frac{dL}{dt} = P(t)L - LW. \tag{4.1.5}$$

Matrix W plays an important part in the analysis of the properties of solutions of linear system (4.1.1), the eigenvalues of W being the characteristic exponents of this system.

4.1.1 Construction of matrices $L(t)$ and W by means of series

Matrices $L(t)$ and W are found on the basis of (4.1.5) in the form of the series

$$L(t) = E + L_1(t) + L_2(t) + \cdots, \tag{4.1.6}$$

$$W = W_1 + W_2 + \cdots, \tag{4.1.7}$$

where E is a unit matrix, and the matrices $L_k(t)$, W_k, $k = 1, 2 \ldots$ satisfy the equations:

$$\begin{aligned} \frac{dL_1}{dt} &= P(t) - W_1, \\ \frac{dL_2}{dt} &= P(t)L_1 - W_2 - L_1W_1, \\ \frac{dL_3}{dt} &= P(t)L_2 - W_3 - L_1W_2 - L_2W_1, \\ &\cdots\cdots\cdots\cdots\cdots \end{aligned} \tag{4.1.8}$$

From these equations we find real 2π-periodic matrices $L_k(t)$ and real constant matrices W_k by the formulae

$$\begin{aligned} W_1 &= \frac{1}{2\pi}\int_0^{2\pi} P(s)\,ds, \qquad L_1(t) = \int_0^t [P(s) - W_1]\,ds, \\ W_2 &= \frac{1}{2\pi}\int_0^{2\pi} [P(s)L_1(s) - L_1(s)W_1]\,ds, \\ L_2(t) &= \int_0^t [P(s)L_1(s) - W_2 - L_1(s)W_1]\,ds, \\ &\cdots\cdots\cdots\cdots\cdots \end{aligned} \tag{4.1.9}$$

Convergence of these series (uniform with respect to t) takes place if the norm $||P(t)||$ of matrix $P(t)$ is comparatively small. According to Erugin [1], convergence is guaranteed if

$$\max_k |\mu_k(t)| < \frac{\ln 2}{2\pi}, \qquad k = 1, 2, \ldots, n, \tag{4.1.10}$$

where μ_k are eigenvalues of the matrix W.

4.1.2 Construction of matrices $L(t)$ and W by means of iterations

Matrices $L(t)$ and W can also be found by means of iterations, using the same equations (4.1.5) (cf. Woodcock, *et al.* [1]). To do this, assume in (4.1.5)

$$L(t) = E + V(t),$$

where E is a unit matrix, and obtain the equation

$$\frac{dV}{dt} = U(t) - W, \tag{4.1.11}$$

where

$$U(t) = P(t)[E + V(t)] - V(t)W. \tag{4.1.12}$$

We will consider matrix W as a constant part of the periodic matrix $U(t)$ and calculate matrices $U(t)$, W, $V(t)$ by means of successive approximations U_k, W_k, $V_k(t)$, $k = 1, 2, \ldots$

The first approximation is:

$$\begin{gathered} U_1(t) = P(t), \qquad W_1 = \frac{1}{2\pi}\int_0^{2\pi} U_1(s)\,ds, \\ V_1(t) = \int_0^{2\pi} g(t,s)[U_1(s) - W_1]\,ds, \end{gathered} \tag{4.1.13}$$

where

$$g(t,s) = \begin{cases} \dfrac{s}{2\pi}, & 0 \le s < t, \\ \dfrac{s}{2\pi} - 1, & t \le s \le 2\pi. \end{cases} \tag{4.1.14}$$

The second approximation is:

$$\begin{gathered} U_2(t) = P(t)[E - V_1(t)] - V_1(t)W_1, \\ W_2 = \frac{1}{2\pi}\int_0^{2\pi} U_2(s)\,ds, \\ V_2(t) = \int_0^{2\pi} g(t,s)[U_2(s) - W_2]\,ds, \end{gathered} \tag{4.1.15}$$

etc.

Let matrix $P(t)$ be representable by the series (polynomials)

$$P(t) = A_0 + \sum_{k\ge 1}(A_k \cos kt + B_k \sin kt). \tag{4.1.16}$$

Then we obtain the following formulae.

The first approximation:

$$U_1(t) = P(t), \quad W_1 = A_0,$$
$$V_1(t) = \sum_{k\geq 1} \frac{1}{k}(A_k \sin kt - \cos kt). \tag{4.1.17}$$

The second approximation:

$$U_2(t) = A_0^{(1)} + \sum_{k\geq 1} \big(A_k^{(1)} \cos kt + B_k^{(1)} \sin kt\big),$$
$$W_2 = A_0^{(1)}, \tag{4.1.18}$$
$$V_2(t) = \sum_{k\geq 1} \frac{1}{k}\big(A_k^{(1)} \sin kt - B_k^{(1)} \cos kt\big),$$

where $A_0^{(1)}$, $A_k^{(1)}$, $B_k^{(1)}$, $k \geq 1$, are coefficients of a Fourier series (polynomials) obtained on the basis of the expression for $U_2(t)$ in (4.1.15), etc.

If matrix $P(t)$ is specified as a Fourier polynomial with numerical coefficients, then the construction of the approximations $U_k(t)$, W_k, $V_k(t)$, as well as the series (4.1.6), (4.1.7) is easy to implement on a computer with the help of software for multiplication, summation and integration of Fourier polynomials. In the general case, when the matrix $P(t)$ is set by a Fourier polynomial with coefficients being letters, then it is expedient to apply the corresponding symbolic computation programs.

The convergence domain in the case of iterations may be wider than in the case of series (4.1.6), (4.1.7), and from the algorithmic point of view the iteration method is preferable. According to Woodcock, *et al.* [1], convergence of iterations is guaranteed if

$$\max_k |\mu_k(t)| < (3 - 2\sqrt{2})\omega, \tag{4.1.19}$$

where $\mu_k(t)$ are eigenvalues of the matrix $P(t)$, and the least period of $P(t)$ is equal to $T = 2\pi/\omega$.

Example Consider the equation

$$\ddot{z} + 0.5\dot{z} + (1 - 0.5\cos\omega t)z = 0,$$

equivalent to the system

$$\dot{x} = P(t)x, \tag{4.1.20}$$

where

$$P(t) = P_0 + P_1 \cos\omega t,$$
$$P_0 = \begin{pmatrix} 0 & 1 \\ -1 & -1/2 \end{pmatrix}, \quad P_1 = \begin{pmatrix} 0 & 0 \\ 1/2 & 0 \end{pmatrix}$$
$$x = (x_1, x_2).$$

In conformity with the described method, we obtain

$$W^{(1)} = P_0, \quad V^{(1)}(t) = \frac{1}{\omega} P_1 \sin \omega t,$$

$$W^{(2)} = W^{(1)}, \quad V^{(2)}(t) = V^{(1)}(t) + \frac{1}{\omega^2} \begin{pmatrix} -1/2 & 0 \\ 1/4 & 1/2 \end{pmatrix} \cos \omega t,$$

$$W^{(3)} = W^{(2)} + \frac{1}{\omega^2} \begin{pmatrix} 0 & 0 \\ -1/8 & 0 \end{pmatrix}, \quad \text{etc.}$$

Let, e.g., $\omega = 10$. Upon the fifth approximation we obtain

$$W = \begin{pmatrix} 0 & 1 \\ -1.001\,297 & -0.500\,006 \end{pmatrix}, \qquad L(t) = E + V(t), \tag{4.1.21}$$

$$V(t) = \sum_{k=1}^{2} A_k \cos k\omega t + B_k \sin k\omega t, \quad E = \begin{pmatrix} 1 & 0 \\ 0 & 1 \end{pmatrix}, \tag{4.1.22}$$

where

$$A_1 = \begin{pmatrix} -0.005\,188 & -0.000\,025 \\ 0.002\,594 & 0.005\,188 \end{pmatrix}, \quad B_1 = \begin{pmatrix} 0.000\,257 & -0.001\,037 \\ 0.050\,907 & -0.000\,257 \end{pmatrix},$$

$$A_2 = \begin{pmatrix} 0.000\,003 & 0 \\ -0.000\,008 & -0.000\,003 \end{pmatrix}, \quad B_2 = \begin{pmatrix} 0 & 0 \\ -0.000\,067 & 0 \end{pmatrix},$$

Here all the coefficients, upon rounding off, are written to an accuracy of 0.000 001, and the other coefficients are supposed to be zero.

4.1.3 Interpolation formulae for Lyapunov transform matrices It is also possible to propose the method of the construction of Lyapunov matrices, based on the numerical integration of system (4.1.1) and on the representation of functions of matrices with the help of Lagrange–Sylvester interpolation polynomials (see Gantmakher [1], and Starzhinsky and Yakubovich [1]). Namely, let the fundamental matrix $X(t)$ of this system be found in this way in the interval $[0, 2\pi]$, i.e. a corresponding table of numerical values of elements of matrix $X(t)$ is constructed. In particular, obtain the numerical matrix $X(2\pi)$, called a monodromy matrix, for the initial system (4.1.1). Then find its eigenvalues $\varrho_1, \ldots, \varrho_n$, called multipliers of this system.

Case (a) There are no negative multipliers. Then the matrix W is calculated by the formula

$$W = \frac{1}{2\pi} \ln X(2\pi), \tag{4.1.23}$$

where $\ln X$ denotes the principal value of the logarithm of X. If all multipliers (complex in the general case) are simple, then the matrix $\ln X(2\pi)$ can be expressed by the finite interpolation formula of Lagrange–Sylvester:

$$\ln X(2\pi) = \sum_{k=1}^{n} \ln \varrho_k \frac{\prod_{j=1,\, j\neq k}^{n} [X(2\pi) - E\varrho_j]}{\prod_{j=1,\, j\neq k}^{n} (\varrho_k - \varrho_j)}, \tag{4.1.24}$$

where E is a unit matrix, and $\ln \varrho_k$, $k = 1, \dots, n$, are the principal values of logarithms of ϱ_k, i.e.

$$\ln \varrho_k = \ln |\varrho_k| + i \arg \varrho_k, \quad (\mathrm{i} = \sqrt{-1}). \tag{4.1.25}$$

The eigenvalues λ_k, $k = 1, \dots, n$, of the matrix W are equal to

$$\lambda_k = \frac{1}{2\pi} \ln \varrho_k \tag{4.1.26}$$

and are called characteristic exponents of the system (4.1.1) (or a solution of that system). Having performed all multiplication operations in (4.1.24), we obtain a matrix $\ln X(2\pi)$ in the form of a n-th power matrix polynomial with respect to $X(2\pi)$ with coefficients depending on $\varrho_1, \dots, \varrho_n$, and $\ln \varrho_1, \dots, \ln \varrho_n$. Since the matrix $X(2\pi)$ is real, then the matrix $\ln X(2\pi)$ found in this way is also real (the imaginary parts of the coefficients of this matrix polynomial are mutually effaced due to the conjugation of complex multipliers).

Having obtained the matrix W, by a similar interpolation formula one can express the matrix $\exp(-Wt)$:

$$e^{-Wt} = \sum_{k=1}^{n} e^{-\lambda_k t} \frac{\prod_{j=1, j\neq k}^{n} [-W - E\lambda_j]}{\prod_{j=1, j\neq k}^{n} (\lambda_k - \lambda_j)}, \tag{4.1.27}$$

where λ_k are eigenvalues of W, calculated in accordance with (4.1.26). This matrix is also real.

If among the multipliers $\varrho_1, \dots, \varrho_n$ there are multiple factors, then the formulae for $\ln X(2\pi)$ and $\exp(-Wt)$ have a more complicated structure. In the general case, if $\lambda_1, \lambda_2, \dots, \lambda_m$ are eigenvalues of the matrix A, and $s_1, s_2, \dots, s_m$ are their multiplicity respectively, then the formula for the function $F(A)$ of that matrix is written as

$$F(A) = \sum_{k=1}^{m} [F(\lambda_k) Z_k^{(1)} + F'(\lambda_k) Z_k^{(2)} + \dots + F^{(s_k - 1)}(\lambda_k) Z^{(s_k)}], \tag{4.1.28}$$

where $Z_k^{(1)}, \dots, Z_k^{(s_k)}$ are matrices independent of $F(A)$. For example, in the case of a second-order matrix with dual eigenvalue $\lambda_1 = \lambda_2$ we have

$$F(A) = EF(\lambda_1) + (A - E\lambda_1) F'(\lambda_1). \tag{4.1.29}$$

In the case of a third-order matrix with eigenvalues λ_1, $\lambda_2 = \lambda_3$ we have

$$F(A) = F(\lambda_1) Z_1 + F(\lambda_2) Z_2 + F'(\lambda_2) Z_3, \tag{4.1.30}$$

where

$$Z_1 = \frac{(A - E\lambda_2)^2}{(\lambda_1 - \lambda_2)^2}, \qquad Z_2 = E - \frac{(A - E\lambda_2)^2}{(\lambda_1 - \lambda_2)^2},$$

$$Z_3 = (A - E\lambda_2) - \frac{(A - E\lambda_2)^2}{(\lambda_1 - \lambda_2)}.$$

But if $\lambda_1 = \lambda_2 = \lambda_3$, then

$$F(A) = EF(\lambda_1) + (A - E\lambda_1)F'(\lambda_1) + \frac{1}{2}(A - \lambda E_1)^2 F''(\lambda_1). \tag{4.1.31}$$

Depending on the nature of the elementary divisor corresponding to multiple factors, secular terms in the given formulae may become zero. In particular, formula (4.1.31) directly corresponds to the case when the third-order matrix A has one elementary divisor $(\lambda - \lambda_1)^3$. If this matrix has two elementary divisors, $(\lambda - \lambda_1)$, $(\lambda - \lambda_1)^2$, then

$$(A - E\lambda_1)^2 = 0, \qquad F(A) = EF(\lambda_1) + (A - E\lambda_1)F'(\lambda_1), \tag{4.1.32}$$

and if matrix A has three elementary divisors $\lambda - \lambda_1$, $\lambda - \lambda_1$, $\lambda - \lambda_1$, then $A - E\lambda_1 = 0$ and

$$F(A) = EF(\lambda_1). \tag{4.1.33}$$

(More detailed information on the representation of the functions $F(A)$ via matrices with multiple eigenvalues can be found in Gantmakher [1].)

Concrete formulae for the above functions will be found, assuming in (4.1.29)–(4.1.33)

$$F(A) = \ln A, \qquad A = X(2\pi),$$

$$F(A) = \exp(-A), \qquad A = \frac{1}{2\pi} \ln X(2\pi).$$

Upon construction of the matrix $\ln X(2\pi)$ we can find the Lyapunov matrix $L(t)$ (or rather a table of values of the elements of this matrix) by the formula

$$L(t) = X(t)e^{-Wt}. \tag{4.1.34}$$

It is easy to see that this formula determines a 2π-periodic real matrix. Using the table of numerical values of its elements, we can find an approximate (but sufficiently accurate) expression of this matrix in the form of a matrix trigonometric polynomial by methods of approximate function theory.

Case (b) There are negative multipliers. Then formulae (4.1.23), (4.1.24), (4.1.34) determine complex matrices W, $L(t)$. To obtain real matrices W, $L(t)$, retaining the formula (4.1.34) for $L(t)$, for the matrix W one should use the formula (see Erugin [1])

$$W = \frac{1}{4\pi} \ln X^2(2\pi). \tag{4.1.35}$$

The eigenvalues of the matrix $X^2(2\pi)$ are equal to the squares $\varrho_1^2, \ldots, \varrho_n^2$ of the eigenvalues $\varrho_1, \ldots, \varrho_n$ of the matrix $X(2\pi)$. If all $\varrho_1^2, \ldots, \varrho_n^2$ are different and nonnegative, then for the construction of the matrix $\ln X^2(2\pi)$ the formula (4.1.24) can be used, replacing $X(2\pi)$ by $X^2(2\pi)$, and ϱ_k by ϱ_k^2, $k = 1, \ldots, n$. Then the matrix $L(t)$ determined in accordance with (4.1.34) is real, but its least period in this case is equal not to 2π, but to 4π, because

$$\ln X^2 \neq 2 \ln X, \quad \exp(-2\pi W) \neq [X(2\pi)]^{-1},$$
$$\exp(-4\pi W) = \exp(-\ln X^2(2\pi)) = [X^2(2\pi)]^{-1}.$$

If among $\varrho_1^2, \ldots, \varrho_n^2$ there are multiple factors, then we can apply formulae of the form (4.1.28). If among $\varrho_1^2, \ldots, \varrho_n^2$ there are negative factors (in the case of purely imaginary multipliers ϱ_k), then the formula (4.1.35) implies the complex matrix W. To obtain a real matrix, the following formula should be applied

$$W = \frac{1}{8\pi} \ln X^4(2\pi). \tag{4.1.36}$$

The eigenvalues of the matrix $X^4(2\pi)$ are equal to $\varrho_1^4, \ldots, \varrho_n^4$. If all of them are nonnegative, then the last formula implies the real matrix W. Matrix $L(t)$ is calculated by the same formula (4.1.34), and it turns out to be real, with least period 8π.

Note that if among $\varrho_1, \ldots, \varrho_n$ there are negative multipliers, then for the construction of matrices W, $L(t)$ the above described methods of series and iterations are not applicable. Indeed, those series and iterations imply the real and 2π-periodic expression for $L(t)$, and in the presence of negative multipliers the real matrix of Lyapunov transform does not have period 2π. Consequently, these series and iterations necessarily diverge.

4.2 Construction of Green and Lyapunov Matrices

Green and Lyapunov matrices are applied in calculation of periodic solutions of the systems

$$\frac{dx}{dt} = P(t)x + f(t), \tag{4.2.1}$$

where $P(t)$ is a periodic matrix, and $f(t)$ is a periodic vector-function. These linear heterogeneous systems are encountered during the construction of periodic solutions of quasi-linear systems

$$\frac{dx}{dt} = P(t)x + F(x, t) \tag{4.2.2}$$

by series or iteration methods.

4.2.1 Noncritical case The noncritical case is when the homogeneous system

$$\frac{dx}{dt} = P(t)x \tag{4.2.3}$$

does not have T-periodic solutions not identically equal to zero, i.e. when among the characteristic exponents of that system there are no purely imaginary exponents, equal or divisible by $\pm(2\pi/T)i$.

Then the periodic solution of system (4.2.1) is representable by the formula

$$x(t) = \int_0^T G(t,s) f(s)\, ds, \tag{4.2.4}$$

where T is the period of matrix $P(t)$ and the function $f(t)$, and $G(t,s)$ is the Green function for the homogeneous system (4.2.3). This function is expressed through the fundamental matrix $X(t)$ of the homogeneous system (4.2.4) normalized at $t = 0$ in the following way:

$$G(t,s) = \begin{cases} [E - X(T)]^{-1} X(t) X^{-1}(s), & 0 \le s < t, \\ [E - X(T)]^{-1} X(t+T) X^{-1}(s), & t \le s \le T, \end{cases} \tag{4.2.5}$$

where E is a unit matrix.

There is also the following variant of formula (4.2.4) for periodic solution of system (4.2.1):

$$x(t) = \int_t^{t+T} G_L(t,s) f(s)\, ds, \tag{4.2.6}$$

where $G_L(t,s)$ is a matrix to be called a Lyapunov matrix. The formula for it will be given below.

Construction of matrices $G(t,s)$, $G_L(t,s)$ in analytic form (approximated, of course) can be performed by means of matrices $L(t)$, W of the Lyapunov transform for the homogeneous system (4.2.3). Indeed, let a transform

$$x = L(t) y \tag{4.2.7}$$

be found, reducing (4.2.3) to the system

$$\frac{dy}{dt} = W y \tag{4.2.8}$$

with a constant matrix W. Then the same transformation applied to the heterogeneous system (4.2.1) reduces this system to the form

$$\frac{dy}{dt} = W y + L^{-1}(t) f(t). \tag{4.2.9}$$

For a periodic solution of this heterogeneous system with constant matrix, we obtain the following formulae in the noncritical case:

$$y(t) = \int_0^T G^0(t,s) L^{-1}(s) f(s)\, ds \tag{4.2.10}$$

or

$$y(t) = \int\limits_{t}^{t+T} G_L^0(t,s) L^{-1}(s) f(s)\, ds, \tag{4.2.11}$$

where $G^0(t,s)$ is a Green matrix equal to

$$G^0(t,s) = \begin{cases} [E - e^{Wt}]^{-1} e^{W(t-s)}, & 0 \le s < t, \\ [E - e^{Wt}]^{-1} e^{Wt} e^{W(t-s)}, & t \le s \le T, \end{cases} \tag{4.2.12}$$

and where $G_L^0(t,s)$ is a Lyapunov matrix equal to

$$G_L^0(t,s) = \left[E - e^{WT}\right]^{-1} e^{WT} e^{W(t-s)}, \quad 0 \le t - s \le T. \tag{4.2.13}$$

Reverting to the periodic solution $x(t)$ of system (4.2.1), we obtain in accordance with (4.2.7)

$$x(t) = \int\limits_{0}^{T} L(t) G^0(t,s) L^{-1}(s) f(s)\, ds \tag{4.2.14}$$

or

$$x(t) = \int\limits_{t}^{t+T} L(t) G_L^0(t,s) L^{-1}(s) f(s)\, ds. \tag{4.2.15}$$

Comparing the last formulae with (4.2.3) and (4.2.5), we obtain the following expressions for matrices $G(t,s)$ and $G_L(t,s)$

$$G(t,s) = L(t) G^0(t,s) L^{-1}(s), \tag{4.2.16}$$

$$G_L(t,s) = L(t) G_L^0(t,s) L^{-1}(s). \tag{4.2.17}$$

4.2.2 Example Consider system (4.1.20). We obtained above expressions (4.1.21) and (4.1.22) for the matrices W, $L(t)$ of this system, which allows us to construct the matrices W and $\exp(Wt)$.

The eigenvalues of matrix W are as follows:

$$\lambda_1, \lambda_2 = c \pm i\mu, \quad c = -0.250\,003, \quad \mu = 0.968\,915,$$

so that the normalized fundamental matrix of this system $\exp(Wt)$ is

$$e^{Wt} = e^{ct} \begin{pmatrix} \cos\mu t - \dfrac{c}{\omega}\sin\mu t & \dfrac{1}{\mu}\sin\mu t \\ \left(-\dfrac{c^2}{\mu} - \mu\right)\sin\mu t & \cos\mu t + \dfrac{c}{\mu}\sin\mu t \end{pmatrix} \tag{4.2.18}$$

or

$$e^{Wt} = e^{ct}\left(E\cos\mu t + M\sin\mu t\right),$$

where E is a unit matrix of order 2×2, and

$$M = \begin{pmatrix} -\dfrac{c}{\mu} & \dfrac{1}{\mu} \\ -\dfrac{c^2}{\mu} - \mu & \dfrac{c}{\mu} \end{pmatrix} = \begin{pmatrix} 0.258\,024 & 1.032\,082 \\ -1.033\,422 & -0.258\,024 \end{pmatrix}. \tag{4.2.19}$$

The period of the sought solution is $T = 2\pi/\mu = 6.484\,764$, so that

$$e^{Wt} = Ee^{cT} = 0.197\,659E, \quad (E - e^{Wt})^{-1} = (1 - e^{cT})^{-1}E = 1.246\,350E,$$

$$(E - e^{Wt})^{-1}e^{WT} = e^{cT}(1 - e^{cT})^{-1}E = 0.307\,042E$$

$$G^0(t,s) = \begin{cases} 1.246\,353e^{c(t-s)}F(t-s), & 0 \le s < t, \\ 0.246\,353e^{c(t-s)}F(t-s), & t \le s \le T, \end{cases} \tag{4.2.20}$$

$$G^0_L(t,s) = 0.246\,353e^{c(t-s)}F(t-s), \quad 0 \le t \le s \le t+T, \tag{4.2.21}$$

where

$$F(t-s) = E\cos\mu(t-s) + M\sin\mu(t-s).$$

Matrices $G(t,s)$, $G_L(t,s)$ will be expressed in accordance with (4.2.16) and (4.2.17) by the formulae

$$G(t,s) = \begin{cases} 1.246\,350e^{c(t-s)}\Phi(t,s), & 0 \le s < t, \\ 0.246\,353e^{c(t-s)}\Phi(t,s), & t \le s \le T, \end{cases} \tag{4.2.22}$$

$$G_L(t,s) = 0.246\,353e^{c(t-s)}\Phi(t,s), \quad 0 \le t \le s \le t+T, \tag{4.2.23}$$

where

$$\Phi(t,s) = L(t)F(t-s)L^{-1}(s) = E\cos\mu(t-s) + L(t)ML^{-1}(s)\sin\mu(t-s). \tag{4.2.24}$$

Multiplication of matrices $L(t)$, $F(t-s)$, $L^{-1}(s)$ is simple in principle, but quite an intricate operation. It is advisable to implement it on a computer, using appropriate software for multiplication of Fourier polynomials with matrix coefficients. For the construction of $L^{-1}(s)$ an iteration method of Newton's type can be used. Assume $L^{-1}(t) = Z$. Then Z satisfies the matrix equation

$$F(Z) \equiv Z^{-1} - L(t) = 0. \tag{4.2.25}$$

Let a zero approximation $Z = Z_0$ be chosen. The next approximation Z_1 is found from the following correlation for $\Delta Z = Z_1 - Z_0$:

$$(Z_0 + \Delta Z)^{-1} - L(t) = 0,$$

where the terms of order higher than one with respect to ΔZ are truncated. We obtain

$$\Delta Z = Z_0(Z_0^{-1} - L(t))Z_0,$$

hence

$$Z_1 = 2Z_0 - Z_0L(t)Z_0. \tag{4.2.26}$$

By a similar formula the successive approximations are found.

4.2.3 Critical cases In critical cases periodic solutions of heterogeneous systems of the form (4.2.1) are also expressed by formulae (4.2.3), (4.2.6), but the matrices $G(t,s)$, $G_L(t,s)$ have a more complicated structure. These matrices are called generalized Green (or Lyapunov) matrices.

By means of formulae (4.2.16), (4.2.17) these matrices can be constructed after their construction for system (4.2.8) with the constant matrix W.

For instance, let a second-order system of the form (4.2.1) with 2π-periodic $P(t)$ and $f(t)$ be considered, and let it turn out after construction of Lyapunov transformation matrices W, $L(t)$ that the eigenvalues of matrix W are equal to $\pm i$. Then it is always possible to find the transform

$$z = Qy \tag{4.2.27}$$

with a constant nonsingular matrix Q that reduces the second-order system (4.2.9) to the system

$$\frac{dz}{dt} = W_1 z + \varphi(t), \tag{4.2.28}$$

where $\varphi(t) = QL^{-1}(t)f(t)$, and the matrix W_1 has the form

$$\begin{pmatrix} 0 & -1 \\ 1 & 0 \end{pmatrix}.$$

The conditions for the existence of periodic solutions of system (4.2.28) are well known (see Grebenikov and Ryabov [1]):

$$\int_0^{2\pi} \varphi_1(t)\cos t\,dt = -\int_0^{2\pi} \varphi_2(t)\sin t\,dt,$$

$$\int_0^{2\pi} \varphi_1(t)\sin t\,dt = \int_0^{2\pi} \varphi_2(t)\cos t\,dt,$$

where $\varphi_1(t)$, $\varphi_2(t)$ are components of the vector $\varphi(t)$. The periodic solution of system (4.2.28) is expressed by the formula

$$z(t) = \int_0^{2\pi} G^0(t,s)\varphi(s)\,ds,$$

where $G^0(t,s)$ is a generalized Green matrix for this system, equal to

$$G^0(t,s) = g(t,s)\begin{pmatrix} \cos(t-s) & -\sin(t-s) \\ \sin(t-s) & \cos(t-s) \end{pmatrix}, \tag{4.2.29}$$

and

$$g(t,s) = \begin{cases} \dfrac{s}{2\pi}, & 0 \le s < t, \\ \left(\dfrac{s}{2\pi} - 1\right), & t \le s \le 2\pi. \end{cases}$$

Periodic solution of the initial system (4.2.1) will be expressed by the formula

$$x(t) = L(t) \int_0^{2\pi} G^0(t,s) Q L^{-1}(s) f(s)\, ds, \tag{4.2.30}$$

so the generalized Green matrix for the initial system is equal to

$$G(t,s) = L(t) G^0(t,s) Q L^{-1}(s). \tag{4.2.31}$$

4.3 Direct Numerical–Analytic Method of Construction of Periodic Solutions

This method is described in the book of Grebenikov and Ryabov [1] basically as applied to quasi-linear equations with a constant matrix of the linear part. That book contains a number of examples. In the case of the construction of periodic solutions by means of iterations with accelerated convergence one has to deal with heterogeneous linear equations with a periodic matrix. In this section we will consider in detail the algorithm of the construction of periodic solutions of such equations.

Let a system of the form (4.2.1) be given, i.e. the system

$$\frac{dx}{dt} = P(t)x + f(t), \tag{4.3.1}$$

where $P(t)$ is a 2π-periodic matrix, and $f(t)$ a 2π-periodic vector. Consider the noncritical case. This fact can be established upon construction of Lyapunov transformation matrices $L(t)$, W and calculation of the eigenvalues of matrix W. Let $P(t)$ and $f(t)$ be presented by Fourier series where we will be restricted to harmonics of order N_0, that is

$$P(t) = \frac{1}{2}A_0 + \sum_{k=1}^{N_0} A_k \cos kt + B_k \sin kt, \tag{4.3.2}$$

$$f(t) = \frac{1}{2}G_0 + \sum_{k=1}^{N_0} G_k \cos kt + H_k \sin kt, \tag{4.3.3}$$

where A_0, A_k, B_k are constant matrices, and G_0, G_k, H_k constant vectors.

We will look for a 2π-periodic solution $x(t)$ in the form of a trigonometric polynomial of sufficiently high order N $(N \geq N_0)$:

$$x(t) = \frac{1}{2}M_0 + \sum_{k=1}^{N} M_k \cos kt + L_k \sin kt, \tag{4.3.4}$$

where M_0, M_k, L_k are constant vectors. Substituting (4.3.4) into (4.3.1) and equating the coefficients at equal harmonics, neglecting harmonics of order higher than N, we will come to a system of algebraic equations with respect to M_0,

M_k, L_k. We would like to propose quite a simple algorithm of the construction of this system.

First of all, consider the algorithm for multiplication of two Fourier polynomials for scalar functions $p(t)$, $\varphi(t)$:

$$\begin{aligned} p(t) &= \frac{1}{2}a_0 + \sum_{k=1}^{N} a_k \cos kt + b_k \sin kt, \\ \varphi(t) &= \frac{1}{2}m_0 + \sum_{k=1}^{N} m_k \cos kt + l_k \sin kt. \end{aligned} \tag{4.3.5}$$

In the product $s(t) = p(t)\varphi(t)$ we will neglect harmonics of order higher than N, so assume that

$$s(t) = \frac{1}{2}g_0 + \sum_{k=1}^{N} g_k \cos kt + h_k \sin kt. \tag{4.3.6}$$

Consider the algorithm simplest in structure, expressing $p(t)$, $\varphi(t)$, $s(t)$ by Fourier polynomials in complex form:

$$p(t) = \sum_{k=-N}^{N} p_k e^{ikt}, \qquad \varphi(t) = \sum_{k=-N}^{N} r_k e^{ikt}, \qquad s(t) = \sum_{k=-N}^{N} s_k e^{ikt}, \tag{4.3.7}$$

where

$$\begin{aligned} p_0 &= \frac{1}{2}a_0, \qquad r_0 = \frac{1}{2}m_0, \qquad s_0 = \frac{1}{2}g_0, \\ p_k &= \frac{1}{2}(a_k \mp ib_k), \qquad r_k = \frac{1}{2}(m_k \mp il_k), \qquad s_k = \frac{1}{2}(g_k \mp ih_k), \end{aligned} \tag{4.3.8}$$

minus corresponds to $k > 0$, and plus to $k < 0$.

Then introduce the vectors

$$\bar{p} = \operatorname{col}(p_{-N}, \dots, p_N), \quad \bar{r} = \operatorname{col}(r_{-N}, \dots, r_N), \quad \bar{s} = \operatorname{col}(s_{-N}, \dots, s_N). \tag{4.3.9}$$

The vector $\bar{s}$ can be expressed as the product

$$\bar{s} = \bar{P}\bar{r}, \tag{4.3.10}$$

where $\bar{P}$ is a matrix of sufficiently simple structure. Write this matrix in the case of $N = 2$ (that this is enough for the structure of this matrix for arbitrary N will become clear):

$$\bar{P} = \begin{pmatrix} p_0 & p_{-1} & p_{-2} & 0 & 0 \\ p_1 & p_0 & p_{-1} & p_{-2} & 0 \\ p_2 & p_1 & p_0 & p_{-1} & p_{-2} \\ 0 & p_2 & p_1 & p_0 & p_{-1} \\ 0 & 0 & p_2 & p_1 & p_0 \end{pmatrix}. \tag{4.3.11}$$

Consequently, the components of the vector $\operatorname{col}(s_{-2}, \dots, s_2)$ are expressed through elements of the matrix $\bar{P}$ and the components of the vector $\operatorname{col}(r_{-2}, \dots, r_2)$

as follows:

$$\begin{aligned} s_{-2} &= p_0 r_{-2} + p_{-1} r_{-1} + p_{-2} r_0, \\ s_{-1} &= p_1 r_{-2} + p_0 r_{-1} + p_{-1} r_0 + p_{-2} r_1 \quad \text{etc.} \end{aligned} \tag{4.3.12}$$

Since p_k and p_{-k}, and r_k and r_{-k} are conjugate complex, s_k and s_{-k} are the same. Upon finding s_0, s_k, s_{-k} obtain in accordance with (4.3.8) the real coefficients g_0, g_k, h_k of the product $p(t)\varphi(t)$ by the formulae

$$g_0 = 2s_0, \quad g_k = 2\operatorname{Re} s_k, \quad h_k = -2\operatorname{Im} s_{-k}. \tag{4.3.13}$$

The complex conjugation of s_k and s_{-k} can be used to check the calculations.

This algorithm of the computation of the product of Fourier polynomials is quite convenient, if the computer program provides for operations with complex numbers. If one is restricted to operations with real numbers, another algorithm can be proposed.

Assume that

$$\bar{p} = p' + ip'', \quad \bar{r} = r' + ir'', \quad \bar{s} = s' + is'', \tag{4.3.14}$$

where $p', p'', \ldots, s''$ are real, expressed through the coefficients of the Fourier polynomials (4.3.5), (4.3.6). In accordance with (4.3.8) for $N = 2$ we obtain

$$2P' = \begin{pmatrix} a_0 & a_1 & a_2 & 0 & 0 \\ a_1 & a_0 & a_1 & a_2 & 0 \\ a_2 & a_1 & a_0 & a_1 & a_2 \\ 0 & a_2 & a_1 & a_0 & a_1 \\ 0 & 0 & a_2 & a_1 & a_0 \end{pmatrix} \tag{4.3.15}$$

$$2P'' = \begin{pmatrix} 0 & b_1 & b_2 & 0 & 0 \\ -b_1 & 0 & b_1 & b_2 & 0 \\ -b_2 & -b_1 & 0 & b_1 & b_2 \\ 0 & -b_2 & -b_1 & 0 & b_1 \\ 0 & 0 & -b_2 & -b_1 & 0 \end{pmatrix} \tag{4.3.16}$$

and

$$\begin{aligned} r' &= \operatorname{col}(m_2, m_1, m_0, m_1, m_2), \quad r'' = \operatorname{col}(l_2, l_1, 0, -l_1, -l_2), \\ s' &= \operatorname{col}(g_2, g_1, g_0, g_1, g_2), \quad s'' = \operatorname{col}(h_2, h_1, 0, -h_1, -h_2). \end{aligned}$$

Formula (4.3.10) will be rewritten as

$$s' + is'' = (P' + iP'')(r' + ir''),$$

so that

$$s' = P'r' - P''r'', \quad s'' = P''r' + P'r'' \tag{4.3.17}$$

and, e.g., for $N = 2$ (we will write the formulae in complete form)

$$
\begin{aligned}
2\,\mathrm{col}\,(g_2, g_1, g_0, g_1, g_2) = {} & \begin{pmatrix} a_0 & a_1 & a_2 & 0 & 0 \\ a_1 & a_0 & a_1 & a_2 & 0 \\ a_2 & a_1 & a_0 & a_1 & a_2 \\ 0 & a_2 & a_1 & a_0 & a_1 \\ 0 & 0 & a_2 & a_1 & a_0 \end{pmatrix} \begin{pmatrix} m_2 \\ m_1 \\ m_0 \\ m_1 \\ m_2 \end{pmatrix} \\
& - \begin{pmatrix} 0 & b_1 & b_2 & 0 & 0 \\ -b_1 & 0 & b_1 & b_2 & 0 \\ -b_2 & -b_1 & 0 & b_1 & b_2 \\ 0 & -b_2 & -b_1 & 0 & b_1 \\ 0 & 0 & -b_2 & -b_1 & b_2 \end{pmatrix} \begin{pmatrix} l_2 \\ l_1 \\ 0 \\ l_1 \\ l_2 \end{pmatrix},
\end{aligned} \tag{4.3.18}
$$

$$
\begin{aligned}
2\,\mathrm{col}\,(h_2, h_1, 0, -h_1, -h_2) = {} & \begin{pmatrix} 0 & b_1 & b_2 & 0 & 0 \\ -b_1 & 0 & b_1 & b_2 & 0 \\ -b_2 & -b_1 & 0 & b_1 & b_2 \\ 0 & -b_2 & -b_1 & 0 & b_1 \\ 0 & 0 & -b_2 & -b_1 & 0 \end{pmatrix} \begin{pmatrix} m_2 \\ m_1 \\ m_0 \\ m_1 \\ m_2 \end{pmatrix} \\
& + \begin{pmatrix} a_0 & a_1 & a_2 & 0 & 0 \\ a_1 & a_0 & a_1 & a_2 & 0 \\ a_2 & a_1 & a_0 & a_1 & a_2 \\ 0 & a_2 & a_1 & a_0 & a_1 \\ 0 & 0 & a_2 & a_1 & a_0 \end{pmatrix} \begin{pmatrix} l_2 \\ l_1 \\ 0 \\ l_1 \\ l_2 \end{pmatrix}.
\end{aligned} \tag{4.3.19}
$$

The structure of the formulae for arbitrary N is obvious.

The symmetry of the components of the vectors s' and s'' can serve to check the calculations.

Using the derived formulae for multiplication of Fourier scalar polynomials, we can write in concrete form the algebraic system for the coefficients M_0, M_k, L_k of the periodic solution $x(t)$ expressed in accordance with (4.3.4).

Represent the matrix $P(t)$ and the vectors $f(t)$, $x(t)$ by complex Fourier polynomials

$$
P(t) = \sum_{k=-N}^{N} P_k e^{ikt}, \qquad f(t) = \sum_{k=-N}^{N} S_k e^{ikt}, \qquad x(t) = \sum_{k=-N}^{N} R_k e^{ikt}, \tag{4.3.20}
$$

where P_k are matrices, and S_k, R_k are vectors expressed in terms of the coefficients A_k, B_k, G_k, H_k, M_k, L_k of the real Fourier polynomials (4.3.2)–(4.3.4) by a relation analogous to (4.3.8).

Now introduce the vectors

$$
w = \mathrm{col}\,(R_{-N}, \dots, R_N), \qquad v = \mathrm{col}\,(S_{-N}, \dots, S_N), \tag{4.3.21}
$$

where the components R_k, S_k are vectors. The system of algebraic equations with respect to vector w will be written in the vector-matrix form as

$$
U_N w = -v, \tag{4.3.22}
$$

where U_k is obtained from $\bar{P}$ in (4.3.10), substituting the scalar constants p_k from P_k and substracting diagonally $\pm ik$, $k = N, N-1, \dots$. For example, for $N = 2$

$$U_N = \begin{pmatrix} P_0 + 2iE & P_{-1} & P_{-2} & 0 & 0 \\ P_1 & P_0 + iE & P_{-1} & P_{-2} & 0 \\ P_2 & P_1 & P_0 & P_{-1} & P_{-2} \\ 0 & P_1 & P_2 & P_0 - iE & P_{-1} \\ 0 & 0 & P_2 & P_1 & P_0 - 2iE \end{pmatrix}. \tag{4.3.23}$$

If instead of system (4.3.1) a first-order scalar equation of the same form is considered, then equations (4.3.22) are the usual system of algebraic linear equations with respect to $2N+1$ scalar unknown $R_{-N}, \dots, R_N$ with a constant matrix with complex scalar elements. Then the direct construction of the inverse matrix U_N^{-1} does not cause difficulties (in the mode of operations with complex numbers). We obtain

$$w = U_N^{-1} v. \tag{4.3.24}$$

Checking the calculation involves checking the complex conjugation of R_k and R_{-k}.

But if (4.3.1) is a system of a certain order, then the elements P_k of matrix U_k are matrices, so that U_k is a block matrix. The structure of an inverse matrix for a block matrix is complicated. For instance (see Bellman [1]),

$$\begin{pmatrix} A_1 & B_1 \\ A_2 & B_2 \end{pmatrix}^{-1} = \begin{pmatrix} (A_1 - B_1 B_2^{-1} A_2)^{-1} & (A_2 - B_2 B_1^{-1} A_1)^{-1} \\ (B_1 - A_1 A_2^{-1} B_2)^{-1} & (B_2 - A_2 A_1^{-1} B_1)^{-1} \end{pmatrix}, \tag{4.3.25}$$

where $A_1, \dots, B_2$ are (nondegenerate) matrices of the same order. Therefore it is advisable to write out completely the rows and columns of matrices P_k and components of vectors v, w, and to replace (4.3.22) by a system of algebraic equations with respect to all scalar components of vectors $R_{-N}, \dots, R_N$.

For example, let (4.3.1) be a second-order system. Then all the P_k are 2×2 matrices, and R_{-N}, R_N are two-dimensional vectors:

$$P_k = \begin{pmatrix} p_k^{(11)} & p_k^{(12)} \\ p_k^{(21)} & p_k^{(22)} \end{pmatrix}, \quad R_k = \begin{pmatrix} r_k^{(1)} \\ r_k^{(2)} \end{pmatrix}, \quad S_k = \begin{pmatrix} s_k^{(1)} \\ s_k^{(2)} \end{pmatrix}. \tag{4.3.26}$$

Then system (4.3.22) can be written as a system with respect to $2(2N+1)$ unknowns $r_0^{(1)}, r_0^{(2)}, r_k^{(1)}, r_k^{(2)}$, $k = 1, 2, \dots, N$.

Introducing the vectors

$$\bar{v} = \mathrm{col}\,(s_{-N}^{(1)}, s_{-N}^{(2)}, \dots, s_N^{(1)}, s_N^{(2)}), \tag{4.3.27}$$

$$\bar{w} = \mathrm{col}\,(r_{-N}^{(1)}, r_{-N}^{(2)}, \dots, r_N^{(1)}, r_N^{(2)}), \tag{4.3.28}$$

rewrite system (4.3.22) in the form

$$\overline{U}_N \bar{w} = -\bar{v}, \tag{4.3.29}$$

where the matrix $\bar{U}_N$ is obtained from U_k upon replacement of P_k by the corresponding matrices. For example, for $N = 2$ we obtain

$$\overline{U}_2 = \begin{pmatrix} p_0^{(11)} + 2i & p_0^{(12)} & p_{-1}^{(11)} & p_{-1}^{(12)} & \cdots \\ p_0^{(21)} & p_0^{(22)} + 2i & p_{-1}^{(21)} & p_{-1}^{(22)} & \cdots \\ p_1^{(11)} & p_1^{(12)} & p_0^{(11)} + i & p_0^{(12)} & \cdots \\ p_1^{(21)} & p_1^{(22)} & p_0^{(21)} & p_0^{(22)} + i & \cdots \\ \vdots & \vdots & \vdots & \vdots & \ddots \end{pmatrix} \tag{4.3.30}$$

Thus we obtain an ordinary system of algebraic linear equations with respect to $4N+2$ unknowns $r_{-N}^{(1)}, r_{-N}^{(2)}, \ldots, r_N^{(1)}, r_N^{(2)}$ with a complex matrix $\overline{U}_N$. Having constructed an inverse matrix by standard methods, we find the vector $\bar{w}$ with the sought components. The requirement for complex conjugation of the components $r_k^{(1)}$ and $r_{-k}^{(1)}$, as well as $r_k^{(2)}$ and $r_{-k}^{(2)}$ serves to check the calculations.

If one is restricted to an operation with real matrices, the following algorithm can be used. Separate the real and imaginary parts in the matrix $\overline{U}_N$ and the vector $\bar{v} = \operatorname{col}(r_{-N}^{(1)}, \ldots, r_N^{(2)})$:

$$\overline{U}_N = C + iD, \quad \bar{v} = v' + iv'',$$

where, owing to the connection between the coefficients of complex and real Fourier polynomials for $f(t)$,

$$v' = \frac{1}{2}\operatorname{col}(G_N^{(1)}, G_N^{(2)}, \ldots, G_0^{(1)}, G_0^{(2)}, \ldots, G_N^{(1)}, G_N^{(2)}),$$
$$v'' = \frac{1}{2}\operatorname{col}(H_N^{(1)}, H_N^{(2)}, \ldots, H_0^{(1)}, H_0^{(2)}, \ldots, -H_N^{(1)}, -H_N^{(2)}).$$

Representing the inverse matrix $\overline{U}_N^{(-1)}$ in the same form $\overline{U}_N^{(-1)} = C_1 + iD_1$, for the real matrices C_1, D_1 we obtain the following expressions:

$$C_1 = (C + DC^{-1}D)^{-1}, \quad D_1 = -(D + CD^{-1}C)^{-1}. \tag{4.3.31}$$

Thus we obtain for the vector $\bar{w}(r_{-N}^{(1)}, r_{-N}^{(2)}, \ldots, r_N^{(1)}, r_N^{(2)})$ the formula

$$\bar{w} = -(C_1 + iD_1)\bar{v} = -C_1 v' + D_1 v'' - i(C_1 v'' + D_1 v'). \tag{4.3.32}$$

The requirement for complex conjugation of $r_k^{(j)}$ and $r_{-k}^{(j)}$, $j = 1, 2$, $k = 1, \ldots, N$, can serve to check the calculations.

Having determined the vectors $R_k = \operatorname{col}(r_k^{(1)}, r_k^{(2)})$, $k = 0, \pm 1, \ldots, \pm N$, we can find the real coefficients M_0, M_k, L_k of the sought polynomials for $x(t)$ by the formulae

$$M_0 = 2R_0, \quad M_k = 2\operatorname{Re}(R_k), \quad L_k = -2\operatorname{Im}(R_k). \tag{4.3.33}$$

4.4 Construction of Periodic Solutions in Hill's Problem of Lunar Motion

Hill [1] considered the equations

$$\begin{aligned} &\frac{d^2x}{d\theta^2} - 2m\frac{dy}{d\theta} + \chi\frac{x}{r^3} = 3m^2x, \\ &\frac{d^2y}{d\theta^2} + 2m\frac{dx}{d\theta} + \chi\frac{y}{r^3} = 0 \end{aligned} \tag{4.4.1}$$

of the plane bounded circular three-body problem of the Earth, Moon and Sun, where x, y are the rectangular coordinates of the Moon in the rotating coordinate system with the center in the Earth, $r = \sqrt{x^2 + y^2}$. In the right-hand members of these equations, in comparison with the exact equations for such a problem, the terms of order x/a', y/a' and higher are discarded (a' is the distance from the Earth to the Sun). The parameter m is equal to

$$m = \frac{n'}{n - n'}, \tag{4.4.2}$$

where n' is the average day angular motion of the Earth around the Sun, and n is the same motion of the Moon around the Earth. Hill assumed $m = 0.0808\,489\,3679$. The parameter χ in (4.4.1) is expressed by the formula

$$\chi = f\frac{m_T + m_L}{(n - n')^2}, \tag{4.4.3}$$

where m_T, m_L are the masses of the Earth and the Moon respectively, and f is the gravitational constant. The variable θ is connected with time t by the formula $\theta = (n - n')(t - t_0)$, where t_0 is the initial time.

Hill constructed the periodic solution of these equations in the form of trigonometric series

$$x(\theta) = \sum_{k=0}^{\infty} A_{2k+1}\cos(2k+1)\theta, \qquad y(\theta) = \sum_{k=0}^{\infty} A'_{2k+1}\sin(2k+1)\theta, \tag{4.4.4}$$

where the constant coefficients are represented by literal power series with respect to m. Hill found those series without clearing up the question of their convergence. The first estimate of the radius of convergence of these series was obtained by Lyapunov [2].

We will consider the methods of numerical–analytic construction of the periodic solution of equations (4.4.1) by means of iteration, and the implementation of these methods for numerical values of m (see Grebenikov and Ryabov [1]). As the initial equations we assume those obtained by Lyapunov from the given equations upon the following change of variables:

$$x + iy = \tilde{a}(1 - p)e^{i\theta}, \quad x - iy = \tilde{a}(1 - q)e^{-i\theta} \quad (i = \sqrt{-1}), \tag{4.4.5}$$

where

$$\tilde{a} = \left(\frac{\chi}{l}\right)^{1/3}, \quad l = 1 + 2m + \frac{3}{2}m^2.$$

Equations with respect to variables p, q are written in the form:

$$\begin{aligned} &\frac{d^2p}{d\theta^2} + 2(1+m)i\,\frac{dp}{d\theta} - \frac{3}{2}l(p+q) = \frac{3}{2}\lambda(q-1)e^{-2i\theta} + lR(p,q), \\ &\frac{d^2q}{d\theta^2} - 2(1+m)i\,\frac{dq}{d\theta} - \frac{3}{2}l(p+q) = \frac{3}{2}\lambda(p-1)e^{2i\theta} + lR(p,q), \end{aligned} \qquad 4.4.6$$

where

$$\begin{aligned} R_1(p,q) &= (1-p)^{-1/2}(1-q)^{-3/2} - 1 - \frac{1}{2}p - \frac{3}{2}q, \\ R_2(p,q) &= R_1(q,p), \quad \lambda = m^2. \end{aligned} \tag{4.4.7}$$

This is close to the resonance case, because for $\lambda = 0$ the homogeneous equations (obtained from (4.4.6) with $m = 0$, $\lambda_1 = 0$, $R_1 \equiv 0$, $R_2 \equiv 0$) have a 2π-periodic solution (the corresponding characteristic equation has a couple of imaginary roots $\pm i$ and a multiple zero root). However the structure of the right-hand members of equations (4.4.6) is such that no supplementary existence conditions for periodic solutions are required, which in resonance cases are usually imposed on the right-hand members of the initial equations.

Write equation (4.4.6) in the following form:

$$D_s(p,q) = f_s(p,q,\theta,\lambda), \quad s = 1,2, \tag{4.4.8}$$

where $D_s(p,q)$ are differentiation operators, and $f_s(p,q,\theta,\lambda)$ are functions corresponding to the right-hand members of equations (4.4.6).

We will calculate the periodic solution of those equations by means of successive approximations $p_j(\theta)$, $q_j(\theta)$, $j = 1,2,\ldots$, assuming that the functions $p_j(\theta)$, $q_j(\theta)$ satisfy the equations

$$D_s(p_j,q_j) = f_s(p_{j-1},q_{j-1},\theta,\lambda), \quad s = 1,2, \tag{4.4.9}$$

and $p_0 \equiv q_0 \equiv 0$.

Consider the auxiliary equations

$$D_s(p,q) = \varphi_s(\theta), \quad s = 1,2, \tag{4.4.10}$$

where the functions $\varphi_1(\theta)$, $\varphi_2(\theta)$ are represented by Fourier polynomials in complex form, and

$$\varphi_1(\theta) = A_0 + \sum_{k=1}^{N}\left(A_k e^{2ik\theta} + A_{-k}e^{-2ik\theta}\right), \quad \varphi_2(\theta) = \varphi_1(-\theta), \tag{4.4.11}$$

where A_0, A_k, A_{-k} are the real coefficients, and N is a sufficiently large integer. With such functions $\varphi_1(\theta)$, $\varphi_2(\theta)$, equations (4.4.10) have a periodic solution of the following form:

$$p(\theta) = M_0 + \sum_{k=1}^{N}\left(M_k e^{2ik\theta} + M_{-k}e^{-2ik\theta}\right), \quad q(\theta) = p(-\theta), \tag{4.4.12}$$

where the coefficients M_0, M_k, M_{-k} are real. These coefficients are expressed by the formulae

$$M_0 = -\frac{A_0}{3l}, \quad M_k = \alpha_k A_k + \delta_k A_{-k}, \quad M_{-k} = \delta_k A_k + \gamma_k A_{-k}, \tag{4.4.13}$$

where

$$\alpha_k = \frac{1}{\Delta_k}\left[-4k^2 + 4(1+m)k - \frac{3}{2}l\right], \qquad \delta_k = \frac{3}{2\Delta_k}l,$$

$$\gamma_k = \frac{1}{\Delta_k}\left[-4k^2 - 4(1+m)k - \frac{3}{2}l\right], \qquad \Delta_k = 4k^2\left[4k^2 - 1 - 2m + \frac{1}{2}m^2\right].$$

The equations for the first approximation p_1, q_1 are as follows:

$$D_1(p_1, q_1) = -\frac{3}{2}\lambda e^{-2i\theta}, \quad D_2(p_1, q_1) = -\frac{3}{2}\lambda e^{2i\theta}. \tag{4.4.14}$$

Since their right-hand members have the form (4.4.11), they have the periodic solution

$$p_1(\theta) = M_1^{(1)} e^{2i\theta} + M_{-1}^{(1)} e^{-2i\theta}, \quad q_1(\theta) = p_1(-\theta), \tag{4.4.15}$$

where

$$M_1^{(1)} = -\frac{9(2+4m+3m^2)m^2}{16(6-4m+m^2)}, \qquad M_{-1}^{(1)} = \frac{3(38+28m+9m^2)m^2}{16(6-4m+m^2)}.$$

If we substitute the obtained expressions for $p_1(\theta)$, $q_1(\theta)$ into the right-hand members of the equations with respect to p_2, q_2, we find that the functions

$$f_s(p_1(\theta), q_1(\theta), \theta, \lambda), \quad s = 1, 2 \tag{4.4.16}$$

may be represented by Fourier polynomials of the form (4.4.11), if we neglect the harmonics of order higher than a certain N. Consequently, we can find a periodic solution of the form

$$p_2(\theta) = M_0^{(2)} + \sum_{k=1}^{N}\left(M_k^{(2)} e^{2ik\theta} + M_{-k}^{(2)} e^{-2ik\theta}\right), \quad q_2(\theta) = p_2(-\theta), \tag{4.4.17}$$

where the coefficients $M_0^{(2)}$, $M_k^{(2)}$, $M_{-k}^{(2)}$ are expressed in conformity with (4.4.13).

For the direct construction of solution (4.4.17) it is necessary to express the function $R_1(p_1(\theta), q_1(\theta))$, and at the same time the function $f_1(p_1, q_1, \theta, \lambda)$ in the first of the equations with respect to $p_2(\theta)$, $q_2(\theta)$ by a Fourier polynomial. Here we can use the following formula:

$$R_1(p_1, q_1) = \frac{3}{4}p_1 q_1 + A + B + \frac{3}{2}q_1 A + \frac{1}{2}p_1 B + AB, \tag{4.4.18}$$

where

$$A = \frac{1\cdot 3}{2\cdot 4}p_1^2 + \frac{1\cdot 3\cdot 5}{2\cdot 4\cdot 6}p_1^3 + \frac{1\cdot 3\cdot 5\cdot 7}{2\cdot 4\cdot 6\cdot 8}p_1^4 + \cdots,$$

$$B = \frac{3\cdot 5}{2\cdot 4}q_1^2 + \frac{3\cdot 5\cdot 7}{2\cdot 4\cdot 6}q_1^3 + \frac{3\cdot 5\cdot 7\cdot 9}{2\cdot 4\cdot 6\cdot 8}q_1^4 + \cdots.$$

Having performed the operations of multiplication, exponentiation of Fourier polynomials for $p_1(\theta)$, $q_1(\theta)$, and the subsequent summation, we can construct for $R_1(p_1(\theta), q_1(\theta))$ a Fourier polynomial of any order N.

When $p_2(\theta)$, $q_2(\theta)$ are found, the approximations $p_3(\theta)$, $q_3(\theta)$, etc. are constructed analogously.

Calculations were done on a computer for different values of m (0.2, 0.35, 0.43, 0.45). In the computational program $N = 24$ was assumed, but in the output of the final results only the Fourier coefficients larger than 10^{-6} in absolute value were retained.

The maximum possible number of iterations was assumed to be $j = 150$. If all Fourier coefficients in two neighboring approximations at a certain step $j < 100$ coincided with an accuracy of 10^{-6}, it was assumed that there was practical convergence of the iteration process; the Fourier coefficients in the last approximation $p_k(\theta)$, $q_k(\theta)$ were printed out.

Table 4.4.1

m	0.2	0.35	0.43
n	5	13	86
M_0	−0.000 712	−0.003 584	−0.004 192
M_1	−0.012 495	−0.052 377	−0.087 296
M_{-1}	0.062 919	0.234 994	−0.384 072
M_2	−0.000 358	−0.005 661	−0.015 325
M_{-2}	−0.000 023	−0.000 610	−0.002 130
M_3	−0.000 013	−0.000 782	−0.003 380
M_{-3}	−0.000 001	−0.000 077	−0.000 369
M_4	−0.000 001	−0.000 122	−0.000 841
M_{-4}		−0.000 009	−0.000 067
M_5		−0.000 020	−0.000 226
M_{-5}		−0.000 001	−0.000 014
M_6		−0.000 004	−0.000 062
M_{-6}		0	−0.000 003
M_7		−0.000 001	−0.000 016
M_{-7}			−0.000 001
M_8			−0.000 005
M_{-8}			0
M_9			−0.000 003
M_{-9}			0
M_{10}			−0.000 001

Practical convergence of the proposed iterations was found for $m \leq 0.43$. For $m = 0.45$ the process explicitly diverges. The results of the calculations (coefficients M_0, M_k, M_{-k} for these values of m) are given in Table 4.4.1. In this table you will also find the number of iterations n required with the given m for the coincidence of the neighboring approximation with an accuracy of 10^{-6}.

Note that the value $m = 0.43$ substantially exceeds the previously obtained (see Grebenikov and Ryabov [1]) theoretical estimates of the domain of series convergence and iterations used for the construction of the Hill's periodic solutions.

4.5 Numerical–Analytic Construction of Mathieu Functions

Mathieu functions, as one knows (see Hayashi [1]), are periodic solutions of the differential equation

$$\frac{d^2 z}{dt^2} + (a + q\cos 2t)z = 0, \tag{4.5.1}$$

where a and q are constants. This equation is called the Mathieu equation. If the constant q taken as a parameter is fixed, then periodic solutions of Mathieu equations exist with certain values of the constant a which depend on q and are called the eigenvalues of this equation or of the corresponding Mathieu functions. These values form the two sequences:

$$a_{cn}, \quad n = 0, 1, 2, \dots \quad \text{and} \quad a_{sn}, \quad n = 1, 2, \dots$$

The so-called n-th order Mathieu cosine functions, denoted by $ce_n(t)$, correspond to the values a_{cn}, and inverting into $\cos nt$ as $t \to 0$. Mathieu n-th order sine functions, denoted by $se_n(t)$, correspond to values a_{sn}, and inverting into $\sin nt$ as $t \to 0$.

Grebenikov and Ryabov [1] and Ryabov [1] considered some variations of iteration algorithms for the construction of Mathieu functions in the form of trigonometric series (practically polynomials). The convergence of those algorithms is proved for the parameter q not exceeding certain limits (different for one or another Mathieu function); and estimations of some of those limits were obtained. Grebenikov and Ryabov [1] give examples of computer implementation of such algorithms in the case of functions $ce_1(t)$, $se_1(t)$. In this section a common algorithm is considered for the construction of Mathieu functions of any order n, and the respective Quick-Basic implementation of this algorithm is described.

4.5.1 Algorithm construction The n-th order sine and cosine Mathieu functions $ce_n(t)$, $se_n(t)$ will be calculated as the periodic solutions of the equation

$$\frac{d^2 z}{dt^2} + (n^2 + qh + q\cos 2t)z = 0, \tag{4.5.2}$$

where q is a specified numerical parameter (> 0), h is a constant found together with the Mathieu functions and determining the eigenvalues $A = n^2 + qh$ of those functions.

Note that with negative values of the parameter q we do not obtain new Mathieu functions, because (see, e.g., Witteker and Watson [1])

$$ce_{2n}(t, -q) = (-1)^n ce_{2n}\left(\frac{\pi}{2} - t,\, q\right),$$

$$se_{2n}(t, -q) = (-1)^n se_{2n}\left(\frac{\pi}{2} - t,\, q\right),$$

$$ce_{2n+1}(t, -q) = (-1)^n se_{2n+1}\left(\frac{\pi}{2} - t,\, q\right),$$

$$se_{2n+1}(t, -q) = (-1)^n ce_{2n+1}\left(\frac{\pi}{2} - t,\, q\right).$$

For the construction of $se_n(t)$, $n = 1, 2, 3, \ldots,$ assume

$$z = \sin nt + x \tag{4.5.3}$$

$$x = \sum_{L=1}^{L0} E_L \sin(n - 2L)t + \sum_{K=1}^{K0} F_K \sin(n + 2K)t, \tag{4.5.4}$$

where E_L, F_K are unknown coefficients, $K0$ is a sufficiently large number (we assumed $K0 = 30$), $L0 = (n-1)/2$ for odd n and $L0 = n/2 - 1$ for even n.

For the construction of $ce_n(t)$, $n = 0, 1, 2, \ldots,$ assume

$$z = \cos nt + x, \tag{4.5.5}$$

$$x = \sum_{L=1}^{L0} E_L \cos(n - 2L)t + \sum_{K=1}^{K0} F_K \cos(n + 2K)t, \tag{4.5.6}$$

where $L0 = (n-1)/2$ for odd n, $L0 = n/2$ for even n, and for even n $E_{n/2}$ should be replaced by $E_{n/2}/2$.

For x in cases (4.5.3), (4.5.5) we correspondingly have the equations

$$\ddot{x} + n^2 x = -q[h \sin nt + \cos 2t \cdot \sin nt + (h + \cos 2t)x], \tag{4.5.7}$$

$$\ddot{x} + n^2 x = -q[h \cos nt + cos2t \cdot \cos nt + (h + \cos 2t)x]. \tag{4.5.8}$$

Substituting (4.5.4), (4.5.6) into these equations and equating the left-hand and right-hand expressions at the cosines and sines of the respective arguments, we obtain algebraic recurrences for F_K, E_L, h. They are as follows.

For the coefficients F_K, $K \geq 1$, of all the functions $ce_n(t)$, $se_n(t)$ we have the relation

$$F_K = G_K(F_{K-1} + 2hF_K + F_{K+1}), \tag{4.5.9}$$

where $G_K = q/(8K(n + K))$, $F_0 = 1$ for $n > 0$ and $F_0 = 2$ for $n = 0$.

If we denote $M_L = -q/(8L(n-L))$, $L = 1,2,3,\ldots$, then for the coefficients E_1, E_2 of the functions $ce_4(t)$, $ce_n(t)$, $se_n(t)$, $N \geq 5$, we have the relation

$$E_1 = M_L(1 + 2hE_1 + E_2). \tag{4.5.10}$$

In the case of functions $ce_2(t)$ and $se_2(t)$ we have

$$E_1 = 2M_1(1 + hE_1) \quad \text{and} \quad \mathrm{E}_1 = 0 \tag{4.5.11}$$

respectively. For $N = 3$ we have

$$E_1 = M_1(1 + 2hE_1 \pm E_1), \tag{4.5.12}$$

where the plus corresponds to the function $ce_3(t)$, and the minus to the function $se_3(t)$. In the case of function se_4 we have the relation

$$E_1 = M_1(1 + 2hE_1). \tag{4.5.13}$$

The constant h is connected with c the oefficients F_1, E_1 by the relations

$$h + \frac{1}{2}F_1 = 0 \quad (\text{for } ce_0(t),\ se_2(t)), \tag{4.5.14}$$

$$h \pm \frac{1}{2} + \frac{1}{2}F_1 = 0 \quad (+ \text{ for } ce_1(t),\ - \text{ for } se_1(t)), \tag{4.5.15}$$

$$h + \frac{1}{2}(F_1 + E_1) = 0 \quad (\text{for } ce_2(t) \text{ and all } ce_n(t),\ se_n(t),\ n > 2). \tag{4.5.16}$$

For the coefficients E_L, $2 \leq L < n/2$ (n — even), $2 \leq L < (n-1)/2$ (n — odd) of functions $ce_n(t)$, $se_n(t)$ we have the relation

$$E_L = M_L(E_{L-1} + 2hE_L + E_{L+1}). \tag{4.5.17}$$

With even n and $L = n/2$ in the case of $ce_n(t)$, $se_n(t)$ we have

$$E_L = 2M_L(E_{L-1} + hE_L) \quad \text{and} \quad E_L = 0 \tag{4.5.18}$$

respectively. With odd n and $L = (n-1)/2$ we have

$$E_L = M_L(E_{L-1} + 2hE_L \pm E_L), \tag{4.5.19}$$

where plus corresponds to the function $ce_n(t)$, and minus to the function $se_n(t)$.

These are the initial relations constituting the basis for the iteration algorithm for the calculation of the value of h and all coefficients F_K, E_L.

For this, one can use the method of simple iterations. Then, for example, in the case of $ce_2(t)$

$$\begin{gathered} F_1^1 = G_1, \quad E_1^1 = 2M_1, \quad h_1 = -(F_1 + E_1)/2, \\ F_K^1 = G_K F_{K-1}^1, \quad K \geq 2, \\ F_1^2 = G_1(1 + 2h_1 F_1^1 + F_2^1), \quad E_1^2 = 2M_1(1 + h_1 E_1^1), \end{gathered} \tag{4.5.20}$$

etc.

If q remains a literal parameter, then simple iterations allow us to obtain approximations F_K^j, E_L^j, h_j, $j = 1, 2, 3\dots$ in analytic form as polynomials of powers of q. However the implementation of these iterations with different numerical values of q has shown that the domain of convergence of such iterations is comparatively small. For example, in the case of $ce_0(t)$ the upper bound of convergence is approximately equal to $4.7 - 4.8$.

Here we build another algorithm (of an irrational type) which is implemented with numeric values of q and has a sufficiently large convergence domain.

4.5.2 Computational layout 1. Calculation at the first step (step counter $IR = 1$) of the first approximation $F_1^{(1)}$, $E_1^{(1)}$, $h^{(1)}$ for F_1, E_1, h by means of formulae (4.5.10) – (4.5.16) provided that $F_2 = E_2 = 0$ (subprogram T1).

2. Calculation at the same step of the first approximation $F_K^{(1)}$, $E_L^{(1)}$, $K \geq 2$, $L \geq 2$, by formulae (4.5.9), (4.5.17) – (4.5.19), where $F_{K+1} = E_{L+1} = 0$ is assumed (subprogram T2).

3. Calculation at the second step ($IR = 2$) of the second approximation $F_1^{(2)}$, $E_1^{(2)}$ and $h^{(2)}$ by formulae (4.5.10) – (4.5.16), where $F_2 = F_2^{(1)}$, $E_2 = E_2^{(1)}$ (subprogram T1).

4. Calculation at the same step of the second approximation $F_K^{(2)}$, $E_L^{(2)}$, $K \geq 2$, $L \geq 2$, by formulae (4.5.9), (4.5.17) – (4.5.19), where $F_{K+1} = F_{K+1}^{(1)}$, $E_{L+1} = E_{L+1}^{(1)}$ (subprogram T3).

5. Calculation at the third step ($IR = 3$) of the third approximation $F_1^{(3)}$, $E_1^{(3)}$, and $h^{(3)}$ by means of formulae (4.5.10) – (4.5.16), where $F_2 = F_2^{(2)}$, $E_2 = E_2^{(2)}$ (subprogram T1), etc.

Subprogram T1 for the calculation of F_1, E_1, h consists of two parts. The first part provides for the calculation of $F_1^{(1)}$, $h^{(1)}$ for $ce_0(t)$, $ce_1(t)$, $se_1(t)$, $se_2(t)$ (then $E_1 = 0$). For example, in the case of $ce_1(t)$ at the first step according to (4.5.9), (4.5.16) we obtain

$$F_2 = 0, \quad F_1^{(1)} = \left(\sqrt{1 + 4G^2} - 1\right)/2G, \quad h^{(1)} = -\frac{1}{2} - \frac{1}{2}F_1^{(1)}, \quad G = \frac{1}{4}q. \tag{4.5.21}$$

Upon calculation of $F_K^{(1)}$, $K \leq 2$, by subprogram T2 we obtain at the second step

$$F_1^{(2)} = \frac{1}{2}\left(\sqrt{1 + 4G^2(1 + F_2^{(1)})} - 1\right), \quad h^{(2)} = -\frac{1}{2} - \frac{1}{2}F_1^{(2)}, \tag{4.5.22}$$

etc. In the cases $ce_0(t)$, $se_1(t)$, $se_2(t)$ the algorithms are similar.

In the cases $ce_2(t)$, $ce_n(t)$, $se_n(t)$, $n \leq 3$, the relations for F_1, E_1, h are more complicated. They can be written in the following general form:

$$E_1 + F_1 = -2h, \tag{4.5.23}$$

$$E_1\left[1 + q_3 - q_1(E_1 + F_1)\right] = -q_4(1 + R), \tag{4.5.24}$$

$$F_1\left[1 + q_2(E_1 + F_1)\right] = q_2(1 + S), \tag{4.5.25}$$

where

$$q_1 = q/(8N-8), \quad q_2 = q(8N+8), \quad S = F_2,$$
$$q_3 = 0 \ \text{ for } \ N = 2 \ \text{ and all } \ N > 3,$$
$$q_3 = q_1 \ \text{ for } \ ce_3(t) \ \text{ and } \ q_3 = -q_1 \ \text{ for } \ se_3(t),$$
$$q_4 = 2q_1 \ \text{ for } \ N = 2 \ \text{ and } \ q_4 = q_1 \ \text{ in the other cases,}$$
$$R = 0 \ \text{ for } \ ce_2(t), \ ce_3(t), \ se_3(t) \ \text{ and } \ R = E_2 \ \text{ in the other cases.}$$

Calculation of F_1, E_1, h provided that R, S are known is achieved by the second part of subprogram T1, namely by subprogram T11 included into T1. At the first step ($IR = 1$) assume $R = S = 0$ and consider relations (4.5.23) – (4.5.25) as a system of algebraic equations with the two unknowns E_1, F_1. With comparatively small values of q (for $q < 2N$) for the calculation of the roots of these equations use the Newton iteration algorithm, assuming the initial approximations

$$E_1^0 = -q_4, \quad F_1^0 = q_2.$$

The maximum allowed number of Newton iterations is assumed to be $J1 = 150$. This process stops if the neighboring approximations for F_1, E_1 coincide with an accuracy of $\alpha = 10^{-10}$; the obtained values are assumed as the first approximation $F_1^{(1)}$, $E_1^{(1)}$, and

$$h^{(1)} = -\left(F_1^{(1)} + E_1^{(1)}\right)/2$$

is calculated.

If after 150 iterations the coincidence of the neighboring approximations for F_1, E_1 is not achieved, then the calculation stops, and a message about the divergence of the process in subprogram T11 appears. With q values exceeding $2N$ the initial values F_1^0, E_1^0 for the Newton approximation are chosen with much more accuracy (otherwise those approximations either converge to an extraneous solution, or diverge). With this purpose, reduce relations (4.5.23) – (4.5.25), assuming that R, S are known, to the cubic equation with respect to h:

$$c_0 h^3 + c_1 h^2 + c_2 h + c_3 = 0, \tag{4.5.26}$$

where the coefficients $c_0, \ldots, c_3$ are expressed through $q_1, \ldots, q_4, R, S$:

$$c_0 = q_1 q_2,$$
$$c_1 = \left[q_2(1+q_3) - q_1\right]/2,$$
$$c_2 = \left[-1 - q_3 - q_2 q_4(1+R) - q_1 q_2(1+S)\right]/4,$$
$$c_3 = \left[q_4(1+R) - q_2(1+q_3)(1+S)\right]/8.$$

For the sought values F_1, E_1 we have the formulae

$$F_1 = \frac{q_2(1+S)}{1-2hq_2}, \qquad E_1 = -\frac{q_4(1+R)}{1+q_3+2hq_1} \tag{4.5.27}$$

Analysis and respective calculations show that for all considered values of N, q this equation has three real roots. It is necessary to calculate the root $h^0 = h^0(q)$

which depends continuously on q and becomes zero as $q \to 0$. To calculate it, apply the Kardano–Gudde formulae, viz, first rearrange (4.5.26) by means of the substitution $h = y - c_1/3c_0$ into the equation

$$y^3 - b_1 y + b_2 = 0, \tag{4.5.28}$$

where b_1, b_2 are expressed through $c_0, \dots, c_3$:

$$b_1 = c_1^2/3c_0^2 - c_2/c_0, \quad b_2 = -c_1c_2/3c_0^2 + 2c_1^3/27c_0^3 + c_3/c_0. \tag{4.5.29}$$

At the first step assume $R = S = 0$. The root y^0 in this case is calculated by the formula

$$y^0 = \sqrt{\frac{b_1}{3}}(-\cos D + \sqrt{3}\sin D), \tag{4.5.30}$$

where

$$D = D_1/3, \text{ if } b_2 < 0, \text{ and } D = D_1 + \pi, \text{ if } b_2 > 0,$$
$$D_1 = \operatorname{arctg} D_0, \quad -\pi/2 < D_1 < \pi/2, \quad D_0 = -2\sqrt{D_{00}}/b_2,$$
$$D_{00} = b_1^3/27 - b_2^2/4.$$

Having found y^0, calculate $h^0 = y^0 = -c_1/3c_0$, and also F_1^0, E_1^0 by formulae (4.5.27) with $R = S = 0$.

Use the values F_1^0, E_1^0 as an initial approximation for the calculation of the roots of equations (4.5.24) – (4.5.25) by Newton's method. The values F_1^0, E_1^0 are quite close to the exact roots $F_1^{(1)}$, $E_1^{(1)}$, therefore it takes quite a few iterations to get the values of the latter with a specified accuracy 10^{-10}.

Assume the obtained $F_1^{(1)}$, $E_1^{(1)}$ as the first approximation of the coefficients F_1, E_1 (calculated at the first step $IR = 1$). Then calculate

$$h^{(1)} = -\big(F_1^{(1)} + E_1^{(1)}\big)/2$$

and $F_K^{(1)}$, $E_L^{(1)}$, $K \geq 2$, $L \geq 2$, in accordance with subprogram T2.

At the second step $(IR = 2)$ the computational algorithm only differs by the assumption $R = E_2^{(1)}$, $S = F_2^{(1)}$ instead of $R = S = 0$.

With $q < 2N$ the approximation $F_1^{(2)}$, $E_1^{(2)}$ is calculated by Newton's method from equations (4.5.24) – (4.5.25) with the initial approximation $F_1^0 = F_1^{(1)}$, $E_1^0 = E_1^{(1)}$, and with $q \geq 2N$ this initial approximation is calculated by means of the cubic equation (4.5.26), etc.

This is the layout of the calculation of the approximations $F_K^{(j)}$, $E_L^{(j)}$, $h^{(j)}$, $j = 1, 2, \ldots$ (The computational algorithm is given in full in the Quick-Basic program described below.)

The maximum quantity $K0$ of the calculated coefficients F_K is fixed, and, as mentioned above, $K0 = 50$ is assumed. In addition, put a boundary on the absolute value of these coefficients and of the coefficients E_L:

$$|F_K| > \alpha = 10^{-10}, \quad |E_L| > \alpha = 10^{-10}.$$

Calculations show that these coefficients decrease in absolute value with an increase of their indices. Therefore only those first coefficients F_K, E_L that satisfy these estimations are calculated. The number of such coefficients obtained during computations are denoted by $K1$, $L1$ respectively. The neighboring approximations for these coefficients and also for h are compared, and if they coincide with an accuracy of $\epsilon = 10^{-8}$, we consider that with the given N, q there is (practical) convergence of the iteration processes, the calculation stops, and the last approximation for F_K, E_L, h is displayed on the monitor and printed out as the sought result.

The maximum allowable number of iterations is assumed as $I\max = 200$. If at step $IR = 200$ the required coincidence of neighboring approximations is not achieved, the calculation stops, and a message about the divergence of the process appears, as well as the results obtained at the last steps.

4.5.3 Quick-Basic program

```
PRINT "Construction of Mathieu functions CE(t),SE(t) of order"
PRINT " N(<=100) for diff.eq. Z'' + (N^2 + QH + Qcos2t) Z = 0"
PRINT " in form of polynomials"
PRINT "...+E(2)cos(N-4)t+E(1)cos(N-2)t+cosNt+F(1)cos(N+2)t+"
PRINT " F(2)cos(N+4)t+..."
PRINT "or"
PRINT "...+E(2)sin(N-4)t+E(1)sin(N-2)t+sinNt+F(1)sin(N+2)t+"
PRINT "F(2)sin(N+4)t+..."
PRINT "for given numerical value of parameter Q."
PRINT "Coefficients E(1),..., E(LO),F(1),...,F(KO) and"
PRINT " constant H are required quantities"
PRINT "Maximal number of coeff.E(L) equals LO,"
PRINT "maximal number of coeff.F(KO) is taken KO."
PRINT "The indicator ICS=1 corresponds to CE(t)"
PRINT "and indicator ICS=0 corresponds to SE(t)."
DEFDBL B-H, M, Q-S, X-Z
DIM F(2, 51), E(2, 51), G(50), H(2), M(50), U(50), V(50)
KO = 50: IMAX = 200: EPS = 1E-08: ALPH = 1E-10
100 :
INPUT ; "N="; N: INPUT ; "  ICS="; ICS: INPUT "    Q="; Q
IF N MOD 2 = 0 THEN
        LO = N / 2
ELSE
         LO = (N - 1) / 2
END IF
FOR K = 2 TO KO
        G(K) = Q / (8 * K * (N + K))
NEXT K
```

```
FOR L = 2 TO L0

      M(L) = -Q / (8 * L * (N - L))
NEXT L
FOR J = 1 TO 2
FOR I = 1 TO 51
F(J, I) = 0: E(J, I) = 0
NEXT I
NEXT J
IR = 1
R = 0: S = 0
GOSUB T1 'Calculation of X,Y,Z for F(1,1),E(1,1), H(1)
F(1, 1) = Y: E(1, 1) = X: H(1) = Z
GOSUB T2 'Calculation of F(1,K), E(1,L), K>1, L>1
NextIteration:
IR = IR + 1
R = E(1, 2): S = F(1, 2)
GOSUB T1 'Calculation of X,Y,Z for F(2,1), E(2,1), H(2)
F(2, 1) = Y: E(2, 1) = X: H(2) = Z
GOSUB T3 'Calculation of F(2,K), E(2,L), K1, L1
W0 = H(2) - H(1)
IF N + ICS > 2 THEN
U1 = F(2, 1) - Q2 * (1 + 2 * H(2) * F(2, 1) + F(2, 2))
END IF
IF N = 2 AND ICS = 1 THEN
   V1 = E(2, 1) + 2 * Q1 * (1 + H(2) * E(2, 1))
ELSEIF N = 3 AND ICS = 1 THEN
   V1 = E(2, 1) + Q1 * (1 + 2 * H(2) * E(2, 1) + E(2, 1))
ELSEIF N = 3 AND ICS = 0 THEN
   V1 = E(2, 1) + Q1 * (1 + 2 * H(2) * E(2, 1) - E(2, 1))
ELSEIF N = 4 AND ICS = 1 THEN
   V1 = E(2, 1) + Q1 * (1 + 2 * H(2) * E(2, 1) + E(2, 2))
ELSEIF N = 4 AND ICS = 0 THEN
   V1 = E(2, 1) + Q1 * (1 + 2 * H(2) * E(2, 1))
ELSEIF N - 4 > 0 THEN
  V1 = E(2, 1) + Q1 * (1 + 2 * H(2) * E(2, 1) + E(2, 2))
END IF
FOR J = 1 TO K1
   U(J) = F(2, J) - F(1, J)
NEXT J
FOR J = 1 TO L1
   V(J) = E(2, J) - E(1, J)
NEXT J
```

```
IF K1 < L1 THEN J1 = L1 ELSE J1 = K1
FOR J = 1 TO J1
     IF ABS(U(J)) > EPS OR ABS(V(J)) > EPS GOTO 200
NEXT J
IF ABS(W0) > EPS GOTO 200
IF ABS(U1) > EPS OR ABS(V1) > EPS GOTO 200
GOTO PrintRes
200 :
IF IR = IMAX GOTO 300
FOR J = 1 TO J1
    F(1, J) = F(2, J): E(1, J) = E(2, J): H(1) = H(2)
NEXT J
GOTO NextIteration
300 :
PRINT "No convergence for Q="; Q; SPC(3); "N="; N
PRINT "ICS="; ICS
LPRINT "No convergence for Q="; Q; SPC(3); "N="; N
LPRINT "ICS="; ICS
PrintRes:
A = N ^ 2 + Q * H(2)
PRINT "N="; N; SPC(3); "ICS="; ICS; SPC(3); "Q="; Q
PRINT "IR="; IR; SPC(3); "K1="; K1; SPC(3); "L1="; L1
LPRINT "N="; N; SPC(3); "ICS="; ICS; SPC(3); "Q="; Q
LPRINT "IR="; IR; SPC(3); "K1="; K1; SPC(3); "L1="; L1
IF N + ICS > 2 THEN
  PRINT "U1="; U1; SPC(1); "V1="; V1; SPC(1); "D00="; D00
LPRINT "U1="; U1; SPC(1); "V1="; V1; SPC(1); "D00="; D00
END IF
PRINT "H="; H(2); SPC(1); "W0="; W0; SPC(1); "A="; A
LPRINT "H="; H(2); SPC(1); "W0="; W0; SPC(1); "A="; A
FOR J = 1 TO K1
PRINT "F("; J; ")="; F(2, J)
PRINT "U("; J; ")="; U(J)
LPRINT "F("; J; ")="; F(2, J)
LPRINT "U("; J; ")="; U(J)
IF J MOD 17 = 0 THEN
   PRINT "press any key"
   AA$ = "":  WHILE AA$ = "": AA$ = INKEY$: WEND
END IF
NEXT J
PRINT "press any key"
AA$ = "": WHILE AA$ = "": AA$ = INKEY$: WEND
FOR J = 1 TO L1
```

```
PRINT "E("; J; ")="; E(2, J)
PRINT "V("; J; ")="; V(J)
LPRINT "E("; J; ")="; E(2, J)
LPRINT "V("; J; ")="; V(J)
IF J MOD 17 = 0 THEN
   PRINT "press any key"
   AA$ = "":  WHILE AA$ = "": AA$ = INKEY$: WEND
END IF
NEXT J
400 :
PRINT "If you want to repeat calculations with others or"
PRINT "with the same parameters N,ICS,Q, then enter 1 else 0"
INPUT "W="; W
IF W = 1 GOTO 100
PRINT "Calculations are ended": END
T1: 'Calculation of X,Y,Z  for F1, E1, H
IF N = 0 THEN
   G = Q / 4
   Y = (SQR(1+(2+S)*G^2)-1)/G: X=0: Z=-Y/2
   RETURN
ELSEIF N = 1 AND ICS = 1 THEN
   G = Q / (16 + Q): X = 0
   Y = (SQR(1+4*(1 + S)*G^2) - 1)/(2*G): Z = -(1 + Y) / 2
   RETURN
ELSEIF N = 1 AND ICS = 0 THEN
   G = Q / 16: X = 0
   Y = (G-1 + SQR((G-1)^2 + 4*(1 + S)*G^2)) / (2 * G)
   Z = (1 - Y) / 2
RETURN
END IF
IF N > 1 THEN
      Q1 = Q / (8 * (N - 1)): Q2 = Q / (8 * (N + 1))
END IF
IF N = 2 AND ICS = 1 THEN
      Q4 = 2 * Q1: Q3 = 0
ELSEIF N = 2 AND ICS = 0 THEN
      G = Q / 24: X = 0
      Y = (SQR(1+4*(1+S)*G^2) - 1) / (2*G): Z = -Y / 2
      RETURN
END IF
IF N > 2 THEN Q4 = Q1
IF N = 3 AND ICS = 1 THEN
     Q3 = Q1
```

```
ELSEIF N = 3 AND ICS = 0 THEN
     Q3 = -Q1
ELSEIF N > 3 THEN
     Q3 = 0
END IF
GOSUB T11
RETURN
T2: 'Calculation of F(1,K),E(1,L), K>1,L>1
FOR K = 2 TO K0
      F(1, K) = G(K) * F(1, K - 1) / (1 - 2 * H(1) * G(K))
NEXT K
IF N = 4 AND ICS = 1 THEN
      E(1, 2) = -Q * E(1, 1) / (16 + H(1) * Q)
ELSEIF N = 4 AND ICS = 0 THEN
      E(1, 2) = 0
END IF
IF N = 5 THEN
     IF ICS = 1 THEN E1 = 1 ELSE E1 = -1
     E(1, 2) = -Q * E(1, 1) / (48 + Q * (2 * H(1) + E1))
END IF
IF N > 5 THEN
     L00 = L0 - 1
     FOR J = 2 TO L00
           E(1, J) = M(J)*E(1, J - 1) / (1 - 2*H(1)*M(J))
     NEXT J
     J = L0
     IF N MOD 2 = 0 THEN
        IF ICS = 1 THEN
           E(1, J) = 2*M(J)*E(1, J - 1)/(1 - 2*H(1)*M(J))
        ELSE
           E(1, J) = 0
        END IF
     ELSE
        IF ICS = 1 THEN E1 = 1 ELSE E1 = -1
        E(1, J) = M(J)*E(1, J - 1)/(1 - M(J)*(2*H(1) + E1))
     END IF
END IF
RETURN
T11: 'Calculation  of X,Y,Z for F1,E1,H
C0 = Q1 * Q2
C1 = (Q2 * (1 + Q3) - Q1) / 2
C2 = (-1 - Q3 - Q2 * Q4 * (1 + R) - Q1 * Q2 * (1 + S)) / 4
C3 = (Q4 * (1 + R) - Q2 * (1 + Q3) * (1 + S)) / 8
```

```
B1 = C1 ^ 2 / (3 * C0 ^ 2) - C2 / C0
B2 = -C1*C2 / (3*C0^2) + (2*C1^3)/(27*C0^3) + C3/C0
D00 = B1 ^ 3 / 27 - B2 ^ 2 / 4
N1 = 2 * N
IF Q < N1 GOTO 500
D0 = -SQR(D00) * 2 / B2
D1 = ATN(D0)
IF B2 < 0 THEN
     D = D1 / 3
ELSE
     D = (D1 + 4 * ATN(1)) / 3
END IF
Z0 = SQR(B1/3)*(-COS(D)+SQR(3)*SIN(D)) - C1/(3*C0)
X1 = -Q4 * (1 + R) / (1 + Q3 + 2 * Q1 * Z0)
Y1 = Q2 * (1 + S) / (1 - 2 * Q2 * Z0)
JT = 1: J1 = 150
GOTO 600
500 :
IF IR = 1 THEN
     X1 = -Q4: Y1 = Q2
ELSE
   X1 = E(1, 1): Y1 = F(1, 1)
END IF
JT = 1: J1 = 150
600 :
F1 = X1 * (1 + Q3) - Q1 * (X1 + Y1) * X1 + Q4 * (1 + R)
F2 = Y1 + Q2 * (X1 + Y1) * Y1 - Q2 * (1 + S)
DF1 = (1 + Q3) * (1 + Q2 * (X1 + 2 * Y1))
DF = DF1 - Q1 * (2 * X1 + Y1) - 2 * Q1 * Q2 * (X1 + Y1) ^ 2
DF2 = (1 + Q2 * (X1 + 2 * Y1)) * F1 + Q1 * X1 * F2
DF3 = (1 + Q3 - Q1 * (2 * X1 + Y1)) * F2 - Q2 * Y1 * F1
X2 = X1 - (DF2 / DF): Y2 = Y1 - (DF3 / DF)
IF ABS(X2 - X1) > ALPH OR ABS(Y2 - Y1) > ALPH GOTO 700
X = X2: Y = Y2: Z = -(X + Y) / 2
RETURN
700 :
IF JT > J1 GOTO 800
X1 = X2: Y1 = Y2
JT = JT + 1
GOTO 600
800 :
PRINT "No convergence of T11 for IR="; IR
PRINT "Press any key"
```

```
AA$ = "": WHILE AA$ = "": AA$ = INKEY$: WEND
GOTO 400
T3: 'Calculation of F(2,K),E(2,L),K=2 to K1,L=2 to L1
FOR K = 2 TO K0
F(2, K) = G(K)*(F(2, K - 1)+F(1, K+1))/(1 - 2*H(2)*G(K))
IF ABS(F(2, K)) < ALPH THEN EXIT FOR
NEXT K
K1 = K - 1
IF N = 4 AND ICS = 1 THEN
    E(2, 2) = -Q * E(2, 1) / (16 + H(2) * Q)
ELSEIF N = 4 AND ICS = 0 THEN
    E(2, 2) = 0
END IF
IF N = 5 THEN
    IF ICS = 1 THEN E1 = 1 ELSE E1 = -1
    E(2, 2) = -Q * E(2, 1) / (48 + Q * (2 * H(2) + E1))
END IF
IF N > 5 THEN
    L00 = L0 - 1
    FOR J = 2 TO L00
      E(2, J) = M(J)*(E(2, J-1)+E(1, J+1))/(1-2*H(2)*M(J))
    NEXT J
    J = L0
    IF N MOD 2 = 0 THEN
         IF ICS = 1 THEN
            E(2, J) = 2*M(J)*E(2, J-1)/(1-2*M(J)*H(2))
         ELSE
            E(2, J) = 0
         END IF
    ELSE
         IF ICS = 1 THEN E1 = 1 ELSE E1 = -1
         E(2, J) = M(J)*E(2, J - 1)/(1 - M(J)*(2*H(2)+E1))
    END IF
END IF
FOR J = 1 TO L0
    I = L0 + 1 - J
    IF ABS(E(2, I)) > ALPH THEN EXIT FOR
NEXT J
L1 = L0 + 1 - J
RETURN
STOP
```

4.5.4 Comments on the program The Quick-Basic program given above implements the algorithm in double precision. In response to a user query the program sets the order N, the criterion $ICS = 1$ or $ICS = 0$ for the functions $ce_N(t)$, se_N respectively, and the numerical value of the parameter q. After execution of the program, the following are displayed on the monitor and printed out (in the case of convergence of iterations):

1) the given values of N, ICS, q;

2) the number IR of iterations allowed for the required accuracy $\alpha = 10^{-10}$ of the coefficients F_K, E_L;

3) the quantities $K1$, $L1$ of the coefficients F_K, E_L exceeding α in absolute value;

4) the residual discrepancies $U1$, $V1$ obtained upon substitution of the final values F_1, F_2, E_1, E_2, h into the initial relations (4.5.9) for $K = 1$;

5) the quantity h, coefficients F_K, E_L, $1 \le K \le K1$, $1 \le L \le L1$ (with double precision), differences $W0$, $U(K)$, $V(K)$ between the last two approximations for h, F_K, E_L respectively, and the quantity $D00$ proportional to the discriminant of the cubic equation (4.5.26).

If $K1 > 17$, the first values to be displayed and printed out are the first 17 coefficients F_K, and then upon pressing any key, the next 17 coefficients, etc. (the maximum allowed number of coefficients F_K is 50). In the same way the coefficients E_L are displayed on the monitor and printed out.

The maximum order N of Mathieu functions, allowed by the program, is 100.

Note that at his own discretion the user can change the program parameters $K0$, $I\max$, $alpha$, eps, as well as the dimension parameter for the calculated coefficients F_K, E_L and the associated quantities G_K, M_L, $U(K)$, $V(L)$.

The implementation of this program shows that the (practical) convergence of the proposed iteration process is ensured in the case of functions $ce_0(t)$, $ce_1(t)$, $se_1(t)$, $se_2(t)$ in a very large domain, at least for $0 < q < 7000$. In the case of the other Mathieu functions the convergence domain is much smaller, though it is large enough, and this domain increases with the increased order N of these functions. For different functions $ce_N(t)$, $se_N(t)$ see the table of upper bounds q_* of the parameter q, for which the number of iterations IR ensuring the prescribed accuracy 10^{-8} tends towards 200.

Table 4.5.1

ce_N	ce_2	ce_3	ce_4	ce_5	ce_6	ce_7	ce_8	ce_9	ce_{10}	ce_{20}	ce_{30}	ce_{40}
q_*	40.5	160	33	59	48	65	67	80	88	182	273	365
se_N		se_3	se_4	se_5	se_6	se_7	se_8	se_9	se_{10}	se_{20}	se_{30}	se_{40}
q_*		30.3	70	40	60	58	72	78	88	182	270	360

An example of results printed out in the case $ce_2(t)$, $q = 32$, is given below.

```
Construction of Mathieu functions
N= 2     ICS= 1     Q= 32
IR= 36     K1= 11     L1= 1
U1=-6.795683E-09  V1= 4.56666E-16
C3= .1160455991877158   D00= .2469067098022291
H= 1.160101914937828D-02  DH=-1.631752510888873D-10
A= 4.371233
F( 1 )= 7.297387560717799           U( 1 )= 9.071099E-09
F( 2 )= 4.303726409970468           U( 2 )= 5.096762E-09
F( 3 )= 1.210210035599587           U( 3 )= 1.367956E-09
F( 4 )= .2064818841575251           U( 4 )= 2.250614E-10
F( 5 )= 2.389046878665688D-02       U( 5 )= 2.531559E-11
F( 6 )= 2.00541015603754D-03        U( 6 )= 2.078218E-12
F( 7 )= 1.27923482678711D-04        U( 7 )= 1.302111E-13
F( 8 )= 6.416610614172347D-06       U( 8 )= 6.435982E-15
F( 9 )= 2.598511597751983D-07       U( 9 )= 2.5746E-16
F( 10 )= 8.67651371123749D-09       U( 10 )= 8.508039E-18
F( 11 )= 2.429387587345403D-10      U( 11 )= 2.359225E-19
E( 1 )=-7.320589599016555           V( 1 )=-8.744748E-09
```

4.6 Algorithm for Construction of Solutions of the Plane Bounded Three-Body Problem

Consider the bounded plane circular problem of three bodies: Sun–Jupiter–asteroid in the nonresonance case. An algorithm is proposed based on iterations with accelerated (quadratic) convergence. These iterations are not connected with successive changes of variables and belong to iterations of an ordinary type, i.e. each subsequent approximation is directly determined from the previous one. At the same time, for comparison of the obtained results, we indicate an algorithm for constructing the solution by means of simple iterations, equivalent in essence to the power series in a small mass.

The proposed algorithms require a huge number of algebraic operations with double Fourier polynomials with the help of appropriate computer program packages.

4.6.1 Initial differential equations of the problem The following equations will be considered as the initial equations of the given problem (see Subotin [1]):

$$\begin{aligned} \frac{dp}{d\theta} &= \mu F_1(p,e,G,L), & \frac{dG}{d\theta} &= 1 - \mu F_3(p,e,G,L), \\ \frac{de}{d\theta} &= \mu F_2(p,e,G,L), & \frac{dL}{d\theta} &= 1 - F_4(p,e,G), \end{aligned} \tag{4.6.1}$$

where θ is the asteroid longitude measured from some constant direction, p the orbital parameter, e the orbital eccentricity, $G = \theta - g$ the true anomaly along the orbit (the difference between θ and the longitude of the orbital perihelion g), $L = \theta - n_j t$ (the difference between θ and the longitude of Jupiter moving uniformly at the angular velocity n_j), and μ the mass of Jupiter. Units of time and mass are chosen so that the gravitational constant $k^2 = 1$, the semiaxis of the circular orbit of Jupiter $a_j = 1$, the angular velocity $n_j = 1$, the sum of the mass of the Sun m_s and the mass of Jupiter μ is equal to 1. Then the functions on the right-hand sides of equations (4.6.1) will be expressed by the following formulae:

$$\begin{aligned} F_1 &= 2r^2 R'_L, \\ F_2 &= \frac{r^2}{p}(2e + 2\cos G - e\sin^2 G)R'_L + r^2 \sin G R_r, \\ F_3 &= \frac{1}{e}\left[\frac{r^2}{p}(2 + e\cos G)\sin G R'_L - r^2 \cos G R'_r\right], \\ F_4 &= \frac{1}{\sqrt{p}} r^2, \end{aligned} \tag{4.6.2}$$

where R is the disturbing function of the problem, equal to

$$R = -\frac{1}{r} - r\cos L + \frac{1}{\Delta}, \tag{4.6.3}$$

r the radius-vector of the asteroid, equal to

$$r = \frac{p}{1 + e\cos G}, \tag{4.6.4}$$

and Δ the distance between the asteroid and Jupiter, equal to

$$\Delta = (1 + r^2 - 2r\cos L)^{1/2}. \tag{4.6.5}$$

The functions $F_1, \dots, F_4$, after calculation of the partial derivatives R'_L, R'_r, can be written in the following form:

$$\begin{aligned} &F_1 = 2pT_1, \quad F_2 = (2e + 2\cos G - e\sin^2 G)T_1 + T_2 \sin G, \\ &F_3 = \frac{1}{e}(2 + e\cos G)\sin G \cdot T_1 - \frac{1}{e}\cos G \cdot T_2, \quad F_4 = p^{3/2}(1 + e\cos G)^{-2}, \end{aligned} \tag{4.6.6}$$

where

$$\begin{aligned} T_1 &= p^2[(1 + e\cos G)^{-3} - A^{-3/2}]\sin L, \\ T_2 &= 1 - p^2(1 + e\cos G)^{-2}\cos L + p^2[(1 + e\cos G)\cos L - p]A^{-3/2}, \\ A &= p^2 + (1 + e\cos G)^2 - 2p(1 + e\cos G)\cos L. \end{aligned} \tag{4.6.7}$$

The relation between θ and time t is expressed by the equation

$$\frac{d\theta}{dt} = p^{-3/2}(1 + e\cos G)^2. \tag{4.6.8}$$

From these expressions it is seen that the functions F_1, F_2 are odd functions of the angular variables G, L, and the functions F_3, F_4 are even, i.e.

$$F_j(p,e,-G,-L) = -F_j(p,e,G,L), \quad j = 1,2,$$
$$F_3(p,e,-G,-L) = F_3(p,e,G,L),$$
$$F_4(p,e,-G) = F_4(p,e,G).$$

4.6.2 Basic equations for the coefficients of the sought solution We find the solution to the initial equations (4.6.1) with given initial values $p(0)$, $e(0)$, $G(0)$, $L(0)$ in the following form:

$$\begin{aligned} p &= U_0 + \sum_{\|k\|=1}^{N} U_k \cos(k,\psi), \qquad & G &= \psi_1 + \sum_{\|k\|=1}^{N} W_K \sin(k,\psi), \\ e &= V_0 + \sum_{\|k\|=1}^{N} V_k \cos(k,\psi), \qquad & L &= \psi_2 + \sum_{\|k\|=1}^{N} S_k \sin(k,\psi), \end{aligned} \tag{4.6.9}$$

where N is a given large enough number, k the integer vector (k_1, k_2), and ψ_1, ψ_2 components of the vector $\psi = (\psi_1, \psi_2)$, and

$$\|k\| = |k_1| + |k_2|, \quad (k,\psi) = k_1\psi_1 + k_2\psi_2,$$
$$\psi_j = \omega_j\theta + \psi_{j0}, \quad j = 1,2, \quad \psi_0 = (\psi_{10}, \psi_{20}).$$

We seek the coefficients $U_0, U_k, \dots S_k$ of the Fourier polynomials, frequencies ω_1, ω_2 and values ψ_{10}, ψ_{20}. Our goal is to find an algorithm for the construction of the solution in the form of (4.6.9) with numerical coefficients. Therefore we fix the numerical values of the mass μ and initial values $p(0)$, $e(0)$, $G(0)$, $L(0)$ (initial value of θ is zero).

Substituting (4.6.9) into (4.6.1), we obtain

$$\begin{aligned} -\sum_{\|k\|=1}^{N} (k,\omega) U_k \sin(k,\psi) &= \mu F_1(p,e,G,L), \\ -\sum_{\|k\|=1}^{N} (k,\omega) V_k \sin(k,\psi) &= \mu F_2(p,e,G,L), \\ \omega_1 + \sum_{\|k\|=1}^{N} (k,\omega) W_k \cos(k,\psi) &= 1 - \mu F_3(p,e,G,L), \\ \omega_2 + \sum_{\|k\|=1}^{N} (k,\omega) S_k \cos(k,\psi) &= 1 - F_4(p,e,G), \end{aligned} \tag{4.6.10}$$

where ω_1, , ω_2 are the components of the vector $\omega = (\omega_1, \omega_2)$ and $(k,\omega) = k_1\omega_1 + k_2\omega_2$. The arguments p, e, G, L of the functions $F_1, \dots, F_4$ in the right-hand members of these relations are represented by polynomials (4.6.9). In addition, we have the relations connecting the initial values $p(0), \dots, L(0)$ and the quantities U_0, V_0, ψ_{10}, ψ_{20}:

$$\begin{aligned} p(0) &= U_0 + \sum_{\|k\|=1}^{N} U_k \cos(k, \psi_0), & G(0) &= \psi_{10} + \sum_{\|k\|=1}^{N} W_k \sin(k, \psi_0), \\ e(0) &= V_0 + \sum_{\|k\|=1}^{N} V_k \cos(k, \psi_0), & L(0) &= \psi_{20} + \sum_{\|k\|=1}^{N} S_k \sin(k, \psi_0). \end{aligned} \tag{4.6.11}$$

Theoretically the functions $F_1, \dots, F_4$ can be expressed by Fourier polynomials (if harmonics of order higher than N are neglected):

$$\begin{aligned} F_j(p, e, G, L) &= \sum_{\|k\|=1}^{N} F_{jk}(U, V, W, S) \sin(k, \psi), j = 1, 2, \\ F_3(p, e, G, L) &= \sum_{\|k\|=0}^{N} F_{3k}(U, V, W, S) \cos(k, \psi), \\ F_4(p, e, G) &= \sum_{\|k\|=0}^{N} F_{4k}(U, V, W) \cos(k, \psi), \end{aligned} \tag{4.6.12}$$

where F_{lk}, $l = 1, 2, 3, 4$, $\|k\| \le N$, are the corresponding Fourier coefficients depending on all U_k, V_k, $0 \le \|k\| \le N$, W_k, S_k, $1 \le \|k\| \le N$. $U, \dots, S$ denote the vectors with components equal to all of $U_k, \dots, S_k$ respectively. In some simple cases it is possible to construct explicit expressions for F_{jk}, but in the case of this particular problem such expression are too complicated, and we will not need them in future.

From (4.6.10) and (4.6.12) follow the equations with respect to the unknowns U_k, V_k, W_k, S_k, $1 \le \|k\| \le N$, ω_1, ω_2:

$$\begin{aligned} -(k, \omega) U_k &= \mu F_{1k}(U, \dots, S), & -(k, \omega) V_k &= \mu F_{2k}(U, \dots, S), \\ \omega_1 &= -\mu F_{30}(U, \dots, S), & (k, \omega) W_k &= -\mu F_{3k}(U, \dots, S), \\ \omega_2 &= 1 - F_{40}(U, V, W), & (k, \omega) S_k &= F_{4k}(U, V, W). \end{aligned} \tag{4.6.13}$$

These equations are supplemented with relations (4.6.11).

For us these algebraic equations are the basic ones. The number $\bar{N}$ of the sought values U_k, $1 \le \|k\| \le N$, (as well as V_k, W_k, S_k) is large enough. For instance, $\bar{N} = 42$ for $\|k\| \le 6$, $\bar{N} = 72$ for $\|k\| \le 8$, $\bar{N} = 110$ for $\|k\| \le 10$. The total number of unknown scalars $U_k, \dots, S_k, \omega_1, \omega_2$, $1 \le \|k\| \le N$, is equal to $4\bar{N} + 2$.

The solutions of these equations will be sought by means of successive approximations.

4.6.3 Construction of a solution by the method of simple iterations

Zero approximation

$$U_k^{(0)},\ V_k^{(0)},\ W_k^{(0)},\ S_k^{(0)},\ \omega_1^0, \omega_2^0$$

corresponds to the known formulae of undisturbed (Keplerian) motion with $\mu = 0$ (see Szebhely [1]). We obtain

$$U_0^{(0)} = p(0) = p_0, \quad V_0^{(0)} = e(0) = e_0,$$
$$G^{(0)} = \psi_1^{(0)} = \theta + \psi_{10}^{(0)}, \quad \psi_{10}^{(0)} = G(0), \quad \omega_1^{(0)} = 1,$$
$$U_k^{(0)} = V_k^{(0)} = W_k^{(0)} = 0, \quad 1 \le ||k|| \le N,$$
$$\frac{dL^{(0)}}{d\theta} = 1 - \frac{p_0^{3/2}}{(1 + e_0 \cos G^{(0)})^2} = 1 - \left(\frac{p_0}{1 - e_0^2}\right)^{3/2} \left[1 + \sum_{j=1}^{\infty} A_j \cos jG^{(0)}\right], \tag{4.6.14}$$

where

$$A_j = (-1)^j 2(1 + j\sqrt{1 - e_0^2})\beta^j,$$
$$\beta = \frac{e_0}{1 + \sqrt{1 - e_0^2}} = \frac{1 - \sqrt{1 - e_0^2}}{e_0}. \tag{4.6.15}$$

Hence

$$L^{(0)} = \psi_2^{(0)} + \sum_{||k||=1}^{N} S_k^{(0)} \sin(k, \psi^{(0)}), \tag{4.6.16}$$

where

$$\psi_2^{(0)} = \omega_2^{(0)}\theta + \psi_{20}^{(0)}, \qquad \omega_2^{(0)} = 1 - \left(\frac{p_0}{1 - e_0^2}\right)^{3/2} = 1 - a_0^{3/2},$$

(a_0 is the semiaxis of the undisturbed orbit of the asteroid),

$$S_k^{(0)} = S_{k_1,k_2}^{(0)} = \begin{cases} 0, & \text{if } k_2 \ne 0, \\ S_n^{(0)}, & \text{if } k_2 = 0, \quad k_1 = n \end{cases} \tag{4.6.17}$$

and $S_n^{(0)}$, $n = 1, 2, 3, \ldots$, are some numbers.

The value $\psi_{20}^{(0)}$ is calculated by the formula

$$\psi_{20}^{(0)} = L(0) - \sum_{n=1}^{N} S_n^{(0)} \sin(n\psi_{10}^{(0)}). \tag{4.6.18}$$

For example, if we assume that $p_0 = 0.8$, $e_0 = 0.1$, then $\omega_2^{(0)} = 0.27359$,

$$\psi_2^{(0)} = \omega_2^{(0)}\theta + \psi_{20}^{(0)}, \quad \psi_{10}^{(0)} = G(0), \quad \psi_1^{(0)} = \theta + \psi_{10}^{(0)}, \quad \omega_1^{(0)} = 1,$$
$$L^{(0)} = \psi_2^{(0)} + 0.14528 \sin\psi_1^{(0)} - 0.00545 \sin 2\psi_1^{(0)} \tag{4.6.19}$$
$$+ 0.00024 \sin 3\psi_1^{(0)} - 0.00001 \sin 4\psi_1^{(0)},$$

Note that with certain specified $p(0)$, $e(0)$ a sharp or even an exact resonance occurs, that is, a close or an exact commensurability of low order between the undisturbed frequencies $\omega_1^{(0)} = 1$ and $\omega_2^{(0)}$. Then the combination $k_1\omega_1^{(0)} + k_2\omega_2^{(0)}$ is close to zero for some small integers k_1, k_2. In this example $\omega_2^{(0)}/\omega_1^{(0)}$ notably differs from 1/3 and $\omega_1^{(0)} - 3\omega_2^{(0)} = 0.17923$ and it can be considered that there are no resonances of fourth and fifth order. At the same time $3\omega_1^{(0)} - 11\omega_2^{(0)} \simeq 0.00949$, so there is proximity to the 14th-order resonance.

The first approximation

$$U_k^{(1)},\ V_k^{(1)},\ W_k^{(1)},\ S_k^{(1)},\ \omega_1^{(1)},\ \omega_2^{(1)},\quad 1 \le \|k\| \le N$$

is expressed in accordance with (4.6.13) by formulae similar to those used in classical analytic theories of celestial mechanics:

$$\begin{aligned}
U_k^{(1)} &= -\frac{\mu}{(k,\omega^{(0)})} F_{1k}(U^{(0)},\dots,S^{(0)}),\\
V^{(1)} &= -\frac{\mu}{(k,\omega^{(0)})} F_{1k}(U^{(0)},\dots,S^{(0)}),\\
W_k^{(1)} &= -\frac{\mu}{(k,\omega^{(0)})} F_{3k}(U^{(0)},\dots,S^{(0)}),\\
\omega_1^{(1)} &= 1 - \mu F_3(U^{(0)},\dots,S^{(0)},\\
\omega_2^{(1)} &= 1 - F_{40}(U^{(0)},V^{(0)},W^{(1)}),\\
S_k^{(1)} &= -\frac{1}{(k,\omega^{(0)})} F_{4k}(U^{(0)},V^{(0)},W^{(1)}),
\end{aligned} \tag{4.6.20}$$

and $U_0^{(1)}$, $V_0^{(1)}$ are calculated by means of (4.6.11).

We propose the following method of calculation of the values of

$$F_{jk}(U^{(0)},\dots,S^{(0)}),\quad j=1,2,3,\quad F_{40}(U^{(0)},V^{(0)},W^{(1)}),\quad F_{4k}(U^{(0)},V^{(0)},W^{(1)}),$$

which does not require knowledge of the analytic expressions of those values as functions U_k, V_k, W_k, S_k.

For example, consider the quantities $F_{1k}(U^{(0)},\dots,S^{(0)})$ which, in accordance with (4.6.6), are Fourier coefficients for the function

$$F_1(p_0,e_0,G^{(0)},L^{(0)}) = 2p_0^3\left[\left(1+e_0\cos G^{(0)}\right)^{-3} - A_0^{-3/2}\right]\sin L^{(0)}, \tag{4.6.21}$$

where in accordance with (4.6.14)–(4.6.18)

$$\begin{aligned}
A_0 &= p_0^2 + (1+e_0\cos G^{(0)})^2 - 2p_0(1+e_0\cos G^{(0)})\cos L^{(0)},\\
L^{(0)} &= \psi_2^{(0)} + \sum_{n=1}^{N} S_n^{(0)} \sin nG^{(0)},\\
G^{(0)} &= \psi_1^{(0)} = \omega_1^{(0)}\theta + \psi_{10}^{(0)},\quad \psi_2^{(0)} = \omega_2^{(0)}\theta + \psi_{20}^{(0)}
\end{aligned} \tag{4.6.22}$$

and p_0, e_0, $S_n^{(0)}$ are some numbers.

By means of the appropriate programs for algebraic operations with Fourier polynomials one can represent $\cos L^{(0)}$ and also $\sin L^{(0)}$ by polynomials of the following form:

$$\cos L^{(0)} = \sum_{\|k\|=0}^{N} [\cos L^{(0)}]_k \cos(k, \psi^{(0)}), \qquad \sin L^{(0)} = \sum_{\|k\|=1}^{N} [\sin L^{(0)}]_k \sin(k, \psi^{(0)}), \tag{4.6.23}$$

where

$$(k, \psi^{(0)}) = k_1 \psi_1^{(0)} + k_2 \psi_2^{(0)}, \quad k_2 = \pm 1$$

and $[\cos L^{(0)}]_k$, $[\sin L^{(0)}]_k$ are numerical coefficients. Then, multiplying the polynomial for $\cos L^{(0)}$ by $\cos G^{(0)}$ and $\cos 2G^{(0)}$, we obtain an analogous polynomial for A_0:

$$A_0 = \sum_{\|k\|=0}^{N} [A_0]_k \cos(k, \psi^{(0)}) \tag{4.6.24}$$

with numerical coefficients $[A_0]_k$. The next important operation is the construction of the Fourier polynomial for $A_0^{-3/2}$. It is expedient to apply Newton's iteration algorithm, assuming $Z = A_0^{-3/2}$. For Z we have the equation

$$Z^{-2} - A_0^3 = 0$$

and the corresponding successive approximations

$$Z_{j+1} = \frac{3}{2} Z_j - \frac{1}{2} A_0^3 Z_j^3, \quad j = 0, 1, 2, \dots \tag{4.6.25}$$

As an initial approximation Z_0, $Z_0 = p_0^2$ can be assumed. Through the operations corresponding to (4.6.25), one can obtain, with prescribed accuracy, the polynomial

$$A_0^{-3/2} = \sum_{\|k\|=0}^{N} [A_0^{-3/2}]_k \cos(k, \psi^{(0)}) \tag{4.6.26}$$

with numerical coefficients $[A_0^{-3/2}]_k$, $\|k\| \leq N$. It is essential to note that we obtained this polynomial, which is equivalent to the Fourier polynomial for the third power of the value inverse to the distance Δ between the asteroid and Jupiter, without the use of the known (see Subotin [1]) series of disturbing function $R(p, e, G, L)$.

The Fourier polynomial for $(1 + e_0 \cos G^{(0)})^{-3}$ can also be constructed by means of Newton iterations, assuming that

$$Z = (1 + e_0 \cos G^{(0)})^{-1}, \quad Z^{-1} - (1 + e_0 \cos G^{(0)}) = 0.$$

For the iterations Z_j we obtain the formula

$$Z_{j+1} = 2Z_j - (1 + e_0 \cos G^{(0)}) Z_j^2, \quad j = 0, 1, 2, \dots \tag{4.6.27}$$

for $Z_0 = 1$. To obtain the coefficients of the polynomial for Z, e.g., for $e_0 = 0.9$ with an accuracy of $2 \cdot 10^{-11}$, it is sufficient to perform eight iterations corresponding to the calculation of terms up to e_0^{256}.

Cubing the polynomial for Z, we obtain the polynomial for $(1+e_0\cos G^{(0)})^{-3}$, and then, multiplying the polynomials for $(1+e_0\cos G^{(0)})^{-3}$, $A_0^{-3/2}$, $\sin L^{(0)}$, we obtain in accordance with (4.6.21) the polynomial for $F_1(p_0, e_0, G^{(0)}, L^{(0)})$:

$$F_1(p_0, e_0, G^{(0)}, L^{(0)}) = \sum_{\|k\|=1}^{N} F_{1k}(U^{(0)}, V^{(0)}, W^{(0)}, S^{(0)}) \sin(k, \psi^{(0)}) \tag{4.6.28}$$

with the sought numerical coefficients $F_{1k}(U^{(0)}, \dots, S^{(0)})$.

In a similar way one can construct Fourier polynomials for the functions $F_j(p_0, e_0, G^{(0)}, L^{(0)})$, $j=2,3$, and obtain the numerical values of $F_{jk}(U^{(0)}, \dots, S^{(0)})$, $\|k\| \leq N$. The first four formulae in (4.6.20) allow us to calculate the numerical values of $U_k^{(1)}$, $V_k^{(1)}$, $W_k^{(1)}$, $\omega_1^{(1)}$ for $1 \leq \|k\| \leq N$.

To find the values of $F_{4k}(U^{(0)}, V^{(0)}, W^{(1)})$, $0 \leq \|k\| \leq N$, it is necessary, in accordance with (4.6.10), to construct a Fourier polynomial for the function

$$F_4(p_0, e_0, \tilde{G}^{(0)}) = p_0^{3/2}(1+e_0\cos\tilde{G}^{(0)})^{-2}, \tag{4.6.29}$$

where

$$\tilde{G}^{(0)} = \psi_1^{(0)} + \sum_{\|k\|=1}^{N} W_k^{(1)} \sin(k, \psi^{(0)}). \tag{4.6.30}$$

Such a construction is performed through operations on Fourier polynomials (4.6.30). As a result, we obtain the polynomial

$$F_4(p_0, e_0, \tilde{G}^{(0)}) = \sum_{\|k\|=0}^{N} F_{4k}(U^{(0)}, V^{(0)}, W^{(1)}) \cos(k, \psi^{(0)}) \tag{4.6.31}$$

with the sought numerical coefficients F_{4k}, $0 \leq \|k\| \leq N$, which allows us, using the last two formulae of (4.6.20), to find $\omega_2^{(1)}$, and $S_k^{(1)}$, $1 \leq \|k\| \leq N$. Then the values of $\psi_{10}^{(1)}$, $\psi_{20}^{(1)}$, $U_0^{(1)}$, $V_0^{(1)}$ are calculated by means of (4.6.11).

Thus we obtain, in accordance with (4.6.9), the first approximation for the initial elements p, e, G, L:

$$\begin{aligned}
p^{(1)} &= U_0^{(1)} + \sum_{\|k\|=1}^{N} U_k^{(1)} \cos(k, \psi^{(1)}), \\
e^{(1)} &= V_0^{(1)} + \sum_{\|k\|=1}^{N} V_k^{(1)} \cos(k, \psi^{(1)}), \\
G^{(1)} &= \psi_1^{(1)} + \sum_{\|k\|=1}^{N} W_k^{(1)} \sin(k, \psi^{(1)}), \\
L^{(1)} &= \psi_2^{(1)} + \sum_{\|k\|=1}^{N} S_k^{(1)} \sin(k, \psi^{(1)}),
\end{aligned} \tag{4.6.32}$$

where

$$\psi_j^{(1)} = \omega_j^{(1)}\theta + \psi_{j0}^{(1)}, \quad j = 1, 2,$$
$$\psi^{(1)} = (\psi_1^{(1)}, \psi_2^{(1)}), \quad (k, \psi^{(1)}) = k_1\psi_1^{(1)} + k_2\psi_2^{(1)}. \tag{4.6.33}$$

Formulae for the second and further approximations are similar to (4.6.20):

$$\begin{aligned}
U^{(j+1)} &= -\frac{\mu}{(k, \omega^{(j)})} F_{1k}(U^{(j)}, \dots, S^{(j)}),\\
V_k^{(j+1)} &= -\frac{\mu}{(k, \omega^{(j)})} F_{2k}(U^{(j)}, \dots, S^{(j)}),\\
W_k^{(j+1)} &= -\frac{\mu}{(k, \omega^{(j)})} F_{3k}(U^{(j)}, \dots, S^{(j)}),\\
\omega^{(j+1)} &= 1 - \mu F_{30}(U^{(j)}, \dots, S^{(j)},\\
\omega_2^{(j+1)} &= 1 - F_{40}(U^{(j)}, V^{(j)}, W^{(j+1)}),\\
S_k^{(j+1)} &= -\frac{1}{(k, \omega^{(j)})} F_{4k}(U^{(j)}, V^{(j)}, W^{(j+1)}),\\
& 1 \le \|k\| \le N, \quad j = 2, 3, \dots
\end{aligned} \tag{4.6.34}$$

The method of calculation of the values of

$$F_{1k}(U^{(j)}, \dots, S^{(j)}), \dots, F_{4k}(U^{(j)}, V^{(j)}, W^{(j)})$$

is the same as in the case of the first approximation. The basic thing here is the construction of polynomials for the functions $F_j(p, e, G, L)$, if p, e, G, L are expressed by known polynomials of the form (4.6.9).

4.6.4 Construction of a solution by the method of iterations with quadratic convergence

Zero approximation

$$U_k^{(0)}, \dots, \omega_2^{(0)}, \psi_1^{(0)}, \psi_2^{(0)}$$

will remain the same as that obtained in Section 4.6.3.

First approximation In conformity with Newton's method of iterations, we assume in the basic algebraic equations (4.6.13)

$$\begin{gathered}
U_k = U_k^{(0)} + u_k, \quad V_k = V_k^{(0)} + v_k, \quad W_k = W_k^{(0)} + w_k, \quad S_k = S_k^{(0)} + s_k,\\
\omega_j = \omega_j^{(0)} + \nu_j, \quad j = 1, 2, \quad \omega = (\omega_1, \omega_2), \quad \nu = (\nu_1, \nu_2),\\
U_k^{(0)} = V_k^{(0)} = W_k^{(0)} = 0, \quad \omega_1^{(0)} = 1,
\end{gathered} \tag{4.6.35}$$

and $\omega_2^{(0)}$, $\psi_{10}^{(0)}$, $\psi_{20}^{(0)}$ are expressed in accordance with (4.6.14) – (4.6.18). After this substitution, equations (4.6.13) can be rewritten in the form:

$$\begin{aligned}
-(k,\, \omega^{(0)} + \nu)(U_k^{(0)} + u_k) &= \mu F_{1k}(U^{(0)} + u,\, \dots,\, S^{(0)} + s),\\
-(k,\, \omega^{(0)} + \nu)(V_k^{(0)} + v_k) &= \mu F_{2k}(U^{(0)} + u,\, \dots,\, S^{(0)} + s),\\
\omega_1^{(0)} + \nu_1 &= 1 - \mu F_{30}(U^{(0)} + u,\, \dots,\, S^{(0)} + s),\\
(k,\, \omega^{(0)} + \nu)(W_k^{(0)} + w_k) &= -\mu F_{3k}(U^{(0)} + u,\, \dots,\, S^{(0)} + s),\\
\omega_2^{(0)} + \nu_2 &= 1 - F_{40}(U^{(0)} + u,\, V^{(0)} + v,\, W^{(0)} + w),\\
(k,\, \omega^{(0)} + \nu)(S_k^{(0)} + s_k) &= -F_{4k}(U^{(0)} + u,\, V^{(0)} + v,\, W^{(0)} + w),\\
&1 \le \|k\| \le N.
\end{aligned} \tag{4.6.36}$$

Linearizing these equations, taking into account that $U_k^{(0)} = V_k^{(0)} = W_k^{(0)} = 0$, we find the sought linear equations with respect to u_k, v_k, w_k, s_k and ν_1, ν_2.

To ease notation we introduce the vectors

$$\begin{aligned}
X &= (U, V, W, S), & x &= (u, v, w, s),\\
\tilde{X} &= (U, V, W), & \tilde{x} &= (u, v, w),\\
F &= (F_1, F_2, F_3), & \tilde{F}_k &= (F_{1k}, F_{2k}, F_{3k})
\end{aligned} \tag{4.6.37}$$

and the same vectors with subscripts or superscripts. In addition, we introduce the following notation

$$\left(\frac{\partial f}{\partial Z_l},\, z_l\right)_m = \sum_{\|l\|=1}^{N} \left(\frac{\partial f}{\partial Z_l}\right)_m z_l, \tag{4.6.38}$$

where $f = f(Z)$ is a vector-function or scalar function of the argument Z with components Z_l, $1 \le \|l\| \le N$, z_l is an analogous vector, $l = (l_1,\, l_2)$ is the vector index; and the subscript m means that all derivatives $\partial f / \partial Z_l$ with respect to the components Z_l are calculated at $Z = Z^{(m)}$.

Then the linearized equations with respect to u, v, w, s, ν_1, ν_2 can be written in the form

$$\begin{aligned}
(k, \omega^{(0)})x_k + \mu\left(\frac{\partial \tilde{F}_k}{\partial X_l},\, x_l\right)_0 &= -\mu(\tilde{F}_k)_0,\\
\nu_1 + \mu\left(\frac{\partial F_{30}}{\partial X_l},\, x_l\right)_0 &= -\mu(F_{30})_0,\\
\nu_2 + \left(\frac{\partial F_{40}}{\partial \tilde{X}_l},\, \tilde{x}_l\right)_0 &= 0,\\
(k, \nu_1)S_k^{(0)} + \left(k, \omega^{(0)}\right) s_k + \left(\frac{\partial F_{4k}}{\partial \tilde{X}_l},\, \tilde{x}_l\right)_0 &= 0,
\end{aligned} \tag{4.6.39}$$

where

$$(\tilde{F}_k)_0 = \tilde{F}_k(U^{(0)}, \dots, S^{(0)}), \quad (F_{30})_0 = F_{30}(U^{(0)}, \dots, S^{(0)}).$$

Analytic expressions for $F_{jk}(U, \dots, S)$, $j = 1, 2, 3$, $F_{4k}(U, V, W)$ as functions of their vector arguments, and for their derivatives are unknown. The calculation of their numerical value with predefined numerical components of vectors U, V, W, S is described above. Here we consider the calculation of the derivatives of F_{jk} with respect to the components of vectors U, V, W, S, which is similar in a sense.

Let there be given an arbitrary smooth function $\Phi(y)$ of the scalar argument y,

$$y = y_0 + \sum_{l=1}^{N} y_l \cos l\psi, \tag{4.6.40}$$

where $y_0 = y(\psi)$ is a known even 2π-periodic function of the argument ψ, and $y_1, y_2, \dots$ are unknown small quantities. This function can be represented by a Fourier series with respect to a Fourier cosine series, therefore, if we are restricted to harmonics of an order not higher than N, then

$$\Phi\Big(y_0 + \sum_l y_l \cos l\psi\Big) = \Phi(y_1, y_2, \dots) + \sum_{l=1}^{N} \Phi_l(y_1, y_2, \dots) \cos l\psi, \tag{4.6.41}$$

where Φ_0, $\Phi_1, \dots$ are functions of the arguments $y_1, y_2, \dots$. Linearizing $\Phi(y)$ with respect to $y_1, y_2, \dots$, we obtain:

$$\begin{aligned}\Phi(y) = {} & [(\Phi_0)_0 + (\Phi_1)_0 \cos\psi + (\Phi_2)_0 \cos 2\psi + \cdots] \\ & + \left[\left(\frac{\partial\Phi_0}{\partial y_1}\right)_0 y_1 + \left(\frac{\partial\Phi_0}{\partial y_2}\right)_0 y_2 + \cdots\right] \\ & + \left[\left(\frac{\partial\Phi_1}{\partial y_1}\right)_0 y_1 + \left(\frac{\partial\Phi_1}{\partial y_2}\right)_0 y_2 + \cdots\right] \cos\psi \\ & + \left[\left(\frac{\partial\Phi_2}{\partial y_1}\right)_0 y_1 + \left(\frac{\partial\Phi_2}{\partial y_1}\right)_0 y_2 + \cdots\right] \cos 2\psi + \cdots,\end{aligned} \tag{4.6.42}$$

where the lower index 0 means that all Φ_j and the derivatives $\partial\Phi_j/\partial y_l$ are calculated for $y_1 = y_2 = \cdots = 0$.

At the same time

$$\Phi\Big(y_0 + \sum_l y_l \cos l\psi\Big) = (\Phi)_0 + \left(\frac{d\Phi}{dy}\right)_0 \sum_l y_l \cos l\psi + \cdots, \tag{4.6.43}$$

where

$$(\Phi)_0 = \Phi(y_0), \quad \left(\frac{d\Phi}{dy}\right)_0 = \left(\frac{d\Phi}{dy}\right)_{y=y_0}.$$

Now we write down the Fourier series for the functions $(d\Phi/dy)_0 \cos l\psi$, $l = 1, 2, 3, \ldots$:

$$\left(\frac{d\Phi}{dy}\right)_0 \cos\psi = \sum_{j\geq 0} A_j^{(1)} \cos j\psi, \quad \left(\frac{d\Phi}{dy}\right)_0 \cos 2\psi = \sum_{j\geq 0} A_j^{(2)} \cos j\psi, \quad \ldots \tag{4.6.44}$$

and obtain:

$$\Phi(y) = \Phi(y_0) + \left(\sum_{j\geq 0} A_j^{(1)} \cos j\psi\right) y_1 + \left(\sum_{j\geq 0} A_j^{(2)} \cos 2\psi\right) y_2 + \cdots$$

Hence it follows that formula (4.6.43) for $\Phi(y)$ can be rewritten in the form

$$\begin{aligned} \Phi(y) = \Phi(y_0) &+ \left[A_0^{(1)} y_1 + A_0^{(2)} y_2 + \cdots\right] \\ &+ \left[A_1^{(1)} y_1 + A_1^{(2)} y_2 + \cdots\right] \cos\psi \\ &+ \left[A_2^{(1)} y_1 + A_2^{(2)} y_2 + \cdots\right] \cos 2\psi + \\ &\ldots\ldots\ldots\ldots\ldots\ldots\ldots\ldots \end{aligned} \tag{4.6.45}$$

Comparing (4.6.42) and (4.6.45), we conclude that the arrays of derivatives (column vectors)

$$\operatorname{col}\left[\left(\frac{\partial\Phi_0}{\partial y_l}\right)_0, \left(\frac{\partial\Phi_1}{\partial y_l}\right)_0, \left(\frac{\partial\Phi_2}{\partial y_l}\right)_0, \ldots\right], \quad l = 1, 2, 3, \ldots$$

are equal to the corresponding arrays of Fourier coefficients

$$\left[A_0^{(l)},\ A_1^{(l)},\ A_2^{(l)}, \ldots\right].$$

The matrix of these derivatives

$$\begin{pmatrix} \left(\frac{\partial\Phi_0}{\partial y_1}\right)_0 & \left(\frac{\partial\Phi_0}{\partial y_2}\right)_0 & \cdots & \left(\frac{\partial\Phi_0}{\partial y_l}\right)_0 & \cdots \\ \left(\frac{\partial\Phi_1}{\partial y_1}\right)_0 & \left(\frac{\partial\Phi_1}{\partial y_2}\right)_0 & \cdots & \left(\frac{\partial\Phi_1}{\partial y_l}\right)_0 & \cdots \\ \left(\frac{\partial\Phi_2}{\partial y_1}\right)_0 & \left(\frac{\partial\Phi_2}{\partial y_2}\right)_0 & \cdots & \left(\frac{\partial\Phi_2}{\partial y_l}\right)_0 & \cdots \\ \vdots & \vdots & \vdots & \vdots & \ddots \end{pmatrix} \tag{4.6.46}$$

coincides with the matrix

$$\begin{pmatrix} A_0^{(1)} & A_0^{(2)} & \cdots & A_0^{(l)} & \cdots \\ A_1^{(1)} & A_1^{(2)} & \cdots & A_1^{(l)} & \cdots \\ A_2^{(1)} & A_2^{(2)} & \cdots & A_2^{(l)} & \cdots \\ \vdots & \vdots & \vdots & \vdots & \ddots \end{pmatrix}, \tag{4.6.47}$$

where the first column consists of Fourier coefficients for the function $(d\Phi/dy)_0 \cos\psi$, the second column consists of Fourier coefficients for the function $(d\Phi/dy)_0 \cos 2\psi$, etc.

The same results will be obtained if $\Phi(y)$ is an odd function y and we consider Fourier sine series instead of (4.6.40), (4.6.4).

Now revert to the basic blocks of the matrix of coefficients of equations (4.6.39):

$$\begin{pmatrix} \mu\left(\dfrac{\partial \tilde{F}_k}{\partial X_l}\right)_0 \\ \mu\left(\dfrac{\partial F_{30}}{\partial X_l}\right)_0 \end{pmatrix}, \qquad \begin{pmatrix} \left(\dfrac{\partial F_{40}}{\partial \tilde{X}_l}\right)_0 \\ \left(\dfrac{\partial F_{4k}}{\partial \tilde{X}_l}\right)_0 \end{pmatrix}. \tag{4.6.48}$$

If the blocks of this matrix are represented with its scalar elements, then the columns in these blocks will be similar to those of matrix (4.6.46), i.e. consisting of the derivatives of the corresponding functions F_j, $j = 1, \dots, 4$ with respect to any component of the vectors U, V, W, S. If the vectors $l = (l_1, l_2)$, $k = (k_1, k_2)$, $1 \le \|k\| \le N$, $1 \le \|l\| \le N$, are renumbered in a certain order, and the vectors $F_{jk}, U, V, W,\ S(F_{j1}, F_{j2}, \dots, U_1, U_2, \dots)$ are also renumbered in the same order, then, e.g. (dropping the subscripts 0),

$$\begin{pmatrix} \dfrac{\partial F_{40}}{\partial \tilde{X}_l} \\ \dfrac{\partial F_{4k}}{\partial \tilde{X}_l} \end{pmatrix} = \begin{pmatrix} \dfrac{\partial F_{40}}{\partial U_1} & \dfrac{\partial F_{40}}{\partial U_2} & \cdots & \dfrac{\partial F_{40}}{\partial V_1} & \dfrac{\partial F_{40}}{\partial V_2} & \cdots & \dfrac{\partial F_{40}}{\partial W_1} & \dfrac{\partial F_{40}}{\partial W_2} & \cdots \\ \dfrac{\partial F_{41}}{\partial U_1} & \dfrac{\partial F_{41}}{\partial U_2} & \cdots & \dfrac{\partial F_{41}}{\partial V_1} & \dfrac{\partial F_{41}}{\partial V_2} & \cdots & \dfrac{\partial F_{41}}{\partial W_1} & \dfrac{\partial F_{41}}{\partial W_2} & \cdots \\ \dfrac{\partial F_{42}}{\partial U_1} & \dfrac{\partial F_{42}}{\partial U_2} & \cdots & \dfrac{\partial F_{42}}{\partial V_1} & \dfrac{\partial F_{42}}{\partial V_2} & \cdots & \dfrac{\partial F_{42}}{\partial W_1} & \dfrac{\partial F_{42}}{\partial W_2} & \cdots \\ \vdots & \vdots & \vdots & \vdots & \vdots & \vdots & \vdots & \vdots & \ddots \end{pmatrix} \tag{4.6.49}$$

Based on what was said regarding matrices (4.6.46) and (4.6.47), we can conclude that the columns of matrix (4.6.49) consist of Fourier coefficients for the functions

$$\left(\frac{\partial F_4}{\partial p}\right)_0 \cos(l_1, \psi), \quad \left(\frac{\partial F_4}{\partial p}\right)_0 \cos(l_2, \psi), \dots,$$
$$\left(\frac{\partial F_4}{\partial e}\right)_0 \cos(l_1, \psi), \quad \left(\frac{\partial F_4}{\partial e}\right)_0 \cos(l_2, \psi), \dots$$

respectively.

So, to calculate all numerical values of the elements of matrix blocks (4.6.48) it is necessary to construct Fourier series for all functions

$$\left(\frac{\partial F_j}{\partial p}\right)_0 \cos(l, \psi), \quad \left(\frac{\partial F_j}{\partial e}\right)_0 \cos(l, \psi), \quad \left(\frac{\partial F_j}{\partial G}\right)_0 \sin(l, \psi), \quad \left(\frac{\partial F_j}{\partial L}\right)_0 \sin(l, \psi),$$
$$1 \le \|l\| \le N.$$

The elements of all these arrays should be neatly renumbered in accordance with the form of (4.6.39).

For instance, let $N = 2$. Then the ordered array of vectors $l = (l_1, l_2)$ is as follows:

$$l_1 = (0,1), \quad l_2 = (1,0), \quad l_3 = (0,2), \quad l_4 = (1,1), \quad l_5 = (1,-1), \quad l_6 = (2,0).$$

The ordered array of vectors $k = (k_1, k_2)$, $\|k\| \le 2$, is similar.

Having constructed the Fourier series for the functions

$$\left(\frac{\partial F_4}{\partial p}\right)_0 \cos(l_1, \psi) = \frac{3}{2}\sqrt{p_0}(1 + e_0 \cos G^0)^{-2} \cos(l_1, \psi),$$
$$\left(\frac{\partial F_4}{\partial e}\right)_0 \cos(l_1, \psi) = -2p_0^{3/2}(1 + e_0 \cos G^0)^{-3} \cos G^0 \cos(l_1, \psi),$$
$$\left(\frac{\partial F_4}{\partial G}\right)_0 \cos(l_1, \psi) = -2p_0^{3/2}(1 + e_0 \cos G^0)^{-3} e_0 \sin G^0 \cos(l_1, \psi),$$

where p_0, e_0 are some numbers, and $G^0 = \psi$, we find the numerical matrix of the derivatives

$$\begin{pmatrix} \left(\frac{\partial F_{40}}{\partial U_1}\right)_0 & \cdots & \left(\frac{\partial F_{40}}{\partial U_6}\right)_0 & \cdots & \left(\frac{\partial F_{40}}{\partial W_1}\right)_0 & \cdots & \left(\frac{\partial F_{40}}{\partial W_6}\right)_0 \\ \vdots & & \vdots & & \vdots & & \vdots \\ \left(\frac{\partial F_{46}}{\partial U_1}\right)_0 & \cdots & \left(\frac{\partial F_{46}}{\partial U_6}\right)_0 & \cdots & \left(\frac{\partial F_{46}}{\partial W_1}\right)_0 & \cdots & \left(\frac{\partial F_{46}}{\partial W_6}\right)_0 \end{pmatrix}$$

where $F_{40}, \ldots, F_{46}$ are the coefficients of Fourier polynomials for $F_4(p, l, G)$:

$$F_4(p, l, G) = F_{40}(U, V, W) + \cdots + F_{46}(U, V, W) \cos(k_6, \psi).$$

So the calculation of all coefficients of the equations system (4.6.39) amounts to the construction of Fourier series of the known functions with their arguments represented by Fourier polynomials; for this purpose it is necessary to perform a large enough number of algebraic operations on the latter.

The next step is to find the solution of the linear algebraic system (4.6.39) with a known numerical matrix of coefficients. This system certainly has high order (with $N = 10$ the order is 442), but in view of the fact that among the matrix elements there are pivotal elements that do not contain the small multiplier μ, one can hope that solving this system with enough accuracy will not present significant difficulties.

As a result of solving the system (4.6.39) we will find

$$x_k^{(0)} = (u_k^{(0)}, v_k^{(0)}, w_k^{(0)}, s_k^{(0)}), \quad \|k\| \le N, \quad \nu_1 = \nu_1^{(0)}, \quad \nu_2 = \nu_2^{(0)},$$

which are corrections to the zero approximation $U_k^{(0)}, \ldots, S_k^{(0)}$, and obtain the first approximation

$$\begin{aligned} &U_k^{(1)} = U_k^{(0)} + u_k^{(0)}, \quad \ldots, \quad S_k^{(1)} = S_k^{(0)} + s_k^{(0)}, \\ &\omega_1^{(1)} = \omega_1^{(0)} + \nu_1^{(0)}, \quad \omega^{(1)} = \omega_2^{(0)} + \nu_2^{(0)} \end{aligned} \tag{4.6.50}$$

to the solution of the basic nonlinear algebraic equations (4.6.13). Then, upon calculation of $U_0^{(1)}$, $V_0^{(1)}$, $\psi_{10}^{(1)}$, $\psi_{20}^{(1)}$, using (4.6.11), we find the first approximation

$$\begin{aligned} p^{(1)} &= U_0^{(1)} + \sum_{\|k\|=1}^{N} U_k^{(1)} \cos(k, \psi^{(1)}), \\ e^{(1)} &= V_0^{(1)} + \sum_{\|k\|=1}^{N} V_k^{(1)} \cos(k, \psi^{(1)}), \\ G^{(1)} &= \psi_1^{(1)} + \sum_{\|k\|=1}^{N} W_k^{(1)} \sin(k, \psi^{(1)}), \\ L^{(1)} &= \psi_2^{(1)} + \sum_{\|k\|=1}^{N} S_k^{(1)} \sin(k, \psi^{(1)}) \end{aligned} \tag{4.6.51}$$

to the solution of the initial differential equations (4.6.1).

Second approximation In equations (4.6.13) assume

$$\begin{aligned} &U_k = U_k^{(1)} + u_k, \quad \ldots, \quad S_k = S_k^{(1)} + s_k, \\ &\omega_1 = \omega_1^{(1)} + \nu_1, \quad \omega_2 = \omega_2^{(1)} + \nu_2 \end{aligned} \tag{4.6.52}$$

and write down the obtained equations with respect to the corrections u_k, v_k, w_k, s_k, ν_1, ν_2 to the second approximation in the form (4.6.36), where the index 0 is replaced by index 1. Linearizing these equations with respect to $u_k, \ldots, \nu_2$, allowing for equations (4.6.39) and notation (4.6.37), (4.6.38), we obtain the following equations:

$$\begin{aligned} (k,\nu)X_k^{(1)} + (k,\omega^{(1)})x_k + \mu\left(\frac{\partial \tilde{F}_k}{\partial X_l}, x_l\right)_1 &= -\mu\Delta_2\left(\tilde{F}_k\right)_1 - (k,\nu^{(0)})x_k^{(0)}, \\ \nu_1 + \mu\left(\frac{\partial F_{30}}{\partial X_l}, x_l\right)_1 &= -\mu\Delta_2(F_{30})_1, \\ \nu_2 + \left(\frac{\partial F_{40}}{\partial \tilde{X}_l}, \tilde{x}_l\right)_1 &= -\Delta_2(F_{40})_1, \\ (k,\nu)S_k^{(1)} + (k,\omega^{(1)})s_k + \left(\frac{\partial F_{4k}}{\partial \tilde{X}_l}, \tilde{x}_l\right)_1 &= -\Delta_2(F_{4k})_1 - (k,\nu^{(0)})s_k^{(0)}, \end{aligned} \tag{4.6.53}$$

where

$$\begin{aligned} \Delta_2(F_k)_1 &= (F_k)_1 - (F_k)_0 - \left(\frac{\partial \tilde{F}_k}{\partial X_l}, x_l\right)_1 x_l^{(0)}, \\ (F_k)_1 &= F_k(U^{(1)}, V^{(1)}, W^{(1)}, S^{(1)}), \end{aligned}$$

and we have similar expressions for $\Delta_2(F_{30})_1, \ldots, \Delta_2(F_{4k})_1$.

Calculation of the numerical values of $(F_1)_1, \ldots, (F_{4k})_1$ and the derivatives $(\partial F_k/\partial X_l)_1, \ldots, (\partial F_{4k}/\partial \tilde{X}_l)_1$ can be performed in a way similar to that proposed above for the construction of the first approximation. These calculations are far more extensive, because the approximations $p^{(1)}(\theta)$, $e^{(1)}(\theta)$ are represented by Fourier polynomials and $G^{(1)} = \psi +$ Fourier polynomial.

Having determined from the algebraic system (4.6.53) the corrections $u_k^{(1)}$, $v_k^{(1)}$, $w_k^{(1)}$, $s_k^{(1)}$, $\nu_1^{(1)}$, $\nu_2^{(1)}$ to the first approximation $U_k^{(1)}, \ldots, \omega_2^{(1)}$, we find the second approximation to the solution of algebraic equations (4.6.13):

$$U_k^{(2)} = U_k^{(1)} + u_k^{(1)}, \quad \ldots, \quad \omega_2^{(2)} = \omega_2^{(1)} + \nu_2^{(1)}.$$

Then, having calculated $U_0^{(2)}$, $V_0^{(2)}$, $\psi_0^{(2)}$, $\psi_0^{(2)}$ by means of (4.6.11), we find the second approximation to the solution of the differential equations (4.6.1)

$$\begin{aligned} p^{(2)} &= U_0^{(2)} + \sum_{\|k\|=1}^{N} U_k^{(2)} \cos(k, \psi^{(2)}), \\ e^{(2)} &= V_0^{(2)} + \sum_{\|k\|=1}^{N} V_k^{(2)} \cos(k, \psi^{(2)}), \\ G &= \psi_1^{(2)} + \sum_{\|k\|=1}^{N} W_k^{(2)} \sin(k, \psi^{(2)}), \\ L &= \psi_2^{(2)} + \sum_{\|k\|=1}^{N} S_k^{(2)} \sin(k, \psi^{(2)}), \end{aligned} \tag{4.6.54}$$

where $\psi^{(2)} = (\psi_1^{(2)}, \psi_2^{(2)})$, $\psi_1^{(2)} = \omega_1^{(2)}\theta + \psi_{10}^{(2)}$, $\psi_2^{(2)} = \omega_2^{(2)}\theta + \psi_{20}^{(2)}$, $(k, \psi^{(2)}) = k_1\psi_1^{(2)} + k_2\psi_2^{(2)}$.

The third and the further approximations are calculated by means of equations only differing from (4.6.53) by the corresponding indices.

Remarks

(1) The proposed iterating is in essence an analytic construction of the solutions, the existence of which at small enough μ and appropriate initial values $p(0)$, $e(0)$ (corresponding to the nonresonance case) is proved in KAM theory. Therefore for small μ and appropriate $p(0)$, $e(0)$ the iterations must converge.

(2) Assume that practical convergence occurs if at some step the differences between the neighboring approximations for $U_k, \ldots, S_k$, ω_1, ω_2 are smaller in absolute values than the predefined small ε.

(3) Varying μ and the initial values $p(0)$, $e(0)$, we can estimate the convergence domain of iterations and their effectiveness.

(4) It is interesting to compare these iterations and simple iterations.

4.7 Numerical–Analytic Implementation of Krylov–Bogolyubov Transform

The key tool for the construction of asymptotic solutions of differential equations regular with respect to μ is the Krylov–Bogolyubov transform. Construction of this transform in explicit form requires very extensive computations, therefore the use of computers with branched mathematical software seems very promising. To illustrate this statement consider a particular but very important class of equations with constant real frequencies.

First suppose that the frequencies are rationally incommensurable (nonresonance case), i.e. let there be a system with slow and fast phase variables

$$\frac{dx}{dt} = \mu X(x, y), \qquad \frac{dy}{dt} = \omega_0 + \mu Y(x, y), \tag{4.7.1}$$

where $x = (x_1, \dots, x_m)$ is the slow variable vector, $y = (y_1, \dots, y_n)$ the fast variable vector, $\omega_0 = (\omega_{10}, \dots, \omega_{n0})$ the constant vector with real, rationally incommensurable components, and X, Y vector-functions of the respective dimensions. Furthermore, let $X(x, y)$ and $Y(x, y)$ be represented by Fourier polynomials of order n with respect to the vector y, and by algebraic power polynomials of degree s with respect to the vector x:

$$X(x, y) = \sum_{0 \le \|k\| \le N} X_k(x) e^{i(k,y)}, \qquad Y(x, y) = \sum_{0 \le \|k\| \le M} Y_k(x) e^{i(k,y)}, \tag{4.7.2}$$

where $k = (k_1, \dots, k_n)$ is an integer vector,

$$(k, y) = \sum_{s=1}^{n} k_s y_s, \quad \|k\| = \sum_{s=1}^{n} |k_s|, \quad k_s = 0, \pm 1, \pm 2, \dots, \pm N.$$

Assume that the algebraic equation

$$X_0(x) = 0 \tag{4.7.3}$$

has the real solution $x = x^{(0)}$, i.e. there exists a constant vector $x^{(0)}$ such that the average value of the function $X(x, y)$ with respect to y is equal to zero.

Find the Krylov–Bogolyubov equation in the form usual for multifrequency systems:

$$x = \bar{x} + \mu u(\bar{x}, \bar{y}), \quad y = \bar{y} + \mu v(\bar{x}, \bar{y}). \tag{4.7.4}$$

The algorithm described in Section 2.4 for locally nonresonance systems is fully applicable to equations (4.7.1); therefore we can write the solutions of partial differential equations determining the transfer functions u and v. Here the existence of a trigonometric change of variables is guaranteed.

The vector-functions $u(\bar{x},\bar{y})$ and $v(\bar{x},\bar{y})$ are expressed by the formulae

$$u(\bar{x},\bar{y}) = \sum_{0\le\|k\|\le N} \frac{X_k(\bar{x})e^{i(k,\bar{y})}}{i(k,\omega_0)},$$
$$v(\bar{x},\bar{y}) = \sum_{0\le\|k\|\le N} \frac{Y_k(\bar{x})e^{i(k,\bar{y})}}{i(k,\omega_0)}, \qquad \|k\| \ne 0. \tag{4.7.5}$$

Due to the rational incommensurability of the components of the vector ω_0, the denominators (k,ω_0) in (4.7.5) do not become zero; these relations have no peculiarities, though for some k the values $|(k,\omega_0)|$ can be small enough.

Instead of the averaged slow variable $\bar{x}$ we introduce a new variable ξ by the formula

$$\xi = \bar{x} - x^{(0)}, \tag{4.7.6}$$

i.e. we construct asymptotic solutions in the neighborhood of $x^{(0)}$. It is easy to derive equations satisfied by the new sought variables ξ and y:

$$\left(\left(E + \mu\frac{\partial u}{\partial \bar{x}}\right), \frac{d\xi}{dt}\right) + \mu\left(\frac{\partial u}{\partial \bar{y}}, \frac{d\bar{y}}{dt}\right) = \mu X\left(x^{(0)} + \xi + \mu u,\, \bar{y} + \mu v\right),$$
$$\mu\left(\frac{\partial v}{\partial \bar{x}}, \frac{d\xi}{dt}\right) + \left(\left(E + \mu\frac{\partial v}{\partial \bar{y}}\right), \frac{d\bar{y}}{dt}\right) = \omega_0 + \mu Y\left(x^{(0)} + \xi + \mu u,\, \bar{y} + \mu v\right), \tag{4.7.7}$$

where E is a unit matrix.

The functions $u(\bar{x},\bar{y})$, $v(\bar{x},\bar{y})$, for which the formulae (4.6.5) have already been written, evidently satisfy the relations

$$\left(\frac{\partial u}{\partial \bar{y}}, \omega_0\right) = X(\bar{x},\bar{y}) - X_0(\bar{x}), \qquad \left(\frac{\partial v}{\partial \bar{y}}, \omega_0\right) = Y(\bar{x},\bar{y}) - Y_0(\bar{x}). \tag{4.7.8}$$

Now using these equalities, equations (4.7.7) can be rewritten as

$$\left(\left(E + \mu\frac{\partial u}{\partial \bar{x}}\right), \frac{d\xi}{dt}\right) + \mu\left(\frac{\partial u}{\partial \bar{y}}, \left(\frac{d\bar{y}}{dt} - \omega_0\right)\right)$$
$$= \mu\left[X\left(x^{(0)} + \xi + \mu u,\, \bar{y} + \mu v\right) - X\left(x^{(0)} + \xi,\, \bar{y}\right) + X_0\left(x^{(0)} + \xi\right)\right],$$
$$\mu\left(\frac{\partial v}{\partial \bar{x}}, \frac{d\xi}{dt}\right) + \left(\left(E + \mu\frac{\partial v}{\partial \bar{y}}\right), \left(\frac{d\bar{y}}{dt} - \omega_0\right)\right)$$
$$= \mu\left[Y\left(x^{(0)} + \xi + \mu u,\, \bar{y} + \mu v\right) - Y\left(x^{(0)} + \xi,\, \bar{y}\right) + Y_0\left(x^{(0)} + \xi\right)\right]. \tag{4.7.9}$$

Since $X_0(x^{(0)}) = 0$, the right-hand member of the first of the relations (4.7.9) can be represented in the form

$$\mu\left[(\bar{H},\xi) + P(\xi,\bar{y},\mu)\right], \tag{4.7.10}$$

where $\bar{H} = \partial X_0(x^{(0)})/\partial x$ is a constant matrix, and the norm of the function $P(\xi,\bar{y},\mu)$ has second order of infinitesimal with respect to $\|\xi\|$ and first order with respect to μ, that is

$$\|P(\xi,\bar{y},\mu)\| \sim \|\xi\|^2 + \mu. \tag{4.7.11}$$

The right-hand member of the second equation of (4.7.9) can be written as

$$\mu\big[Y_0(x^{(0)}) + Q(\xi, \bar{y}, \mu)\big], \tag{4.7.12}$$

where the norm of the function $Q(\xi, \bar{y}, \mu)$ has first order of infinitesimal with respect to $||\xi||$ and μ, i.e.

$$||Q\,(\xi, \bar{y}, \mu)|| \sim ||\xi|| + \mu. \tag{4.7.13}$$

In accordance with the expressions for X, Y, u, v the functions P and Q are representable by n-tuple Fourier series of the form (4.7.2) with respect to $\bar{y}$, with their coefficients being power series with respect to the components of the vector ξ. We approximate P and Q by Fourier polynomials with respect to y of order N_0 with coefficients that are algebraic polynomials of order s_0 with respect to the components of the vector ξ. Jacobi matrices are also representable by similar Fourier polynomials, but with matrix coefficients:

$$P_1 = \frac{\partial u\,(\bar{x}, \bar{y})}{\partial \bar{x}}, \qquad P_2 = \frac{\partial u\,(\bar{x}, \bar{y})}{\partial \bar{y}}, \qquad Q_1 = \frac{\partial v\,(\bar{x}, \bar{y})}{\partial \bar{x}}, \qquad Q_2 = \frac{\partial v\,(\bar{x}, \bar{y})}{\partial \bar{y}}. \tag{4.7.14}$$

Now considering the relations (4.7.9) as linear algebraic equations with respect to $d\xi/dt$, $d\bar{y}/dt$, we find

$$\begin{aligned} \frac{d\xi}{dt} &= \mu(\bar{H}, \xi) + \mu\bar{F}(\xi, \bar{y}, \mu), \\ \frac{d\bar{y}}{dt} &= \omega_0 + \mu Y_0(x^{(0)}) + \mu\Phi(\xi, \bar{y}, \mu), \end{aligned} \tag{4.7.15}$$

where $\bar{F}$ and $\bar{\Phi}$ are representable by Fourier polynomials with respect to $\bar{y}$ (of order N_0) with coefficients in the form of algebraic polynomials of powers of s_0 with respect to the components of the vector ξ. Here we have the estimations

$$\mu\left\|\bar{F}\right\| \sim \mu\left\|\xi\right\|^2 + \mu^2, \quad \mu\left\|\bar{\Phi}\right\| \sim \mu\left\|\xi\right\| + \mu^2. \tag{4.7.16}$$

The system of differential equations (4.6.15) determines the new unknown functions ξ and $\bar{y}$. It is essential that in the first of these equations the terms present in the vector-function $\mu\bar{F}$ have coefficients of infinitesimal order at least μ^2. Therefore the linear part of this equation, with respect to ξ, has constant coefficients corresponding to the matrix $\mu\bar{H}$ with an accuracy to terms of order μ. Furthermore, in the right-hand member of the second equation of (4.7.15) the variable summands included in the functions $\mu\Phi$ have order μ^2 or higher. It is this analytic structure of the right-hand members of equations with new unknown variables that is the goal to be achieved by means of transformation (4.7.4). If we had applied such a transform to system (4.7.15) once more, that is if instead of ξ, $\bar{y}$ we had introduced the variables ξ_1, $\bar{y}_1$ by the same formulae, we would have come to similar equations, where:

1) the linear part of the first equation, with respect to ξ_1, has constant coefficients with an accuracy to terms of order μ^2;

2) in the right-hand member of the second equation all variables depending on $\bar{y}_1$ and independent of ξ_1 have order μ^3 or higher.

From the point of view of calculations, the difficulty of the operations on the transformation of system (4.7.1) into system (4.7.15) lies in the fact that in all the constructed Fourier polynomials the analytic dependence on ξ is retained. In other words, we deal with polynomials of the form

$$\sum_{0\le\|k\|\le N_0} S_k(\xi)e^{i(k,y)}, \tag{4.7.17}$$

where the coefficients $S_k(\xi)$ are expressed by power series (in practice, algebraic polynomials of some finite power) with respect to the components of vector ξ:

$$S_k(\xi) = \sum_{0\le\|r\|\le s_0} h^{(k)}_{r_1,\dots,r_m}\xi_1^{r_1}\dots\xi_m^{r_m}. \tag{4.7.18}$$

The coefficients $h_r^{(k)}$ are specified numbers, and they are involved in computer operations. One has to constantly separate their real and imaginary parts.

So the full layout of the algorithm for the construction of transform (4.7.4) for any specified numeric value of μ in the nonresonance case consists of the following blocks.

1. Retaining the previous notation, write the multifrequency system (4.7.1) as

$$\frac{dx}{dt} = X(x,y), \qquad \frac{dy}{dt} = \omega_0 + \mu Y(x,y), \tag{4.7.19}$$

i.e. we imply that the numerical value of μ is specified and fixed. The functions X, Y are expressed by Fourier polynomials of the form (4.7.2) with the coefficients X_k, Y_k specified in the form of algebraic polynomials with respect to the components of the vector x.

2. Formulae determining the change of variables can be written in the form

$$x = \bar{x} + u(\bar{x},\bar{y}), \quad y = \bar{y} + v(\bar{x},\bar{y}), \quad \bar{x} = x^{(0)} + \xi, \tag{4.7.20}$$

where the functions u and v are expressed by means of polynomials (4.7.5).

3. Fix the numbers N_0 and s_0, as well as the admissible error for the computation of the coefficients.

4. Solve the functional equation $X_0(x) = 0$ and find the real solution $x = x^{(0)}$.

5. Having performed the change $x = x^{(0)} + \xi$, find, using relations (4.7.5), the Fourier polynomials of the form (4.7.17) for the functions $\mu u\left(x^{(0)}+\xi,\bar{y}\right)$, $\mu v\left(x^{(0)}+\xi,\bar{y}\right)$, denoting them by $\tilde{u}(\xi,\bar{y})$, $\tilde{v}(\xi,\bar{y})$.

6. In series (4.7.2) for the functions X, Y, perform the change of variables

$$x = x^{(0)} + \xi + \tilde{u}(\xi,\bar{y}), \quad y = \bar{y} + \tilde{v}(\xi,\bar{y}).$$

As a result obtain:

$$X\left(x^{(0)}+\xi+\tilde{u},\,\bar{y}+\tilde{v}\right)=\sum_{0\le\|k\|\le N}X_k\left(x^{(0)}+\xi+\tilde{u}\right)e^{i(k,\bar{y})}e^{i(k,\tilde{v})},$$
$$Y\left(x^{(0)}+\xi+\tilde{u},\,\bar{y}+\tilde{v}\right)=\sum_{0\le\|k\|\le N}Y_k\left(x^{(0)}+\xi+\tilde{u}\right)e^{i(k,\bar{y})}e^{i(k,\tilde{v})}. \tag{4.7.21}$$

The coefficients X_k, Y_k are algebraic polynomials with respect to the vector x. Then, as a result of successive mathematical operations on polynomials of the form (4.7.17), we will express the coefficients X_k, Y_k as polynomials of a similar structure, but of order N_0 and with coefficients that are algebraic polynomials of degree s_0 with respect to the new vector ξ.

7. Find Fourier polynomials for the exponents $e^{i(k,\tilde{v})}$ for different vectors k, using, e.g., Taylor series

$$e^{i(k,\tilde{v})}\cong 1+\sum_{j=1}^{\alpha}\frac{\left[i\,(k,\tilde{v})\right]^j}{j!}. \tag{4.7.22}$$

The superscript of the sum is chosen so as to achieve the specified accuracy for all coefficients. Then construct polynomials of the form (4.7.17) for vector-functions (4.7.21), and then also for functions (4.7.10) and (4.7.12).

8. After that construct Fourier polynomials of the form (4.7.17) for the coefficients $d\xi/dt$, $d\bar{y}/dt$ in the left-hand members of system (4.7.9), using the "machine" differentiation program (if available) for the calculation of partial derivatives of vector-functions $u(\bar{x},\bar{y})$, $v(\bar{x},\bar{y})$ from the arguments $\bar{x}$, $\bar{y}$.

9. Considering the system (4.7.9) as an algebraic system with respect to $d\xi/dt$, $d\bar{y}/dt$, find its inverse matrix, and then write the equations (4.7.15) that are necessary for the further calculations, in the form

$$\frac{d\xi}{dt}=\left(\bar{H},\xi\right)+\bar{F}\left(\xi,\bar{y}\right),$$
$$\frac{d\bar{y}}{dt}=\omega+\bar{\Phi}(\xi,\bar{y}). \tag{4.7.23}$$

The new frequency vector

$$\omega=\omega_0+\mu Y_0\left(x^{(0)}\right) \tag{4.7.24}$$

and functions $\bar{F},\bar{\Phi}$ are expressed by means of a Fourier polynomial of the form (4.7.17), i.e.

$$\bar{F}\left(\xi,\bar{y}\right)=\sum_{0\le\|k\|\le N_0}F_k(\xi)e^{i(k,\bar{y})},\qquad \bar{\Phi}\left(\xi,\bar{y}\right)=\sum_{0\le\|k\|\le N_0}\Phi_k(\xi)e^{i(k,\bar{y})},$$

where F_k, Φ_k are algebraic polynomials of degree s_0 with respect to the components of vector ξ.

10. Then, using numerical, analytic or some "hybrid" method, solve the Cauchy problem for system (4.7.23) with the initial conditions determined through

solving the following functional equations with respect to ξ_0, $\bar{y}_0$:

$$\tilde{u}\,(\xi_0, \bar{y}_0) + \xi_0 + x^{(0)} = x_0, \quad \tilde{v}\,(\xi_0, \bar{y}_0) + \bar{y}_0 = y_0, \tag{4.7.25}$$

where x_0, y_0 are the initial values for the initial system (4.7.1).

Note that a similar algorithm can be applied in the case of rationally commensurable frequencies $\omega_{10}, \omega_{20}, \ldots, \omega_{n0}$, and also in the case of the variable frequency vector $\omega(x)$.

4.8 Comments and References

This chapter does not pretend to be a complete and consistent development of algorithms and their numerical–analytic implementation for the problems of resonance analytic dynamics, described by nonlinear differential equations. It is a kind of introduction to such a development. But in any case, examples such as the construction of Lyapunov and Krylov–Bogolyubov transforms, the construction of Hill's solutions and Mathieu functions prove the ample efficiency of numerical–analytic methods and the enlarged possibility of implementation of the corresponding algorithms by these methods compared with purely analytic technique. The described version of numerical–analytic methods, mainly providing for the implementation of algorithms by way of operations on the appropriate structures with numerical (but not literal) coefficients, in a number of cases results in simple enough schemes of calculation that require modest computer aids (e.g., Hill solutions and Mathieu functions). Certainly, in many problems, e.g. in the three-body problem described above, implementation of algorithms requires computers with large enough memory and operating speed. This is mainly due to the fact that we propose to focus on iterations with quadratic convergence. It is this computational process that allows us to surmount to a certain extent the small denominators problem that is typical for problems of resonance analytic dynamics. We would also note that for the further development of the proposed numerical–analytic methods one should use modern computer analytic calculus packages like MAPLE, MATHEMATICA 3.0 (see Wolfram [1]) and other similar software.

Section 4.1 For the derivation of algorithms for the Lyapunov transformation the material described in the papers Erugin [1], Lyapunov [1], Woodcock, *et al.* [1], Starzhinski and Yakubovich [1] is used.

Section 4.2 Algorithms in the construction of Green and Lyapunov matrices were derived in accordance with Lyapunov [1], Grebenikov and Ryabov [1, 4].

Section 4.3 The derived algorithms are new.

Section 4.4 The described method of the construction of periodic Hill solutions conforms to Grebenikov and Ryabov [4].

Section 4.5 The described full algorithm for the construction of Mathieu functions and the computer program are new.

Section 4.6 The shortened version of the described algorithm for the construction of solutions to the three-body problem was given in Ryabov [3, 4].

Section 4.7 The described algorithm corresponds to that from (Grebenikov and Ryabov [1].

References

Akoulenko, L. D. and Chernousko, F. L.

[1] Averaging method in optimal control problems. *Journal of Computational Mathematics and Mathematical Physics* **15**(4) (1975) 20–28. [Russian]

Amirkhanov, I. V., Zhidkov, E. P. and Zhidkova, I. E.

[1] Influence of resonances $2\nu_z + 2\nu_x = 3$, $3\nu_z - \nu_x = 2$, $4\nu_z + \nu_x = 4$ on amplitudes of betatron oscillations. *Preprint, Joint Institute for Nuclear Research*, Doubna, 1988. [Russian]

[2] Averaged equations for the nonlinear resonance $3\nu_x = 2$ in higher approximations by the method of Krylov–Bogolyubov. *Preprint, Joint Institute of Nuclear Research*, Doubna, 1988. [Russian]

Arnaud, J. F.

[1] Progress of Intelsat. *JEEE Int. Conf. Commun.*, Vol. 4, Philadelphia, Pa., 1968.

Arnold, V. I.

[1] Small denominators and stability problem in classical and celestial mechanics. *Russian Mathematical Surveys* **18**(6) (1963) 92–191. [Russian]

[2] *Mathematical Methods of Classical Mechanics.* Moscow: Nauka, 1974. [Russian]

[3] Applicability conditions and error estimate for the speedup method for the systems passing through resonances in the process of evolution. *Dokl. Acad. Nauk SSSR* **161**(1) (1965) 9–12. [Russian]

Aubin, J. P. and Ekland, I.
[1] *Applied Nonlinear Analysis.* Moscow: Mir, 1988.
Avramchuk, N. A. and Klikh, Yu. A.
[1] Solution of the method of averaging of one optimal control problem. *Space Research* **9**(2) (1971) 211–218. [Russian]

Bakhvalov, N. S.
[1] *Numerical Methods.* Moscow: Nauka, 1973. [Russian]

Barton, D.
[1] On literal developments of the theory with the aid of computer. *Astron. Journ.* **72** (1967) 1281–1288.

Basic REFAL
[1] *Basic REFAL, its realization on computers.* Moscow, Funt of Programms and Algorithms, 1977. [Russian]

Bellman, R.
[1] *Introduction into Theory of Matrices.* Moscow: Nauka, 1969.

Bogolyubov, N. N.
[1] *On some Statistical Methods in Mathematical Physics.* Kiev: Publ. of the Ukraine Academy of Sciences, 1945. [Russian]

Bogolyubov, N. N. and Mitropolsky, Yu. A.
[1] *Asymptotic Methods in Nonlinear Oscillation Theory.* Moscow: Nauka, 1963. [Russian]

Bogolyubov, N. N., Mitropolsky, Yu. A. and Samoilenko, A. M.
[1] *Method of Accelerated Convergence in Nonlinear Mechanics.* Kiev: Naukova Dumka, 1969. [Russian]

Bogolyubov, N. N. and Zubarev, D. N.
[1] Method of asymptotic approximation for systems with revolving phase, and its application to charged particles motion in magnetic field. *Ukr. Mat. Zhurnal* **VII**(1) (1955) 5–17. [Russian]

Chernousko, F. L.
[1] On resonance in essentially nonlinear system. *J. of Comput. Math. and Math. Phys.* **3**(1) (1963) 131–144. [Russian]

Dekhtyarenko, V. M.
[1] Iteration algorithm of construction of periodic solutions of a plane circular bounded three-body problem. *Mathematical Issues of Nonlinear Analysis in Dynamic Systems Control.* Moscow: Publ. Moscow State University, 1985. [Russian]

Delaunay, C.

[1] Théorie du mouvement de la lune. *Mémoires de l'Académie des Sciences de la France* **1** (1860), **2** (1867). [French]

Deprit, A., Henrard, J. and Rom, A.

[1] Analytical Lunar Ephemeris. I. Definition of the main problem. *Boeing Sci. Res. Lab.*, 1970.

Duboshin, G. N. (Ed.)

[1] *Reference Manual on Celestial Mechanics and Astrodynamics.* Moscow: Nauka, 1976. [Russian]

Erugin, N. P.

[1] Reducible systems. *Proceedings of the Steklov Institute of Mathematics*, V. 12, Leningrad – Moscow: Publ. of the Academy of Sciences of the USSR, 1946. [Russian]

Filatov, A. N.

[1] *Asymptotic Methods in the Theory of Differential and Integral Equations.* Tashkent: FAN, 1974. [Russian]

Gantmacher, F. R.

[1] *Theory of Matrices.* Moscow: Nauka, 1967. [Russian]

Gikhman, I. I.

[1] Apropos of one theorem of N. N. Bogolyubov. *Ukr. Math. Journ.* **4**(2) (1952) 215–219. [Russian]

Godunov, S. K. and Ryaben'ky, V. S.

[1] *Difference Schemes.* Moscow: Nauka, 1973. [Russian]

Gomtzyan, P. Tz. and Grebenikov, E. G.

[1] Algorithm of inversion of the first integral of the three-body problem differential equations averaged by the scheme of N. D. Moiseev. *Numerical Methods of Solving Boundary and Initial Problems for Differential Equations.* Moscow: Publ. Moscow State University, 1986, 3–10. [Russian]

Grebenikov, E. A.

[1] *Introduction into Resonance Systems Theory.* Moscow: Publ. Moscow State University, 1987. [Russian]

[2] Method of construction of trigonometric theories of resonance asteroids motion. *Astron. Zhourn.* **47**(2) 1970 43–140. [Russian]

[3] *Averaging Method in Applications.* Moscow: Nauka, 1986. [Russian]

Grebenikov, E. A., Isayeva F. I. and Prikhodko, V. A.

[1] Asymptotic and numerical approximated solutions of equations of the motion of asteroids of Hecuba group. *Preprint, Institute of Theoretical and Experimental Physics.* Moscow, 1975. [Russian]

Grebenikov, E. A., Mironov, S. V. and Kiosa, M. N.

[1] *Numerical-Analytic Method of the Study of Regularly Disturbed Multifrequency Systems.* Moscow: Publ. Moscow State University 1986. [Russian]

Grebenikov, E. A. and Popova, N. I.

[1] Justification of averaging method for one system of differential equations in the resonance case. *Preprint. Institute of Theoretical and Experimental Physics,* Moscow, 1976. [Russian]

Grebenikov, E. A. and Ryabov, Yu. A.

[1] *Constructive Methods in the Analysis of Nonlinear Systems.* Moscow: Mir, 1983. [Russian]

[2] *New Qualitative Methods in Celestial Mechanics.* Moscow: Nauka, 1971. [Russian]

[3] *Resonances and Small Denominators in Celestial Mechanics.* Moscow: Nauka, 1978. [Russian]

[4] *Metoda Usrednienia w Mechanice Nieliniowej.* Warszawa: Panstwowe Wydawnictwo Naukowe, 1982 [Polish]

Grebenikov, E. A. and Tysynbayev, B.

[1] Construction and solution of standard systems of differential equations describing the oscillations of a long viscous-elastic plate. *Preprint, Institute of Theoretical and Experimental Physics,* Moscow, 1978, No. 39. [Russian]

Grebenikov, E. A., Zhuravlev, S. G. and Ryabov, Yu. A.

[1] Conditionally periodic solutions of canonical differential equations system at sharp resonance. *Preprint, Institute of Theoretical and Experimental Physics,* Moscow, 1976. [Russian]

Hale, J. K.

[1] *Oscillations in Nonlinear Systems.* Moscow: Mir, 1967.

Hansen, P. A.

[1] Austinandersetzung einer zweckmassigen Methode zur Berechnung der absoluten Stoerungen der kleinen Planeten. Abh. I (1857), II (1859), III (1861), Leipzig. [German]

Hayashi, Ch.

[1] *Nonlinear Oscillations in Physical Systems.* Moscow: Mir, 1967.

Hearn, A. C.
[1] *Reduce-2. User's Manual.* University of Utah, VCP-19. 1973.
[2] *Reduce-2. Implementation Guide for Standard Lisp of IBM 360/370.* University of Utah, No. TR 5, 1979.

Hill, G. W.
[1] Researches in the Lunar theory. *J. Math. Works.* **1** (1905) 284–335.

Karaganchu, V. Kh.
[1] *Construction of Analytic Theory of Motion of Planets of Hestia Family.* Thesis of Diss. Paper, Moscow, 1970. [Russian]

Kauderer, G.
[1] *Nonlinear Mechanics.* Moscow: Inostrannaya Literatura, 1961.

Klikh, Yu. A.
[1] Averaging methods in boundary problems of maximum principle. *Proc. of the VIII Intern. Conf. on Nonlin. Osc.*, Prague, 1978, 151–157.
[2] On resonance phenomena in time-optimal problems close to linear ones. *Problems of Analytic Mechanics, Stability Theory and Control Theory.* Moscow: Nauka, 1975, 164–168. [Russian]

Krasnoselsky, M. A. and Krein, S. G.
[1] On averaging principle in nonlinear mechanics. *Russian Mathematical Surveys* **10** 3(65) (1955) 147–152. [Russian]

Kuz'min, A. V.
[1] Operations on polynomials and rational functions in SASM system. Algorithms of celestial mechanics. *Preprint, Institute of Theoretical Astronomy at the Academy of Sciences of the USSR*, Leningrad, 1981, No. 34. [Russian]

Lance, J. N.
[1] *Numerical Methods for High Speed Computers.* Moscow: Inostrannaya Literatura, 1962.

Lanczos, K.
[1] *Practical Methods of Applied Analysis.* Moscow: Physmatgiz, 1961.

Le Verrier, U. J.
[1] Researches astronomique. *Ann. Observ. Imper.* **I** (1855); **II** (1856), Paris. [French]

Levitan, B. M.
[1] *Almost-Periodic Functions.* Moscow: Gostekhizdat, 1953. [Russian]

Lichtenstein, L.

[1] Zur Maxwelschen Theorie der Saturnringe. *Mathem. Zeitschr.* **17** (1923) 62–110. [German]

Loumpov, V. I., Rostovtsev, V. A., and Fedorova, R. N.

[1] Problem-oriented versions of analytic calculus systems and their creation. In: *Analytic Calculus on Computers and their Application in Theoretical Physics*, Doubna: Joint Institute of Nuclear Research, 1983. [Russian]

Lyapunov, A. M.

[1] *General Problem of Stability Theory.* Moscow: Gostekhizdat, 1950. [Russian]

[2] On the series proposed by Hill for representation of Moon's motion. In: *Coll. Works*, V. 1, Moscow: Publ. of the Academy of Sciences of the USSR, 1956, 418–446. [Russian]

Malkin, I. G.

[1] *Some Problems of Nonlinear Oscillation Theory.* Moscow: Gostekhizdat, 1956. [Russian]

Mandelstam, L. I., and Papaleksi, N. D.

[1] On the grounds of one method of approximate solution of differential equations. *Journal of Experimental and Theoretical Physics Letters* **4**(2) (1934) 117–122. [Russian]

Markeev, A. P. and Sokol'sky, A. G.

[1] Some computational algorithms of Hamiltonian systems normalization. *Preprint, Institute of Problems of Mechanics, Akad. Nauk SSSR*, Moscow, 1976, No. 31. [Russian]

Martynyuk, A. A.

[1] *Stability Analysis: Nonlinear Mechanics Equations.* Amsterdam: Gordon and Breach, 1995.

[2] *Qualitative Methods in Nonlinear Dynamics. Novel Approaches to Liapunov's Matrix Functions.* New York: Marcel Dekker, 2002.

Mikhlin, S. G.

[1] *Variational Methods in Mathematical Physics.* Moscow: Physmatgiz, 1957. [Russian]

Mironov, S. V.

[1] Reducing method of hidden periodicity detection. *Mathematical Issues of Nonlinear Analysis and Dynamic Systems Control.* Moscow: Publ. Moscow State University, 1985, 56–70. [Russian]

Mitropolsky, Yu. A.

[1] *Averaging Method in Nonlinear Mechanics.* Kiev: Naukova Dumka, 1971. [Russian]

[2] Slow processes in nonlinear oscillating systems with many degrees of freedom. *Prikl. Mat. Mekh.* **14**(2) (1950) 139–170. [Russian]

[3] On construction of an asymptotic solution of perturbed Klein equations. *Ukr. Mat. Zhurn.* **47**(3) (1995) 1209–1216. [Russian]

[4] *Nonlinear Mechanics: Asymptotic Methods.* Kiev: Institute of Mathematics, National Academy of Sciences of Ukraine, 1995. [Russian]

[5] *Nonlinear Mechanics: One-frequency Oscillations.* Kiev: Institute of Mathematics, National Academy of Sciences of Ukraine, 1997. [Russian]

[6] On construction of an asymptotic solution of perturbed Bretherton equation. *Ukr. Mat. Zhurn.* **50**(1) (1998) 58–73. [Russian]

Mitropolsky, Yu. A. and Limarchenko, O. S.

[1] On the problem of asymptotic approximation of the slow processes in nonlinear dispersive media. *Ukr. Mat. Zhurn.* **50**(3) (1998) 357–371. [Russian]

Mitropolsky, Yu. A. and Moseenkov, B. I.

[1] *Asymptotic Solutions of Partial Differential Equations.* Kiev: Vischa shkola, 1976. [Russian]

[2] Lectures on application of asymptotic methods to partial differential equations solving. *Preprint, Institute of Mathematics, Ukraine Academy of Sciences*, Kiev, 1968. [Russian]

Mitropolsky, Yu. A. and Nguen Van Dao

[1] Introduction to Asymptotic Methods of Nonlinear Dynamics. (to appear)

Moiseev, N. D.

[1] On some basic simplified schemes of celestial mechanics, obtained through averaging of a bounded circular three-point problem. In: *Proc. of the P. K. Sternberg State Astronomical Institute*, V. 15B, part 1, Moscow, 1945, 75–99. [Russian]

[2] *Asymptotic Methods of Nonlinear Mechanics.* Moscow: Nauka, 1971. [Russian]

Naife, A. Kh.

[1] *Perturbation Methods.* Moscow: Mir, 1976.

Neistadt, A. I.

[1] Estimations in Kolmogorov's theorem on preservation of conditionally periodic motions. *Prikl. Mat. Mekh.* **45** (1981) 1016–1025. [Russian]

Pavelle, R. and Wang, P.

[1] Maxima from F to G. *J. of Simb. Computations* **I** (1985) 69–100.

Plotnikov, V. A.
[1] *Asymptotic Methods in Optimal Control Problems.* Training manual. Odessa: Publ. of Odessa State Unversity, 1976. [Russian]

Poincaré, A.
[1] *Lectures on Celestial Mechanics.* Moscow: Physmatgiz, 1965.
[2] *New Methods of Celestial Mechanics.* Moscow: Nauka, 1971.

Pontryagin, L. S., Boltyansky, V. G., Gamkrelidze, R. V. and Mischenko, E. F.
[1] *Mathematical Theory of Optimal Processes.* Moscow: Nauka, 1969. [Russian]

Popova, N. I.
[1] *Justification and Estimates of Errors of Some Averaging Schemes for Resonance Systems of Ordinary Differential Equations.* Thesis of Diss. Paper, Moscow, 1981. [Russian]
[2] On averaging over fast variables in multifrequency systems permitting resonances. *Preprint, Institute of Theoretical and Experimental Physics*, Moscow, 1977. [Russian]

Prikhodko, V. A.
[1] Construction of explicit solutions of three-body problem by the method of semianalytic programming. *Preprint, Institute of Theoretical and Experimental Physics*, Moscow, 1973, No. 60. [Russian]

Rozanov, Yu. A.
[1] *Stationary Random Processes.* Moscow: Physmatgiz, 1963. [Russian]

Ryabov, Yu. A.
[1] Estimate of convergence radius for the expansion into power series of parameter, and algorithms for Mathieu functions construction. *Diff. Uravn.* **15** (1979) 1993–2003. [Russian]
[2] The method for construction of semianalytical periodic and quasi-periodic solutions in the theory of non-linear oscillations by means of electronic computer. *Proc. of the IV Conf. on Non-Lin. Osc.* Prague, 1968, 231–236.
[3] Analytic-numerical solutions of restricted non-resonans planar three-body problem. In: *Dynamics, Ephemerids and Astrometry in Solar System*, Amsterdam: Kluwer Academic Publishers, 1966, 289–292.
[4] Analytic-numerical solutions of the restricted three-body problem. *Romanian Astronomical J.* **6**(1) (1997) 53–59.
[5] Numeric-analytical construction of Mathieu functions. *Mathematica Bohemica* **124**(1) (1999) 15–28.

Samoilenko, A. M. and Ronto, N. I.
[1] *Numerical-Analytic Methods of the Study of Boundary Problems Solutions.* Kiev: Naukova Dumka, 1985. [Russian]

Schwartz, L.
[1] *Analysis.* Moscow: Mir, 1972.
[2] *Mathematical Methods for Physical Science.* Moscow: Mir, 1965.

Shebekhey, V.
[1] *Theory of Orbits.* Moscow: Nauka, 1982.

Shilov, G. E.
[1] *Mathematical Analysis. Special Course.* Moscow: Physmatgiz, 1961. [Russian]

Smirnov, V. I.
[1] *Course of Higher Mathematics.* V. 2. Moscow: Mir, 1965. [Russian]

Sokolsky, A. G. and Khovansky, S. A.
[1] Normalization programmes for Hamiltonian systems with three degrees of freedom. Moscow Aviation Institute. Moscow, 1981, Dep. in All-Russian Scientific and Technical Information Institute 4.08.81, No. 3883-81.

Starzhinsky, V. M.
[1] *Applied Methods of Nonlinear Oscillations.* Moscow: Nauka, 1977. [Russian]

Stepanov, V. V.
[1] *Course of Differential Equations.* Moscow: Nauka, 1968. [Russian]

Subbotin, M. F.
[1] *Introduction into Theoretical Astronomy.* Moscow: Nauka, 1986. [Russian]

Ter Haar, D.
[1] *Elements of Hamiltonian Mechanics.* Moscow: Nauka, 1974.

Tikhonov, A. N. and Samarsky, A. A.
[1] *Equations of Mathematical Physics*, 5th ed. Moscow: Gostekhizdat, 1977. [Russian]

Tschigolev, B. M.
[1] *Mathematical Processing of Observation Results.* Moscow: Nauka, 1973. [Russian]

Van der Pol, B.
[1] *Nonlinear Theory of Electric Oscillations.* Moscow: Svyazisdat, 1935. [Russian]

Verhulst, F.
[1] *Nonlinear Differential Equations and Dynamical Systems.* Berlin: Springer-Verlag, 1990.

Vladimirov, V. S.
[1] *Equations of Mathematical Physics.* Moscow: Nauka, 1988. [Russian]

Volosov, V. M.

[1] *Asymptotic Methods of the Study of Nonlinear Waves in Stratified Medium with Applications to the Theory of Internal Waves in Oceans.* Moscow: Publ. Moscow State University, 1972. [Russian]

[2] Averaging in ordinary differential equations systems. *Russian Mathematical Surveys* **17**(6) (1962) 3–126. [Russian]

Volosov, V. M. and Morgunov, B. I.

[1] *Averaging Method in Nonlinear Oscillation Systems Theory.* Moscow: Publ. Moscow State University, 1971. [Russian]

Whittaker, E. T. and Watson, G. N.

[1] *A Course of Modern Analysis.* P. 2. Moscow: Physmatgiz, 1963.

Wolfram, St.

[1] *The Mathematica Book,* 3rd ed. Cambridge: Cambridge University Press, 1996.

Woodcock, D., Elton, D. and Davis, R. J.

[1] An examination of an iterative procedure for determining the characteristic exponents of linear differential equation with periodic coefficients. London, 1972.

Yakubovich, V. A. and Starzhinsky, V. M.

[1] *Linear Differential Equations with Periodic Coefficients and its Applications.* Moscow: Nauka, 1972. [Russian]

Zharkov, V. N. (Ed.)

[1] *Flows and Resonances in the Solar System.* Moscow: Mir, 1975.

Zhuravlev, S. G.

[1] Analytic theory of daily satellite motion. Part 1. Preliminary orbit. *Optimal Control Problems,* Issue 1. (Collected Articles). Perm, 1972, 68–84. [Russian]

[2] Disturbances of orbit elements of a synchronous satellite moving in a noncentral Earth's gravitation field. *Bulletin of Moscow State University. Series: Physics and Astronomy* (2) (1968) 105–110. [Russian]

[3] Motion of a daily satellite along orbits of great bent in the gravitation field of the Earth, the Moon and the Sun. In: *Collected Works of Post-graduates of the University of Peoples' Friendship. Mathematics, Mechanics and Physics* (Ed.: by Grebenikov E. A.), No. 7, 1970, 193–201. [Russian]

Zigmund, A. M.

[1] *Trigonometric Series.* Moscow: Mir, 1965.

Zubov, V. I.

[1] *Theory of Oscillations.* Moscow: Vyschaya Sckola, 1979. [Russian]

Index